USE OF BIOMARKERS FOR
ENVIRONMENTAL QUALITY ASSESSMENT

Use of Biomarkers for Environmental Quality Assessment

Editors

Laurent Lagadic, Thierry Caquet,
Jean-Claude Amiard and François Ramade

Science Publishers, Inc.

Enfield (NH), USA   Plymouth, UK

This edition is only for sale in North America

Internet Site: *http://www.scipub.net*

SCIENCE PUBLISHERS, INC.
Post Office Box 699
Enfield, New Hampshire 03748
United States of America

ISBN 1-57808-095-9

Library of Congress Cataloging-in-Publication Data

Utilisation de biomarqueurs pour la surveillance de
    la qualité de l'environnement. English. Use of
    biomarkers for environmental quality assessment/
    editors, Laurent Lagadic ... [et al.].
        p. cm.
    Includes bibliographical references and index.
    ISBN 1-57808-095-9
    1. Environmental quality—Evaluation. 2. Biochemi-
    cal markers. I. Lagadic, Laurent. II. Title

    GE140.U75 2000
    363.7'063—dc21

                                                00-032229

Published by arrangement with Technique et Documentation, Lavoisier, Paris,
France

Aidé par le ministère français chargé de la culture.
Published with the support of the French Ministry of Culture.

Translation of: ***Utilisation de biomarqueurs pour la surveillance de la qualité
de l'environnement*** 1998 © Technique et Documentation, Lavoisier, Paris, France
ISBN 2-7430-0230-1. **Updated by the authors for the English edition in 1999.**

Published by Science Publishers, Inc., Enfield, NH, USA
Printed in India

Preface

During the past twenty years, national and international programmes of environmental quality assessment have been implemented, and some of them have included the measurement of biomarkers. These tools were transferred from the laboratory, where their response to a single contaminant was calibrated, to the field, where contaminants are numerous and where many factors may exert a confusing effect on their response. Thus, in some cases, the results obtained were difficult to analyse and the real value of biomarkers in environmental risk assessment became questionable.

Among the numerous publications on biomarkers, only a few have considered the difficulties that arise from their practical use. The objective of this book was to give the opportunity to several research teams that have been involved in the implementation of these programmes to make a critical analysis of the use of biomarkers in this context. The analysis has been based on their own experience and on that of other groups. French, Canadian, Belgian and British scientists have been involved in a common approach to critically assess the actual operationality of biomarkers and to propose improvements in practical strategies for their *in situ* use.

This work is primarily designed for any person or organization in charge of assessment of quality of natural resources and of pollution prevention. It will be useful for scientists in public and private organizations. It also constitutes a reference work for university and high school students.

Contents

Chapter 2

Chapter 3

Chapter 4

Chapter 5

Chapter 7

Biomarkers of Exposure of Birds and Small Mammals to
Pollutants 139
J.-L. Rivière, M.-O. Fouchécourt and C.H. Walker

Chapter 8

Biomarkers of Exposure of Terrestrial Plants to Pollutants:
Application to Metal Pollution 169
J. Vangronsveld, M. Mench, B. Mocquot and H. Clijsters

Chapter 9

Endocrine Biomarkers: Hormonal Indicators of Sublethal
Toxicity in Fishes 187
A. Hontela

Chapter 10

Chapter 11

J. Pellerin-Massicotte and R. Tremblay

Chapter 12

*M. Amichot, J.-B. Bergé, A. Cuany, N. Pasteur, D. Pauron and
M. Raymond*

Chapter 13

Consequences of Individual-Level Alterations on Population
Dynamics and Community Structure and Function 269
Th. Caquet and L. Lagadic

Introduction

Biomarkers as Tools for Environmental Quality Assessment

J.-C. Amiard, Th. Caquet and L. Lagadic

A major problem at the end of the 20th century is the preservation of environmental health. The release of natural or synthetic substances is one of the most important factors in the degradation of the biosphere by human activities. The first harmful effects began when agricultural and animal breeding activities altered natural ecosystems (Dorst, 1965). These effects, at first qualitatively and quantitatively limited, rapidly became predominant with the extension of areas that were intensively exploited. Although the occurrence of pollution can be linked to the development of the earliest civilizations (according to recent analyses, the development of metallurgical activities during the Romen Empire led to copper and lead pollution that extended up to the arctic regions; Nriagu, 1983; Hong et al., 1996), it is mainly since the 19th century that the impact of human activities on the environment increased, in association with population increase and the major technological upheavals of the Industrial Revolution: increase in energy needs and raw materials (activities of extraction and transport), continuously growing use of fossil fuels (source of atmospheric pollution), and development and intensification of numerous activities that caused direct or indirect pollution (such as chemical industries and industrialized agriculture).

One of the essential characteristics of man-made pollution is the dispersal, wilful or otherwise, of certain substances (pesticides, hydrocarbons, etc.) or elements (metals, for example) that can contaminate various parts of the biosphere (atmosphere, hydrosphere, lithosphere), even in areas very far from where they were initially released. Presently, there are no ecosystems free from traces of human activity, because even areas unknown to human habitation are contaminated by pollutants carried by the movements of air masses or by marine and oceanic currents.

1. WHY DO WE MONITOR ENVIRONMENTAL QUALITY?

Since at least the 16th century, people have been interested in the quality of their environment. In a work published in 1567, Paracelse described the symptoms of arsenic and mercury poisoning in miners (Borm and Henderson, 1996). The first studies of environmental health were aimed only at preserving human health. Also, the first works were concerned with the risks of exposure for workers in local industries (Gallo and Doull, 1991; Lauwerys, 1990, 1991; Borm and Henderson, 1996). The environments studied were gradually diversified, extending especially to urban sites and suburban areas, but the preservation of human health remained the chief preoccupation. Presently, studies concerning occupational risks and evaluating the impact of pollution on the health of human populations are still qualitatively and quantitatively the most important (Amdur, 1991; Gallo and Doull, 1991; Lauwerys, 1990, 1991; Lu and Dourson, 1992; Borm and Henderson, 1996; Kavlock et al., 1996).

In fact, only at the beginning of the 20th century were qualitative and quantitative criteria proposed to evaluate the state of ecosystems with greater precision. This evaluation was considered, at least at first, strictly from an anthropocentric point of view, the environment being considered only as a collection of resources to be exploited by and for humans. However, the concept of resource has significantly evolved. While this term was at first closely linked to the idea of food resource, its use was progressively extended to raw materials, renewable or otherwise, and it now covers concepts as diverse as those of genetic or medicinal, or even aesthetic or recreational resources. This recent evolution of the concept of resource linked to the environment has been reinforced by the emergence of the concept of conservation of biodiversity and by the intensification of measures to protect or restore landscapes and areas felt to be ecologically unique (reserves, natural parks, etc.).

Much more recently, it has become clear that in the long term the health of human populations depends to a large extent on the quality of their environment, that term being considered in its most extensive sense. In acting on physical mediums (particularly air and water) as well as on biological mediums, i.e. living organisms, man-made pollutions have a definite impact on human health, and perhaps even on human evolution. Recent studies seem indeed to show that certain environmental contaminants can affect the reproductive capacities of animal species, the most vulnerable being those, like humans, having a low biotic potential (Bergeron et al., 1994; Guillette et al., 1994; Jobling et al., 1995).

2. HOW DO WE MONITOR ENVIRONMENTAL QUALITY?

The numerous methods for monitoring environmental health can be divided into two distinct categories: (1) the detection of pollutants and their quantification, in physical and biological mediums, and (2) the evaluation of the effects of pollution on living organisms, either at the individual level or at the level of populations and/or communities. These methods have been in use for several years, and it is clear that none of them is exclusive and none can by itself provide reliable and complete information on the state of the environment. In other words, it is the judicious combination of these different approaches that enables us to precisely evaluate the state of environments and organisms living in them (Fig. 1).

2.1 Detection and Quantification of Pollutants: Chemical Analysis

Methods of chemical analysis, currently highly improved and often expensive, enable us generally to measure the extent and level of

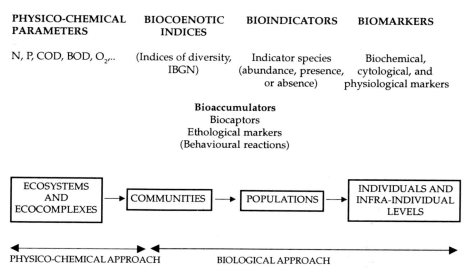

Fig. 1. Different types of measurable parameters in the physical medium and at different levels of biological organization to evaluate the state of the environment (after Amiard, 1994)

environmental contamination, by using a limited number of samples taken from various matrices (water, soil, sediments, plants, animals, etc.). However, apart from the fact that the use of such methods is not always possible because of the properties of the substances being studied, no method of analysis can enable us to quantify simultaneously all the contaminants present in a sample, especially because of the diversity of these contaminants (even without taking into account their products of degradation or transformation). Moreover, the analytical techniques do not allow us to evaluate the effects of the presence of the contaminants on living organisms or on the health of the ecosystem. The bioavailability and toxicological and ecotoxicological effects of certain contaminants may vary according to their chemical form (metal species or ionizable organic molecules, for example), because of interactions between contaminants (synergy or antagonism) or between contaminants and other abiotic or biotic components (reversible processes of adsorption-desorption, bioturbation, etc.), or even according to the organism under consideration, its life stage, or its physiological condition.

2.2 The Biological Approach

Methods based on qualitative and quantitative observations of living organisms in their natural environments were soon considered as potential means of compensating for the deficiencies of chemical analysis. In the beginning of the 20th century, Kolkwitz and Marson (1908, 1909) proposed the use of communities of aquatic organisms as indicators of water quality in rivers through the saprobes method.

At present, two complementary approaches, based on the study of living organisms, are used:

—The study of certain species or groups of species, the presence (or absence) and/or the abundance of which provide information about environmental quality; these methods are sometimes linked to the use of mathematical descriptors of the structure of communities to which these **bioindicators** belong.

—The measurement, in individuals from a natural environment, of molecular, biochemical, cellular, or physiological parameters, grouped under the term **biomarkers**. A biomarker is a change that can be observed and/or measured at the molecular, biochemical, cellular, physiological, or behavioural level and that reveals the present or past exposure of an individual to at least one polluting chemical substance (Lagadic et al., 1997a).

Behaviour is a particular type of biomarker. It is traditionally included in the category of physiological biomarkers but one must keep in mind that a direct link may exist between the behaviour of an organism in response to the presence of a xenobiotic and the fact that it is or is not present in the site studied (escape response). In fact, although behavioural responses themselves have a larger meaning as biomarkers, especially when they can be observed *in situ*, the absence of a bioindicator (i.e., a bioindicator species) in a given site may be due either to the disappearance of that species following the death of individuals that were earlier present, or to a behavioural reaction (escape) of the latter.

2.2.1 Some Bioindicators for Evaluation of the Ecological Health of Environments

The term 'bioindicator' has been defined and used in various ways and some authors even used it to designate biomarker-like descriptors (Blandin, 1986). For the sake of clarity, we recommend the definition of Guelorget and Perthuisot (1984), who consider bioindicators as 'species or groups of species that, by their presence and/or their abundance, are significant of one or several properties of the ecosystem to which they belong.'

Numerous works have been published about the use of bioindicators to evaluate the quality of natural ecosystems. It would be well beyond the scope of this introductory chapter even to list the different taxonomic groups that have been used as bioindicators and to explain the appropriate methodologies in terms of sampling and processing of data. Various publications and/or books synthetize these works (see, for example, Blandin, 1986; Hellawell, 1986; Haslam, 1990; Kramer, 1994).

In considering bioindicators for environmental health assessment, particular attention should be given to two types of species, **bioaccumulator species** and **sentinel species**, because they enable us to make a link between bioindicators and biomarkers.

Because of their mode of life and/or their physiological and metabolic characteristics, **bioaccumulator species** have the ability to accumulate certain contaminants directly from the ambient medium or indirectly (through food, for example) up to levels clearly higher than the level of contamination of the physical compartment (water, sediments, atmosphere, etc.). Two processes are responsible for the bioaccumulation of xenobiotics in organisms: bioconcentration that is a direct transfer from the surrounding medium (water, sediments, suspended particles, etc.) and by biomagnification that results from the transfer of chemicals through the food webs (Ramade, 1979; Amiard and Amiard-Triquet, 1980). Accumulation of pollutants in bioaccumulator organisms enables us to detect and quantify the pollutants by chemical and/or physical analysis in samples of such species. Consequently, bioaccumulator species are frequently used in monitoring the contamination of natural ecosystems by micropollutants, especially through biomonitoring networks. Marine and freshwater bivalve molluscs are excellent indicators of contamination because of their strong capacity for bioconcentration of xenobiotics. Thus they are widely used for the biomonitoring of heavy metals and organic products of industrial origin (PAH, PCB, PCDD, PCDF, and especially organochlorine pesticides). Biomonitoring networks of coastal waters are usually based on marine mussels, alone or in association with other species (such as oysters). Mussel Watch in the United States (Goldberg et al., 1978, 1983; Bayne, 1989) and Réseau National d'Observation (RNO) of the IFREMER in France (RNO, 1981; Cossa, 1989; Joanny et al., 1993) are two such programmes. Among their other characteristics, bioaccumulator species are in general sessile species, are abundant at least locally, are relatively resistant to the toxins being studied, and have a life span long enough to allow them to integrate variations in the quality of the environment.

Sentinel species constitute a particular type of indicator organisms. They include all living species that can be used as indicators of the presence and toxicity of at least one contaminant, and that enable us to evaluate the potential effects of this contaminant on human and/or environmental health (Lower and Kendall, 1990). Sentinel species live on the site under study, being present either spontaneously or because they were deliberately introduced (in cages, for example). Individuals can thus be sampled according to a definite plan and be used for various analytical and biochemical measurements (Riviere, 1993). These species have the same advantages as bioaccumulator species. Furthermore, a detailed knowledge of their physiology is needed for interpreting the variations in the biomarkers being measured.

Finally, apart from bioindicator species, there is a definite advantage in studying biomarkers in **species at risk**, i.e., species indispensable to the proper functioning of the ecosystem (pollinators, for example; Paine, 1966) or economically advantageous (marketable species), and in **target species** (crop pests, disease vectors, etc.) deliberately exposed to certain biocides. The latter are studied in order to improve the efficiency of treatments as the evolution of genetic characteristics of species has made them more tolerant, or even resistant.

In terms of sensitivity of species to xenobiotics, there are different possibilities (Fig. 2):

— the species at risk are species not targeted by xenobiotics; they are most often sensitive and have poor capacity to adapt;

— the target species of certain xenobiotics, because of a significant capacity for adaptation, are most often resistant (or at least mildly sensitive) or have the potential to become resistant;

— the sentinel species and species used as biological models may be, depending on the case, resistant or sensitive to xenobiotics.

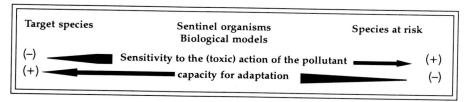

Fig. 2. Differences in sensitivity and capacity for adaptation characterizing target species, species at risk, and sentinel species

2.2.2 Biomarkers as Tools for Evaluating the Health of Individuals

Experiments under controlled conditions have widely demonstrated that biomarkers can be used to evaluate the exposure of individuals to xenobiotics and, in some cases, the effects of the latter on the structures and vital functions of the organism (Huggett et al., 1992; Lagadic et al., 1997b). These experimental bases have enabled us to consider the use of biomarkers in the natural environment, especially in environmental quality assessment programmes (McCarthy and Shugart, 1990; Peakall, 1992; Peakall and Shugart, 1993). Biomarkers are indeed elements for biological evaluation since they can provide information on the condition of individuals. According to the traditional approach of bioindication, bioindicator species can, through the abundance of individuals, give an account of the quality of the medium. However, even if they are present in the medium, their condition may be affected by pollutants. Some biomarkers may indicate lesions or physiological dysfunctions in these individuals, as well as in other species not usually used as bioindicators. These biomarkers can also enable us to evaluate the chances these individuals have of reproducing, which is fundamental to the preservation of the species.

In comparison to other methods of biomonitoring, the use of biomarkers presents the advantage of an integrated evaluation in time and in space of bioavailable pollutants, not only in terms of presence, but also in relation to the effects of these products on animal, plant, and microbial populations. From a temporal point of view, biomarkers may help to reconstitute a dynamic picture of variations in the quantities of bioavailable pollutants, considering either the original molecules or their products of degradation. This type of information is particularly important when, for example, the processes of degradation make metabolites more bioavailable, and thus potentially more dangerous, than the original molecules. Biomarkers can also reveal the exposure of individuals to compounds like organophosphorus pesticides, which are metabolized rapidly and accumulate very little in organisms. They are therefore the sign of a past exposure to pollutants that may have affected individuals and their offspring. The development of resistance to pesticides in insects perfectly illustrates this type of situation.

Biomarkers can give an indication of the distribution of contaminants in the environment. Depending on their habitat and their position in the food web, the species that are the source of biomarkers can indicate how pollutants are diluted in the medium or, on the other hand, are concentrated in certain parts of the ecosystem.

Despite these advantages, the major handicap in the use of biomarkers in the natural medium remains the interference from other factors of the environment. In experiments under controlled conditions, the influence of factors other than contamination by pollutants may be easily reduced or at least characterized with satisfactory precision. In the natural medium, different factors may give rise to responses of biochemical and physiological parameters used as biomarkers. Thus, climatic conditions, physico-chemical characteristics of the medium, genetic characteristics of the species, relations between individuals and/or species, or interactions between pollutants themselves, which are found in combination in the environment, may singularly complicate the interpretation of responses of biomarkers. These factors, sometimes grouped under the expression 'confusing factors', make comparison of environments delicate and the choice of natural reference sites problematic. The use of biomarkers to follow the evolution of environmental quality necessitates that the ecosystems be characterized precisely in physico-chemical as well as ecological terms. As a matter of fact, the knowledge of factors responsible for the natural variation of biomarkers is indispensable for distinguishing between a signal of disturbance caused by pollutants and the 'background noise' due to natural fluctuations of biomarkers. It seems unreal, at a time when natural ecosystems are contaminated by products of human activities to a varying extent but universally, to find a site free of pollution that could serve as an 'absolute' reference. The use of relative references, based on the comparison between sites with different contamination levels, or the use of organisms maintained under controlled conditions (laboratory strains, animals caged in the medium, etc.), are means of reducing the problem, but, as shown by studies presented in this work, necessitate the greatest caution in interpreting responses of biomarkers.

3. USE OF BIOMARKERS IN DIFFERENT TYPES OF ECOSYSTEMS

The biological oneness of the living world makes possible the use of the same types of biomarkers in organisms belonging to very different taxonomic groups in the animal and plant kingdoms. The possibility of measuring biomarkers that fulfil similar functions in very different organisms gives us in theory access to all parts of the biosphere in order to detect the presence of pollutants and, in some cases, to assess their effects on animals and plants.

This book proposes to present concrete cases of use of biomarkers in different types of ecosystems. The presentation, which begins with a brief summary of the biological bases at the origin of the emergence of the concept of biomarkers in ecotoxicology (Narbonne; chapter 1), focuses on two points corresponding respectively to terrestrial and aquatic ecosystems. The biomarkers measured in aquatic animals, vertebrates and invertebrates, are virtually the best known today. Above all, they enable us to assess the quality of marine environments, especially for the purpose of monitoring (Michel et al., chapter 2; Burgeot and Galgani, chapter 3; Cosson and Amiard, chapter 5; Galgani and Bocquene, chapter 6; Pellerin-Massicotte and Tremblay, chapter 11), but they are also widely used in continental aquatic ecosystems, lakes and rivers (Flammarion et al., chapter 4; Hontela, chapter 9; Spear and Bourbonnais, chapter 10). The use of biomarkers for monitoring the health of terrestrial ecosystems relies on animals (Rivière et al., chapter 7; Spear and Bourbonnais, chapter 10) and plants (Vangronsveld et al., chapter 8). Most of these biomarkers enable us to demonstrate that the individuals in which they are measured have been exposed to pollutants that may have affected their physiology. In the double perspective of a long-term assessment and a sustainable conservation of the quality of ecosystems, it is useful to examine the consequences of these biochemical and/or physiological alterations on the progeny of individuals exposed. The case of development of resistance to insecticides in insect populations (Amichot et al., chapter 12) and the analysis of structural and functional alterations of communities and ecosystems (Caquet and Lagadic, chapter 13) enable us to deal with the significance of responses of biomarkers for the highest levels of organization.

Each chapter takes into account criteria of applicability of biomarkers that contribute to the validation of their use *in situ*. Some of these biomarkers, for example EROD (ethoxyresorufin O-deethylase) activity, are already validated to a great extent in different types of natural mediums but many methodological aspects must be taken into account in rationalizing and promoting their use. For each biomarker, the precision of its measurement, the reliability of the response, the limit of validity, the dependence on environmental factors, the cost, and the facility of use are evaluated through studies presented in the different chapters. The present tendency to use biomarkers for the monitoring of environmental health is mainly based on these methodological considerations and the relevance of the information provided by the different biomarkers on the health of individuals and the quality of the environment in which they live.

REFERENCES

Amdur M.O. (1991). Air pollutants. In: Casarett and Doull's Toxicology—The basic science of poisons, 4th ed. Amdur M.O., Doull J. and Klassen ED (eds.). Pergamon Press, New York, pp. 854–871.

Amiard J.C. (1994). Surveillance de la qualité des milieux aquatiques par les biomarqueurs. *IREPOLIA, La Chronique*, 9:2.

Amiard J.C. and Amiard-Triquet C. (1980). Le transfert des polluants radioactifs dans les chaînes alimentaires aquatiques. *Année Biol.*, 19:117–146.

Bayne B.L. (1989). Measuring the biological effects of pollution: the Mussel Watch approach. *Wat. Sci. Tech.*, 21:1089–1100.

Bergeron J.M., Crews D. and McLachlan J.A. (1994). PCBs as environmental estrogens: turtle sex determination as a biomarker of environmental contamination. *Environ. Health Persp.*, 102:780–781.

Blandin P. (1986). Bioindicateurs et diagnostic des systèmes écologiques. *Bull. Écol.*, 17:215-307.

Borm P.J.A. and Henderson P.Th. (1996). Occupational toxicology. In: Toxicology, principles and applications. Niesink R.J.M., De Vries J. and Hollinger M.A. (eds.). CRC Press, Boca Raton, pp. 1141–1180.

Cossa D. (1989). A review of the use of *Mytilus* spp. as quantitative indicators of cadmium and mercury contamination in coastal waters. *Oceanol. Acta*, 12:417–432.

Dorst J. (1965). *Avant que Nature Meure*. Delachaux et Niestlé, Neuchâtel.

Gallo M.A. and Doull J. (1991). History and scope of toxicology. In: Casarett and Doull's Toxicology—The basic science of poisons, 4th ed. Amdur M.O., Doull J. and Klassen C.D. (eds.). Pergamon Press, New York, pp. 3–11.

Goldberg E.D., Koide M., Hodeg V. et al. (1983). U.S. Mussel Watch: 1977-1978 results on trace metals and radionuclides. *Estuar. Coastal Shelf. Sci.*, 16:69–93.

Goldberg E.D., Bowen V.T., Farrington J.W. et al. (1978). The Mussel Watch. *Environ. Conserv.*, 5:101–125.

Guelorget O. and Perthuisot J.P. (1984). Indicateurs biologiques et diagnose écologique dans le domaine paralique. *Bull. Écol.*, 15:67–76.

Guillette L.J. Jr., Gross T.S., Masson G.R. et al. (1994). Developmental abnormalities of the gonad and abnormal sex hormone concentrations in juvenile alligators from contaminated and control lakes in Florida. *Environ. Health Persp.*, 102:68–688.

Haslam S.M. (1990). *River pollution. An ecological perspective*. Bellhaven Press, London.

Hellawell J.M. (1986). *Biological indicators of freshwater pollution and environmental management*. Elsevier Applied Science Publishers, London.

Hong S., Candelone J., Patterson C.C. and Boutron C.F. (1996). History of ancient copper smelting pollution during Roman and medieval times recorded in Greenland Ice. *Science*, 272:246–249.

Huggett R.J., Kimerle R.A., Mehrle P.M. Jr. and Bergam H.L. (eds.) (1992). *Biomarkers. Biochemical, physiological and histological markers of anthropogenic stress*. SETAC Special Publications Series, Lewis Publishers, Boca Raton.

Joanny M., Belin C., Claisse D. et al. (1993). *Qualité du milieu marin littoral*. Ifremer, Brest.

Jobling S., Reynolds T., White R. et al. (1995). A variety of environmentally persistent chemicals, including some phthalate plasticizers, are weakly estrogenic. *Environ. Health Persp.*, 103:582-587.

Kavlock R.J., Daston G.P., Derosa C. et al. (1996). Research needs for the risk assessment of health and environmental effects of endocrine disruptors: a report of the U.S. EPA-sponsored workshop. *Environ. Health Persp.*, 104, suppl. 4:715–740.

Kolkwitz R. and Marson M. (1908). Ökologie der pflanzliehen Saprobien. *Ber. Dt. Bot. Ges.*, 26:505–519.

Kolkwitz R. and Marson M. (1908). Ökologie der tierischen Saprobien. *Intern. Rev. Hydrobiol.*, 2:126–152

Kramer K.J.M. (ed.) (1994). *Biomonitoring of coastal waters and estuaries.* CRC Press, Boca Raton.

Lagadic L., Caquet Th. and Amiard J.C. (1997a). Biomarqueurs en écotoxicologie: Principes et définitions. In: Biomarqueurs en écotoxicologie: Aspects Fondamentaux. Lagadic L., Caquet Th. Amiard J.C. and Ramade F. (eds.). Masson, Paris, pp. 1–9.

Lagadic L., Caquet Th., Amiard J.C. and Ramade F. (eds.) (1997b) *Biomarqueurs en ecotoxicologie: Aspects Fondamentaux.* Masson, Paris.

Lauwerys R.R. (1990). *Toxicologie industrielle et intoxications professionnelles.* Masson, Paris.

Lauwerys R.R. (1991). Occupational toxicology. In: Casarett and Doull's Toxicology—The basic science of poisons, 4th ed. Amdur M.O., Doull J. and Klassen C.D. (eds.). Pergamon Press, New York, pp. 947–969.

Lower W.R. and Kendall R.J. (1990). Sentinel species and sentinel bioassay. In: Biomarkers of environmental contamination. McCarthy J.F. and Shugart L.R. (eds.). Lewis Publishers, Boca Raton, pp. 309–331.

Lu F.C. and Dourson M.L. (1992). Safety/risk assessment of chemicals. *J. Occup. Med. Toxicol.*, 1:321–335.

McCarthy J.F. and Shugart L.R. (eds.). (1990). *Biomarkers of environmental contamination.* Lewis Publishers, Boca Raton.

Nriagu J.O. (1983). *Lead and lead poisoning in antiquity.* Wiley, New York.

Paine R.T. (1966). Food web complexity and species diversity. *Am. Nat.* 100:65–76.

Peakall D.B. (1992). *Animal biomarkers as pollution indicators.* Chapman & Hall, London.

Peakall D.B. and Shugart L.R. (eds.) (1993). *Biomarkers: Research and Application in the assessment of environmental health.* NATO ASI Series, serie: Cell Biology, vol. 68. Springer Verlag, Berlin.

Ramade F. (1979). *Écotoxicologie*, 2nd ed., Masson, Paris.

Rivière J.L. (1993). Les animaux sentinelles. *Courrier de l'Environnement de l'INRA*, 20: 59–67.

RNO (1981). *Synthèse des travaux de surveillance 1975-1979 de Réseau National d'Observation de la Qualité du Milieu Marin.* Centre National pour l'Exploitation des Océans et Ministère de l'Environnement.

1

History—Biological Basis of the Use of Biomarkers in Ecotoxicology

J.-F. Narbonne

1. ENVIRONMENTAL MONITORING

Ecotoxicology evolved in the early 1960s when the effects of environmental contamination by polluting products of industrial and agricultural origin became apparent (Ramade, 1992). It was during this period that mercury, cadmium, and PCB poisoning affected human populations in Japan. Similarly, in the North American Great Lakes region, toxic effects due to DDT were observed in birds and fish (Peakall, 1992). Finally, from 1962 to 1979, more than fifty accidents caused by oil tankers were recorded world-wide, each leading to the dispersal of, on the average, 10,000 tonnes of crude oil in the sea, and causing profound changes in marine ecosystems (Lacaze, 1980).

In the face of extensive contamination of the biosphere by ever-growing quantities of pollutants from various sources, it seems essential to try to protect the environment by taking immediate regulatory measures, such as a ban or limit on the use of certain compounds. But to evaluate the consequences of such measures, and be able to identify new contaminations, it is necessary to monitor different parts of the biosphere. In the early 1970s, programmes to monitor the environment were developed, based on chemical analysis of major contaminants (PAHs, PCBs, heavy metals, organochlorine pesticides) in different mediums (water, sediments, soil, organisms). The doses occurring in the biological matrices indicate the bioavailability of contaminants present in the physical medium.

Analyses done in water revealed, however, that the concentrations measured were located, for many weakly soluble compounds, at the detection thresholds of the equipment. On the other hand, when the contaminants were measured in bivalve molluscs, such as mussels, their quantification was much easier and the results were more significant. Such animals have a considerable capacity for bioaccumulation of pollutants because of their filtering mode of feeding. They are more sedentary and easy to study. Numerous species of molluscs have a long life cycle (several months to a few years), which allows us to study variations in the contamination of a given site.

The Mussel Watch programme, in the United States, was begun on this basis (Bayne, 1989). This programme, developed towards the late 1970s under the aegis of the Scripps Institution of Oceanography and the EPA, consists of the measurement of principal contaminants in mussels sampled periodically at more than a hundred points along the coast. Similarly, in 1972, the Réseau National d'Observation de la qualité du milieu marin (RNO) was created in France by the IFREMER and the Ministry of the Environment (Claisse, 1989). More recently, the Conseil International pour l'Éxploration des Mers (CIEM) and the Commission Internationale pour l'Éxploration de la Mer Méditerranée (CIESM) organized sampling campaigns in the context of a monitoring network in the North Sea and the Mediterranean Sea.

2. THE CONCEPT OF BIOMARKER

The monitoring of chemical contamination in an ecosystem does not enable us to assess its impact on the organisms, populations, and communities. In terms of sublethal effects, the response of organisms to contamination can only be evaluated by the measurement of biological, physiological, or biochemical parameters, according to an approach similar to that used for medical diagnostics in human or veterinary clinical toxicology. If numerical ecology detects changes occurring at the level of communities present in an ecosystem, it is necessary, from an ecotoxicological point of view, to have available early indications revealing a risk of modification in the ecosystem.

Most indices available in the 1960s to evaluate and predict harmful effects on organisms in case of contamination of their medium of life were based on laboratory tests aiming at estimation of the ecotoxicity of chemical compounds that could contaminate the environment. These tests were limited either to evaluation of lethal toxicity or to the measurement of non-specific physiological parameters (blood parameters of energy metabolism, for example). Such indices, which were not directly linked to one or more functional and/or structural alterations in the exposed animals, had only a limited credibility as indicators of environmental contamination.

From the 1970s, the considerable development of molecular toxicology allowed rapid progress in the understanding of toxicity mechanisms of xenobiotics, mainly in mammalian models such as the rat. Subsequently, for many pollutants, relatively sensitive and specific biochemical effects became apparent in species of ecotoxicological interest, such as birds (see chapters 5, 7, and 10), fishes (see chapters 2–6, 9–11), and molluscs (see chapters 2, 5, 6, and 11). Some examples of biochemical processes linked to the presence of pollutants are inhibition of acetylcholinesterase activity by organophosphorous pesticides and carbamates (Coppage and Braidech, 1976), the induction of cytochrome P450 monooxygenases by liposoluble contaminants such as PAHs (Payne, 1976) and PCBs (Narbonne and Gallis, 1979), the inhibition of dehydratase activity of delta-amino-levulinic acid by lead (Hodson et al., 1980), and induction of metallothionein synthesis following exposure to salts of metals such as copper or cadmium (Roch et al., 1982).

The possibilities of using these markers in the study of environmental contamination are extensively developed in this work. Among the biochemical indices, the induction of metallothioneins by metals and that of monooxygenases by liposoluble compounds represent an adaptive response to the presence of pollutants. Detoxication mechanisms often come into play before the manifestation of toxic effects strictly speaking and may thus provide sensitive and, to some extent, specific indicators. In parallel, methods have been developed to measure genotoxic effects (Randerath et al., 1981) and immunotoxic effects (Vos, 1980).

It is thus that, in the early 1980s, the notion of biomarker took shape, designating molecular, biochemical, histological, or physiological changes in organisms that could be used to estimate either the exposure to contamination in their medium of life, or the effects induced by the pollution. Although it does not refer to a new concept (Koeman et al., 1993), the notion of biomarker is relatively recent, and the evolution of its use as a tool to monitor and evaluate environmental health is closely linked to progress in our knowledge of molecular toxicity mechanisms of pollutants in different animal and plant species in the ecosystem.

3. HISTORICAL STAGES IN THE DEVELOPMENT OF BIOMARKERS FOR ENVIRONMENTAL MONITORING

In parallel with the evolution of the scientific context, the development of biomarkers in environmental monitoring was, in part, followed by structures responsible for the management of problems linked to environmental contamination and, in part, stimulated by the implementation of research programmes by national and international organizations. Under the impetus of such organizations, there has been a rapid implementation and development of means of surveillance based on the *in situ* use of biomarkers in many species, particularly in marine environments.

Some important steps in the development of biomarkers for monitoring of environmental health in France and other countries are mentioned below:

1978: First international conference at Copenhagen on the feasibility of monitoring the biological effects of pollution, under the aegis of the International Council for the Exploration of the Sea.

1980: UNESCO produced a first report entitled 'Monitoring Biological Variables Related to Marine Pollution'.

1981: Launching of the PRIMA programme (Pollutant Response in Marine Animals) in the United States under the aegis of the National Science Foundation–Office for the International Decade of Ocean Exploration, consisting in a multidisciplinary study of the development and evaluation of biological indices of adverse effects induced in marine animals by low levels of pollution. (This programme involves meetings of PRIMO (Pollutant Response in Marine Organisms), which have taken place every two years alternately in Europe and the United States.)

1982: Launching of a programme on Molecular Mechanisms of Cell Injury and Toxicity in Marine Molluscs, by three European laboratories (IMER Plymouth, University of Genes, University of Bordeaux), supported by the DG XII of the CCE. Launching of several research programmes and therefore several laboratories (INRA centres in Versailles and Lyon, University of Bordeaux) on activities of biotransformation in birds, fishes, and molluscs, by the Méthodologies en Écotoxicologie commission of the Ministry of Environment.

1983: Organization of the International Symposium on Ecotoxicological Testing for the Marine Environment, by the CEE and the International Academy of Environmental Safety, in Gand (Belgium), in the course of which J.F. Payne presented the advantages of including the study of monooxygenases of fish in monitoring programmes (Persoone et al., 1983).

1986: Organization of a Workshop on Biological Effects of Pollutants at Oslo (Norway) by GEEP (Group of Experts on the Effects of Pollutants), under the mandate of the Intergovernmental Oceanographic Commission (IOC) of UNESCO, with the aim of field-testing methods for detecting the effects of pollutants by molecular approaches and by investigations at the cellular and physiological level. The originality of this work consisted in matching biomarkers measured in animals on natural sites and in mesocosms having gradients of pollution, with characteristic chemical and ecological data of these environments (Bayne et al., 1988). This workshop enabled the validation of certain parameters of the system of biotransformation as biomarkers in fishes and molluscs.

1987: On the basis of the Oslo meeting, creation of the Groupe Interface Chimie-Biologie des Écosystèmes Marins (GICBEM) combining teams of the Universities of Bordeaux, Nice, and Marseille, supported at first by the Institut Océanographique de Monaco, then by the CCE and the IFREMER (see chapter 2).

1987: Launching of the Programme de Toxicologie de l'Environnement by the European Science Foundation.

1989: Organization of Biomarkers Workshop, the Eighth Pellston Workshop, at Keystone (Colorado, USA) by the Society of Environmental Toxicology and Chemistry (SETAC) (Huggett et al., 1992).

1990: Following a symposium of the American Chemical Society, publication of results of the evaluation of different types of biomarkers by organizations responsible for monitoring the marine environment in the United States, Canada, the Netherlands, Germany, Great Britain, France, and the Scandinavian countries (McCarthy and Shugart, 1990).

1991: Organization of a workshop on Strategy for Biomarker Research and Application in the Assessment of Environmental Health at Texel (Netherlands), under the aegis of NATO (Peakall and Shugart, 1993).

1992: Organization of a workshop on Use of Biomarkers in Assessing Health and Environmental Impacts of Chemical Pollutants at Luso (Portugal), under the aegis of NATO (Travis, 1993).

1992: Organization of a Training Workshop on the Techniques for Monitoring Biological Effects of Pollutants in Marine Organisms at the University of Nice (France), under the aegis of the FAO, IOC, and PNUE (Mediterranean Action Plan Technical Report Series, 1993).

1994: Organization of workshop on Interpretation and Evaluation of Biomarker Responses to Anthropogenic Pollution: Consequences for Environmental Damage, at the Research Institute for Soil Sciences and Agricultural Chemistry of Budapest (Hungary), during which an assessment was made of the results of the Working Group on Assessment of Effects on Ecosystems of the Scientific Advisory Committee on Environmental Toxicology of the Programme de Toxicologie de l'Environnement implemented by the European Science Foundation in 1987 (Peakall, 1994).

1994: Launch of the BIOMAR programme, supported by the DG XII of the CCE and aiming to develop fundamental aspects and applications of biomarkers in marine environments, in order to evaluate their use in the Mediterranean, Atlantic, North Sea, and Baltic Sea.

1995: Organization of a conference on Air Toxics: Biomarkers in Environmental Applications, in Houston (Texas, USA) (Environmental Health Perspectives, 1996).

1995: Launching of the programme 'Évaluation de l'état de santé écologique des hydrosystèmes par l'utilisation de variables biologiques', by the Ministry of Environment and the GIP Hydrosystèmes, in which some research activities concerned biomarkers.

1996: Launching of the Programme National d'Écotoxicologie (PNETOX), by the Ministry of Environment, one objective of which is the validation of tools, including biomarkers, and methodologies for ecotoxicological risk evaluation.

1997: Organization of the workshop Biomarkers: A Pragmatic Basis for Remediation of Severe Pollution in Eastern Europe, at Cieszyn (Poland), under the aegis of NATO (Peakall, 1997).

At present, several laboratories of the world study biomarkers that can be used in ecotoxicology. The use of some of them as biomarkers of exposure and, in certain cases, of toxic effects, is now in the transfer stage for evaluation of environmental contamination. Present developments concern mainly, on one hand, the study of biomarkers in new sentinel species in aquatic or terrestrial ecosystems, and, on the other hand, methods using organisms deliberately placed in the medium being monitored (programmes of IFREMER, Agence de l'eau Rhône-Méditerranée-Corse, University of Bordeaux/Agence de l'eau Adour-Garonne, etc.). These two types of approaches are developed in the following chapters.

REFERENCES

Bayne B.L. (1989). Measuring the biological effects of pollution: The Mussel Watch approach. *Wat. Sci. Tech.*, 21:1089–1100.

Bayne B.L., Clarke K.R. and Gray J.S. (1988). *Marine Ecology, Progress Series*, vol. 43 (1–3). Inter-Research, Germany.

Claisse D. (1989). Chemical contamination of French coasts. The results of a ten-year Mussel Watch. *Mar. Pollut. Bull.*, 20:523–528.

Coppage D.L. and Braidech T. (1976). River pollution by anticholinesterase agents. *Water Res.*, 10:19–24.

Environmental Health Perspectives (1996). Conference on air toxics: Biomarkers in environmental applications. *Environ. Health Persp.*, 104, suppl. 5.

Hodson P.V., Blunt B.R. and Whittle D.M. (1980). Biochemical monitoring of fish blood as an indicator of biologically available lead. *Thalasia Jugosl.*, 16:389–396.

Huggett R.J., Kimerle R.A., Mehrle P.M. Jr. and Bergam H.L. (eds.) (1992). *Biomarkers. Biochemical, physiological, and histological markers of anthropogenic stress.* SETAC Special Publications Series. Lewis Publishers, Boca Raton.

Koeman J.H., Köhler-Gunther A., Kurelee B. et al. (1993). Applications and objectives of biomarkers research. In: *Biomarkers. Research and applications in the assessment of environmental health.* Peakall D.B. and Shugart L.R. (eds.). NATO Advanced Science Institutes Series, vol. II 68. Springer Verlag, Berlin, Heidelberg, pp. 1–13.

Lacaze J.C. (1980). *La pollution pétrolière en milieu marin. De la toxicologie à l'écologie.* Collection Écologie Appliquée et Sciences de l'Environnement. Masson, Paris.

McCarthy J.F. and Shugart L.R. (eds.) (1990). *Biomarkers of environmental contamination.* Lewis Publishers, Boca Raton.

Narbonne J.F. and Gallis J.L. (1979). *In vitro* and *in vivo* effect of pehoclor DP6 on drug metabolizing activity in mullet liver. *Bull. Environ. Contam. Toxicol.*, 23:344–348.

Payne J.F. (1976). Field evaluation of benzo (a) pyrene hydroxylase induction as a monitor for marine petroleum pollution. *Science*, 191:945–946.

Peakall D.B. (1992). *Animal Biomarkers as Pollution Indicators.* Chapman & Hall, London.

Peakall D.B. (1994). The role of biomarkers in environmental assessment. (1) Introduction. *Ecotoxicology*, pp. 157–160.

Peakall D.B. (1997). *Biomarkers: A pragmatic basis for remediation of severe pollution in Eastern Europe.* Kluwer Academic Publishers, Dordrecht.

Peakall D.B. and Shugart L.R. (eds.) (1993). *Biomarkers. Reserach and application in the assessment of environmental health.* NATO Advanced Science Institutes Series, vol. II 68. Springer Verlag, Berlin, Heidelberg.

Persoone G., Jaspers E. and Claus C. (1983). *Ecotoxicological testing for the marine environment*, vol. 1. State University of Ghent, Ghent.

Ramade F. (1992). *Précis d'écotoxicologie.* Collection d'écologie, vol. 22, Masson, Paris.

Randerath K., Reddy M. and Gupta R.C. (1981). ^{32}P-postlabeling analysis for DNA damage. *Proc. Nat. Acad. Sci. USA*, 78:6126–6129.

Roch M., McCarter J.A., Matheson A.T. et al. (1982). Hepatic metallothionein in rainbow trout (*Salmo gairdneri*) as an indicator of metal pollution in the Campbell River system. *Can. J. Fish. Aquat. Sci.*, 39:1596–1601.

Travis C.C. (ed.) (1993). *Use of biomarkers in assessing health and environmental impacts of chemical pollutants.* NATO Advanced Science Institutes Series, vol. A 250. New York, Plenum Press.

Vos J.G. (1980). Immunotoxicity assessment: Screening and function studies. *Arch. Appl. Toxicol. Suppl.* 4:95–108.

2

Biochemical Indicators of Pollution in Coastal Ecosystems: Experience of the Groupe Interface Chimie-Biologie des Ecosystemes Marins (GICBEM)

X. Michel, J.-F. Narbonne, P. Mora, M. Daubèze, D. Ribera, M. Lafaurie, H. Budzinski and Ph. Garrigues

INTRODUCTION

To preserve the equilibrium of a natural environment in front of the dynamics of human development has become a necessity recognized by all at the turn of this century, which has known an exceptional industrial growth. In the face of pollution caused by economic development, the priority objectives are, on one hand, to maintain the cycles of basic elements of life and, on the other hand, to protect the renewable resources of our environment.

In this context, the basic themes of research applied to the study of environmental pollution revolve around four major axes:

—the reporting and inventory of chemical pollution;
—understanding of the biogeochemical processes of transport, dispersal, and distribution of pollutants in the environment;
—understanding of the mechanisms of disturbance in contaminated organisms and, beyond that, of the repercussions for ecosystems and for human beings;
—the definition and improvement of tools necessary for the evaluation and monitoring of environmental health.

Regarding this last point, three types of approaches can be considered in an attempt to evaluate the biological impact of pollution relatively quickly:

—biological toxicity tests on samples taken *in situ* (sediments, water, extracts);
—measurement of biomarkers in indigenous organisms;
—measurement of biomarkers in organisms transplanted in the sites studied.

It is in this perspective that the Groupe Interface Chimie-Biologie des Ecosystèmes Marins (GICBEM) was constituted in late 1986 with the objective of studying the fundamentals (comparative toxicology) and applications (research of biomarkers of pollution) of the response of marine organisms to the presence of contaminants. The principal objective was to improve techniques to measure sublethal, short-term effects of pollutants on exposed organisms and to research the possible correlations between variations of biological parameters and contaminant levels in the environment.

The purpose of this research is to put forward, to environmental management authorities and to industries, (1) a scientific expertise on the problems of pollution of specific aquatic ecosystems and (2) the tools necessary for the early detection of sublethal biological effects of pollution in the field of surveillance programmes.

In the first part of this chapter, we will describe the experimental programme followed for the campaigns completed in 1989 and 1991 in the Mediterranean Sea. The results obtained will be presented in the second part. In the last part, we discuss the results of a preliminary experiment in the Arcachon basin on mussels transplanted from a less polluted site into a more polluted site.

1. GICBEM MISSIONS IN THE NORTHWESTERN MEDITERRANEAN SEA

1.1 Study Sites

The region in which GICBEM is involved is the northwest of the western Mediterranean Sea. From 1987 to 1991, ten missions were conducted in this zone with the help of vessels from the Musée Oceanographique of Monaco, CNRS, and IFREMER. The choice of sites and contaminants studied took into account especially the data of the Réseau National d'Observation (RNO). During the July 1989, November 1989, and July 1991 campaigns, the results of which are presented here, 15 sites were selected along the coast, from the Gulf of Fos to the Italian Riviera and

Corsica (Fig. 1). These sites comprised very mildly polluted stations (Corsican coast), moderately contaminated stations (continental sites with swimming activities; Portofino, Cannes/Fourmigues, Porquerolles), and sites with significant industrial and port activities (Fos/Carteau, Marseilles/Planier, Toulon/Lazaret, Genes/Nervi). The results of the first campaigns conducted in 1987 and 1988 were the subject of several publications (Raoux et al., 1989; Ribera et al., 1989; Garrigues et al., 1990, 1993; Narbonne et al., 1991; Laufaurie et al., 1992; Raoux and Garrigues,

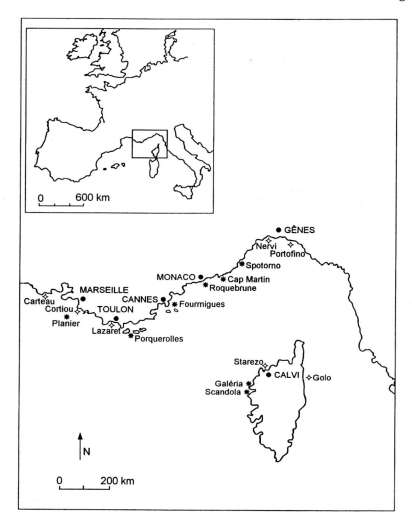

Fig. 1. Sampling sites of GICBEM missions in 1989 and 1991 in the northwestern Mediterranean Sea

1993) and gave us the experience needed to carry out the 1989 and 1991 campaigns.

1.2 Contaminants

Two groups of contaminants were studied by GICBEM, the polycyclic aromatic hydrocarbons (PAHs) and the polychlorobiphenyls (PCBs).

The PAHs, non-substituted aromatic hydrocarbons composed of several aromatic nuclei, are found in the environment primarily because of oil tanker activities and incomplete combustion of fossil fuels.

By virtue of their physico-chemical characteristics, the PAHs are stable. Moreover, certain members of this group of hydrocarbons, such as benzo(a)pyrene (BaP), have a strong genotoxic potential. They are among the most thoroughly studied environmental contaminants because they are among the most persistent and the most toxic. In the GICBEM campaigns, 12 priority PAHs were selected: phenanthrene, anthracene, fluoranthene, pyrene, benzo(a)anthracene, chrysene, perylene, benzo(a)pyrene, benzo(k)fluoranthene, dibenzo(a,h)anthrene, benzo(g,h,i)perylene, and indeno(1,2,3-c,d)pyrene.

PCBs are a particularly important group of environmental contaminants because of their ubiquity, stability, and potential to be bioaccumulated, with possible consequences for the health of organisms. Since the presence of these substances was reported in fish extracts and sea birds in the 1960s, many analytical studies detected PCBs in the entire biosphere, including the polar regions. Despite bans on the manufacture and use of PCBs, pollution by these substances is always present. Thus, in 1981, the RNO indicated a particularly serious PCB contamination in the Seine estuary, the mean concentration of PCB in mussels (2.8 mg/kg dry weight) being four times as high as that measured in other French estuaries (Chevreuil and Granier, 1992). In the earlier research of GICBEM, 7 PCB congeners were studied (CB 28, 52, 101, 118, 153, 138, and 180), being the most representative of the industrial combinations used in France (DP 5 and 6). From 1992, the analysis was extended to all the major congeners (20 to 22 compounds).

PAH and PCB concentrations were determined in samples of sediments taken by divers in the study sites, at a depth of 5 to 15 m. After the plant debris was removed as needed, two or three samples of the superficial layer of the sediments (around 2–3 cm) were taken, mixed, and stored at −20°C. After separation of plant debris by flotation, the sediments were sieved, the fine fractions (< 63 μm) being thus separated from coarser fractions, and dried. The 12 selected PAH were separated and quantified by high pressure liquid chromatography (HPLC) after extraction according to the method of Soclo et al. (1986). After hexane extraction (Soxhlet), the PCBs were purified by passage through a capillary column coupled with an electron capture detector (Monod and Arnoux, 1978).

1.3 Source Species for Biomarkers

The sentinel species in which different biomarkers were studied were chosen on the basis of the following conventional criteria:

—they have a wide geographic distribution in the zone under study;
—they are sedentary;
—they are representative of the coastal ecosystem under consideration; and
—they are easy to capture.

Two species were chosen, a fish, the comber (*Serranus cabrilla*), and the mediterranean mussel (*Mytilus galloprovincialis*).

The comber was selected because of its sedentary nature and simultaneous functional hermaphroditism, which limits, for many parameters, individual variations linked to sex. The individuals were caught by line fishing and dissected as soon as they were out of the water. After dissection, the liver was frozen at −180°C within 5 minutes after capture, at most. The livers were divided to obtain 4 samples of 2 to 5 livers each, and microsomal fractions were prepared from these in the laboratory.

The mussels, particularly the two dominant species of the French coasts, *Mytilus galloprovincialis* and *M. edulis*, answered all the criteria mentioned above and, in addition, had a capacity to accumulate the contaminants, which made these organisms very useful in programmes for monitoring pollution of coastal ecosystems of the Mussel Watch type, such as the RNO in France (Bayne, 1989; Claisse, 1989). Apart from their use as 'integrators' of contaminants present in the marine environment, the mussels are frequently used as a source of biomarkers, the biotransformation systems and toxicity mechanisms of contaminants being well known in these bivalves (Narbonne and Michel, 1997). They are sedentary, quite easy to collect and handle, easily used in transplantation experiments, and widely distributed along the entire coastal ecosystem. In the Mediterranean, *M. galloprovincialis* is the dominant species, especially near rearing sites, although *M. edulis* is sometimes present (Fischer et al., 1987).

The mussels were collected by divers. During the 1989 missions, the individuals were sorted according to sex, which was empirically determined by the colour of the mantle. The animals were then dissected in such a way as to eliminate the byssus, the foot, the crystalline stylet, the retractor muscles of the foot and the byssus, and the adductor muscles of the shell. The samples, numbering 5 per site, were made up of one male and one female and preserved in liquid nitrogen. The subcellular fractions of whole mussels in which various biomarkers were measured (post-mitochondrial fraction, cytosol, and microsomes) were prepared extemporaneously. During the 1991 campaign, the methodology for sample preparation changed because of the difficulties encountered in determining the sex of the animals and because of the limited number of individuals collected in some sites. In particular, the measurements of benzo(a)pyrene hydroxylase (BaPH) were done on microsomes prepared from the hepatopancreas of mussels, this organ being the principal seat of biotransformation activity of xenobiotics in bivalve molluscs. Each sample was made up of 5–10 digestive glands, which limited variability linked to individuals and to sex.

1.4 Biomarkers Studied

Regarding the comber, two enzymatic activities of phase I characteristic of cytochrome P4501A1, ethoxyresorufin-o-deethylase (EROD) and benzo(a)pyrene hydroxylase (BaPH) were measured by the fluorimetric techniques described by Burke and Mayer (1974) and Dehnen et al. (1973), respectively.

In the mussel, several biochemical markers have been studied during campaigns, the results of which are presented. They include biomarkers of enzymatic activities of phase I and II, of neurotoxicity, and of oxidative stress (Table 1).

Benzo(a)pyrene oxidase (BPO) activity is a phase I activity that takes into account the capacity of microsomes to metabolize [^{3}HBaP]. The measurement of this activity is based on a radiometric technique that measures total metabolites of marked BaP. It is quite hard in practice, which led us to improve the measurement of another phase I activity,

Table 1. Biomarkers studied in *Mytilus galloprovincialis* during the 1989 and 1991 GICBEM campaigns in the western Mediterranean Sea

Biomarker (sample)	1989 campaigns	1991 campaign
Benzo(a)pyrene oxidase (BPO) (microsomes)	+	
BaPH (microsomes)		+
Cytochrome P450 concentration (microsomes)	+	
Epoxide hydrolase (EH) (microsomes)		+
Lipid peroxides (TBARs, MDA) (cytosol and microsomes)		+
Acetylcholinesterases (AChE) (post-mitochondrial fraction)	+	+

BaPH, in mussel for the 1991 campaign (Michel et al., 1994). A phase II activity, epoxide hydrolase (EH), was tested during the 1989 campaigns. As for BPO, a radiometric technique was used to measure the metabolites of $[7.^3H]$ styrene oxide.

Hydrophobic organic chemical contaminants and heavy metals can cause significant peroxidative phenomena in the mussel (Livingstone et al., 1990). These may especially bring about a deterioration of membranes, leading to the formation of lipid peroxides. The presence of lipid peroxides in the microsomes or cytosol of mussel was evaluated by measurements of substances reacting with thiobarbituric acid (TBARs) according to the method described by Buege and Aust (1978). Malonedialdehyde (MDA), one of several products resulting from the peroxidation of lipids, is one of the main compounds reacting with this acid to give a pink colour. Different studies have shown that the TBAR concentration can be either increased, or reduced following exposure to certain contaminants, and depending on the product, its concentration, and the duration of exposure (Wenning and Di Giulio, 1988; Ribera et al., 1991; Labrot et al., 1996). In the present state of our understanding, TBARs have potential as a biomarker capable of revealing the exposure of compounds modifying the oxido-reduction equilibrium of the cell.

Acetylcholinesterase (AChE) activities have been used as a biomarker of neurotoxicity. These activities, which have been thoroughly studied in insects, are the subject of active research in molluscs. AChEs are especially sensitive to carbamate or organophosphorus pesticides but also to certain heavy metals. This activity has been measured on the post-mitochondrial fraction of whole mussels by the spectrophotometric technique described by Ellman et al. (1961).

The quality of information given by each of these parameters has been evaluated at each step by the calculation of the intra-site variability of results and of the corresponding coefficient of discrimination. The latter corresponds to the relation between the range of observations (highest value to lowest value) and the mean difference. The higher the coefficient, the more relevant the biomarker is considered.

2. RESULTS OF GICBEM CAMPAIGNS 1989 AND 1991 IN THE WESTERN MEDITERRANEAN SEA

2.1 The 1989 Campaigns

2.1.1 The July 1989 Campaign

Table 2 shows the principal results obtained during this campaign. Large differences were reported among the sites studied.

2.1.1.1 Concentrations of Contaminants

The presence of PAH in the sediments is often associated with the presence of PCB. This was the case especially for the seriously polluted sites of Planier, Carteau, and Lazaret. Similarly, PCB concentrations were generally low in the sites mildly polluted by PAH, such as Galeria and Scandola in Corsica, or the swimming station of Portofino in Italy. Nevertheless, the correlation between these two parameters was not statistically significant. It emerges from these analyses that the sites close to urban, port, and industrial centres such as Carteau (tanker terminal of Fos, Rhône estuary), Lazaret (Toulon harbour), Planier (Marseilles), Nervi (Genes), and Roquebrune (Monaco) are the most heavily polluted. The Corsican sites, on the other hand, are generally much less contaminated by PAH and PCB than the continental coastal sites. Because of this, they can be considered sites of reference with regard to these two contaminants.

2.1.1.2 Biomarkers

The early GICBEM missions seem to reveal the existence of a positive correlation between EH activity and PAH concentration in sediments (Ribera et al., 1989). Moreover, laboratory experiments have shown that EH activity is significantly induced in mussels exposed to BaP or to congeners of PCB (Michel et al., 1993). The measurements done *in situ* in July 1989 on a larger number of stations did not indicate a significant correlation between this activity and the other parameters measured, with the exception of a slight positive correlation with cytochrome P450 concentration. The link between these two parameters can be explained by the complementary character of corresponding enzymatic activities, EH following the oxidative process begun by the monooxygenases with cytochrome P450.

Many authors have demonstrated that, in fish, the rate of cytochrome P450, measured with techniques using anti-P4501A1 antibodies of fish (ELISA test), has significant correlation with PAH concentrations in sediments (Van Veld et al., 1990, Goksoyr, 1991; Collier et al., 1992). The absence of correlation observed during this campaign is due to the very wide range of cytochrome P450 contents measured at each site (intra-site variation greater than 50%), attributable possibly to the methods of measurement used. The research carried out especially by the laboratory of D. Livingstone (Plymouth Marine Laboratory) on the characterization of cytochrome P450 in mussel will eventually enable us to obtain specific anti-cytochrome P450 antibodies of mussel, which would greatly improve the precision of these measurements.

BPO activity is the more reliable measurement used to evaluate phase I activities of metabolism of xenobiotics in the mussel (Livingstone and Farrar, 1984). Earlier research has shown the existence of a positive correlation between BPO activity measured in the mussel and PAH concentration in sediments. Those results were confirmed during this campaign, in which samples were taken at a much larger number of stations (Fig. 2). Only the Golo site stands out, with a high activity and a rather wide intra-site variability (30%), even though it is apparently a mildly polluted site. This site seems to be a passage for cargo ships and oil tankers. Because of this, a limited amount of pollution by hydrocarbons cannot be excluded. Moreover, in relation to other Corsican sites, the sediments of Golo show a larger PAH concentration (Table 2). This result, among others, shows the advantage of using biological markers that respond to a source of contamination not detected by chemical measurement.

TBAR concentrations measured in mussel are positively correlated with PCB concentration in sediments ($r = 0.89$, $p < 0.01$; Fig. 3). An intra-site variation less than 25% has enabled us to obtain for this parameter a coefficient of discrimination of 4. These results will have to be confirmed in other missions. However, the results obtained for the two stations of Cortiou and Carteau are completely out of this range

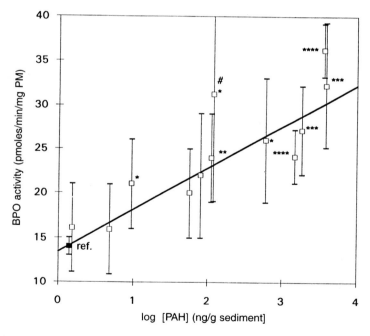

Fig. 2. Correlation between the log of PAH concentration in sediments and the BPO activity measured in microsomes of mussel sampled in different sites along the French and Italian Mediterranean coast in July 1989 ($n = 5$, $\pm$ mean difference, significantly different from the reference site (ref.) at *$p < 0.05$, **$p < 0.02$, ***$p < 0.01$, ****$p < 0.001$; $r = 0.92$; #: point of Golo, excluded from the correlation)

Table 2. PAH and PCB concentrations in sediments and BPO, EH, and AChE activities, and P450 and MDA concentrations measured in microsomes and the S9 of mussels taken from along the French and Italian coasts of the Mediterranean sea during the GICBEM campaign of July 1989

Stations	PAH ng/g sediment dry	PCB ng/g sediment dry	BPO pmoles/mn /mg PM	TBARs nmoles/mg PM	EH pmoles/mn /mg PM	P450 pmoles /mg PM	AChE nmoles/mn /mg P
Planier	1 564	9.6	24±3	3.3±0.2	5.0±2.6	46±23	3.6±1.3
Carteau	1 924	15	27±5	1.4±0.6	3.3±1.0	54±15	3.5±0.9
Cortiou	9.52	4.9	21±5	3.4±0.8	6.1±0.8	36±25	2.0±0.3
Lazaret	3 942	35.2	32±8	3.2±0.5	6.9±1.3	64±23	3.0±0.8
Porquerolles	111	12.8	24±5	2.6±0.4	6.1±1.7	105±33	3.4±0.7
Fourmigues	84	6.7	22±7	2.5±0.2	6.8±1.9	64±5	3.1±0.5
Roquebrune	641	7	26±7	2.6±0.5	5.2±2.9	37±14	2.9±0.7
Nervi	3 732	2.2	36±6	1.3±0.3	3.7±1.7	77±21	1.1±0.4
Portofino	59	2.5	20±5	1.1±0.2	6.2±0.8	42±16	3.4±0.7
Golo	107	4.3	32±11	2.0±0.2	8.3±1.5	97±15	—
Starezo	4.7	8.4	16±5	2.6±0.9	5.5±1.8	67±23	—
Scandola	1.4	3.5	14±1	1.7±0.5	5.4±1.8	52±32	6.1±1.5
Galéria	1.5	3.5	16±5	1.5±0.5	10.2±2.2	115±48	—

Coefficient of correlation between the different chemical and biochemical markers

	PAH	PCB	BPO	TBARs	EH	P450	AChE
PAH		0.44	**0.92**	0.4	-0.5	-0.19	-0.54
PCB			0.14	**0.89**	-0.13	-0.05	0.08
BPO				0.19	-0.46	-0.04	**-0.74**
TBARs					-0.13	-0.19	0.04
EH						**0.58**	0.13
P450							-0.11

BPO, benzo(a)pyrene oxidase; MDA, malonedialdehyde; EH, epoxide hydrolase; AChE, acetylcholinesterase; PM, microsomal proteins of whole mussel; P, proteins of S9 of whole mussel. n = 5, ± mean difference; the calculations of correlation were done with the log PAH and PCB concentrations; the values in bold face indicate significant correlations at $p < 0.05$.

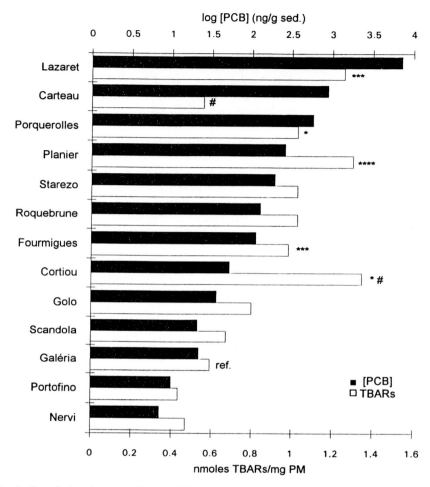

Fig. 3. Correlation between the log PCB concentration in sediments and TBAR concentration measured in microsomes of mussels sampled in different sites along the French and Italian Mediterranean coasts in July 1989 (n = 5, ± mean difference, significantly different from the reference site (ref.) at *p < 0.05, **p < 0.02, ***p < 0.01, ****p < 0.001; r = 0.92; #: points of Carteau and Cortiou, excluded from the correlation)

(Table 2, Fig. 3). We have no particular explanation for these results, except that there are often multiple pollutions observed for the same site, sometimes antagonistic or synergistic, according to the parameter considered. At Cortiou, where the treatment plant of the city of Marseilles discharges effluents, PAH and PCB contamination is relatively low. However, this does not exclude the possibility of a contamination by metals or pesticides such as lindane, dieldrin, or parathion, which are known to induce oxidative stress in the molluscs (Boryslawskyj et al., 1988; Wenning and Di Giulio, 1988). The Durance canal, which supplies Marseilles with water, may carry such pollutions resulting from agricultural activities in the basins toward which it drains. The very low AChE activities measured in the post-mitochondrial fractions of whole mussels sampled in this site may confirm this hypothesis (Fig. 4A, Table 2). In fact, certain pesticides, especially carbamates and organophosphorus pesticides, are inhibitors of AChE activity. Another plausible hypothesis lies in the existence of pollution by iron used in physicochemical treatment processes used in the treatment plant. Romeo et al. (1994) observed a high concentration of iron in tissues of fish taken from this zone. Iron can inhibit AChE activities and induce oxidative stress in exposed animals. A research programme on the impact of the Cortiou treatment plant, headed by the Agence de l'Eau Rhône-Méditerranée-Corse, is in the process of being set up. It will certainly give interesting information on the possible pollution carried by this discharge.

Table 2 and Figure 4 present the results of AChE activities measured on the post-mitochondrial fraction of whole mussels. The activities measured were relatively homogeneous in all the stations, around 3 nmoles/min/mg of proteins. Only three stations—Scandola, Cortiou, and Nervi—were out of range. If Scandola is considered as a reference site, the mussels sampled in all the other sites show a significantly inhibited AChE activity, indicating a generally higher level of pollution on the continental coast. However, the difficulty of interpretation lies in the fact that only one reference is available, this parameter not having been measured in the other Corsican stations. On the other hand, mussels sampled in the sites of Nervi (Genes) and Cortiou (Marseilles) show an AChE activity significantly lower than that measured in the other continental sites. In the case of Cortiou, we have already suggested certain hypotheses. In the case of Nervi, intense industrial and port activities generate multiple pollutions that can easily explain these results.

2.1.2 The November 1989 Campaign

Table 3 shows the results of the November 1989 campaign. Even though the trends observed during the preceding mission were found again, the number of stations studied (5) seemed to be insufficient for a statistically reliable analysis. It should be noted that the values obtained for the parameters linked to metabolism of xenobiotics (BPO, EH and P450) are generally lower than those measured in July. This phenomenon could be linked to seasonal variations in the general metabolism of this mollusc.

Table 3. PAH and PCB concentrations in sediments and BPO and EH activities, and P450 and MDA concentrations measured in microsomes and the S9 of mussels taken from along the French and Italian coasts of the Mediterranean during the GICBEM campaign of November 1989

Stations	PAH ng/g sediment dry	PCB ng/g sediment dry	BPO pmoles/mn /mg PM	TBARs nmoles/mg P	EH pmoles/mn /mg PM	P450 pmoles/mg PM
Lazaret	4745	35.2	20±5 ***	1.9±0.2 —	3.7±0.9 *	55±13 —
Fourmigues	24.1	6.7	13±1 **	1.3±0.4 —	3.1±1.0 **	46±18 —
Roquebrune	250	7	15±6 —	0.7±0.1 *	3.4±0.9 *	49±15 —
Galéria	2	3.5	13±2 —	1.4±0.4 —	6.7±2.6 —	69±11 —
Scandola (a)	1.5	3.5	11±1	1.4±0.5	4.8±0.5	50±25 —
Coefficient of correlation between the different chemical and biochemical markers						
PAH	**0.94**		**0.95**	0.2	−0.64	−0.26
PCB			**0.96**	0.5	−0.54	−0.18
BPO				0.42	−0.38	0.02
TBARs					0.18	0.28
EH						**0.89**

BPO, benzo(a)pyrene oxidase; MDA, malonedialdehyde; EH, epoxide hydrolase; AChE, acetylcholinesterase; PM, microsomal proteins of whole mussel; P, proteins of S9 of whole mussel. $n = 5$, ± mean difference; the calculations of correlation were done with the log PAH and PCB concentrations; the values in bold face indicate significant correlations at $p < 0.05$.
The values of PCB concentrations are those of the July campaign. (a) reference station for Student's t test, $*p < 0.05$; $**p < 0.02$; $***p < 0.01$; $****p < 0.001$.

2.2 The 1991 Campaign

The results of GICBEM's July 1991 sampling campaign on eight points of the coast of the northwestern Mediterranean are reported in Table 4.

2.2.1 Concentrations of Contaminants

The results obtained show that the Corsican site of Galeria (near Calvi) is by far the least polluted by PAH. At the other extreme, sites such as the Martin Cape and Roquebrune (Monaco) or the Planier lighthouse (Marseilles) are among the most contaminated. Generally, the continental coastal sites are much more polluted than the sites in Corsica, which remain relatively well preserved. It must be emphasized, however, that the site of Scandola (a natural reserve) showed a considerable increase in PAH concentration in sediments relative to results obtained in 1989.

2.2.2 Biomarkers

2.2.2.1 AChE Activity in Mussel

The results obtained for AChE activities measured in the post-mitochondrial fraction of whole mussels show levels of activities comparable to those of the July 1989 mission (Fig. 4). As in 1989, the values of activities measured in animals from Corsica (Galeria and Scandola) were significantly higher than those obtained for animals from the continental coast. The data from Planier, Porquerolles, and Fourmigues (animals found on rocks) were of the same order of magnitude as those of 1989. The lowest levels of activity were measured in Fourmigues, Roquebrune (animals found on buoys), and Spotorno. At Fourmigues, the activities were lower in animals found on a buoy than in animals found on rocks. Such differences were not observed for BaPH activities (Table 4). The differences in AChE activities could be due to factors independent of any pollution (different medium of life), or even to possible contaminants found on the buoys themselves (iron, preservative paints). At Spotorno, where PAH contamination is considerable, an increase in BaPH activities and inhibition in AChE activities was observed in mussels. On the other hand, no increase in phase I enzymes (BaPH and EROD) was observed in comber. To all appearances, the data available on this site were insufficient but seem to indicate the presence of contaminants other than PAH. This situation emphasizes the advantage of systematically implementing a more complete, multi-marker approach, on a biological rather than a chemical plan.

2.2.2.2 Phase I Enzymatic Activities in Mussel and Comber

Although measurement of BPO activity is a good biomarker for contamination by contaminants of the PAH type, it nevertheless remains

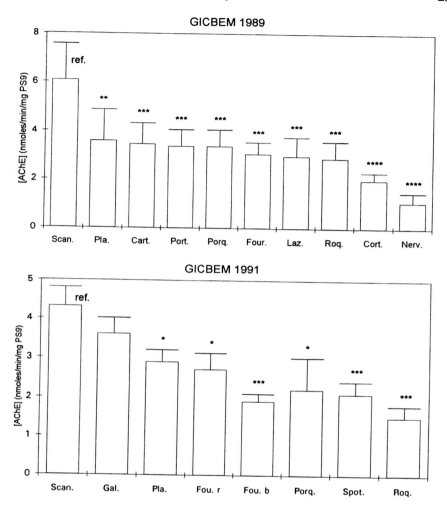

Fig. 4. AChE activities measured in the post-mitochondrial fraction of whole mussels sampled in different sites along the French and Italian Mediterranean coasts in 1989 and 1991 (n = 5, ± mean difference, significantly different from the reference site (ref.) at *p < 0.05, ** p < 0.02, ***p < 0.01, **** p < 0.001; r = 0.92; #: point of Golo, excluded from the correlation) (see Table 2 for the names of the stations)

a difficult technique to carry out (handling of radiolabelled molecules, extractions). For this reason, BaPH activity, which is simpler to measure, has a great advantage. BaPH activity has been tested on samples collected in the 1991 campaign of GICBEM (Table 4). The results of the correlation between PAH correlation in the sediment and BaPH activities measured in mussels and combers are presented in Figure 5. These two parameters are closely correlated, for comber as well as for mussel. At Scandola, all the phase I enzymatic activities (BaPH in comber and

Table 4. PAH and PCB concentrations in sediments and BPO, EH, and AChE activities measured in cellular fractions of mussel and comber taken from along the French and Italian coasts of the Mediterranean Sea during the GICBEM campaign of July 1991

Stations	PAH ng/g sediment dry	PCB ng/g sediment dry	BaPH mussel pmoles/mn/g hepato	EROD comber pmoles/mn/g liver	BaPH comber pmoles/mn/g liver	AChE mussel pmoles/mg PS9
Spotorno	1773	12	9.1±1.8	#546±91	#505±129	2.1±0.3
Cap Martin	1664	6.3	NM	1086±455	1371±384	NM
Planier	3692	109.2	8.2±0.9	756±196	1283±301	2.9±0.3
Roquebrune	2600	3.4	7.7±0.7	NM	NM	1.5±0.3
Scandola	120	1.8	7.5±1.4	895±383	910±454	4.3±0.5
Porquerolles	16	1.9	5.8±1.2	393±98	787±225	2.6±0.4
Fourmigues (r)	83	7.6	6.2±1.9	590±121	772±202	2.7±0.4
Fourmigues (b)	–	–	5.7±1.2	–	–	1.9±0.2
Galéria	18	2	3.8±0.3	478±218	467±116	3.6±0.4

Coefficient of correlation between the different chemical and biochemical markers

	PAH	PCB	BPH mussel	EROD comber	BaPH comber	AChE mussel
PAH		0.67	0.85	0.77	0.93	-0.43
PCB			0.52	0.25	0.63	-0.26
BPH mussel				0.77	0.95	-0.27
EROD comber					0.78	0.49
BaPH comber						-0.09

BaPH activity was measured on microsomal proteins of comber (*Serranus cabrilla*) liver or mussel (*Mytilus galloprovincialis*) hepatopancreas; AChE activity was measured in S9 of whole mussels (PS9 = proteins of S9). EROD activity was measured on microsomes of comber liver; the calculations of correlation were done with the log PAH and PCB concentrations; the values in bold face indicate significant correlations at $p < 0.05$.
#: value excluded from the correlation calculation.
At Fourmigues, the mussels were taken from rocks (r) and from a buoy (b). NM: not measured.

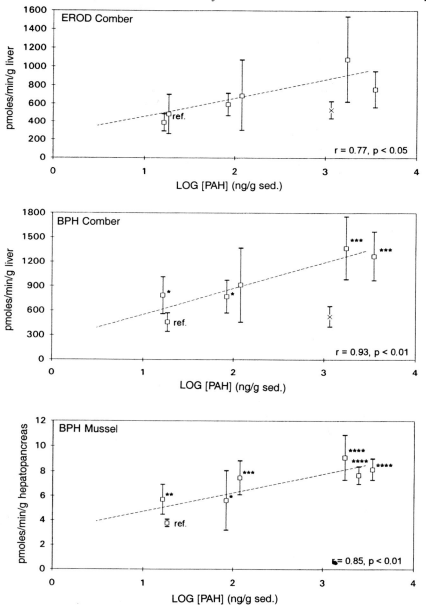

Fig. 5. Correlation between the log PAH concentration in sediments and the BaPH and EROD activity measured in microsomes of digestive glands of mussel (BaPH only) and livers of fish (*Serranus cabrilla*) sampled in different sites along the French and Italian Mediterranean coasts during the GICBEM campaign in 1991 (n = 5 for mussel and n = 4 for fish, ± mean difference, significantly different from the reference site (ref. = Galeria) at *p < 0.05, **p < 0.02, ***p < 0.01, ****p < 0.001; r = 0.92; x: point of Spotorno, excluded from the correlation).

mussel, EROD in comber) measured in 1991 were higher than in 1989 and preceding campaigns. During the earlier missions, in the relatively well-preserved Corsican sites, increases were observed in these biomarkers, which were attributed to limited pollution by hydrocarbons that came very likely from the degassing of oil tankers and that have so far not left traces of contamination in the sediments (Narbonne et al., 1991). The hypothesis of PAH pollution coming from degassing leading to a more marked contamination of this site (increase in PAH in the sediment) is unlikely because that would involve the entire coast, including the nearby site of Galeria. Another hypothesis on the origin of this pollution is that it arises from the particularly high traffic on this site of tourist boats, which are used continuously during the summer, and by yachts that are anchored here.

The BaPH activity in mussels has the best correlation overall with PAH concentration in sediments, in spite of levels much lower than in the comber. EROD activity in the latter species is much less correlated with the presence of PAH, in contrast to the general indications of authors working on fishes (Collier et al., 1992; Stegeman and Lech, 1992).

3. DISCUSSION

The sensitivity of biomarkers of pollution can only be evaluated in relation to chemical indicators of the pollution. PAH concentration in the sediments is frequently used as an indicator of contamination by hydrophobic organic compounds and it is often correlated with parameters relative to phase I enzymatic activities and DNA alteration. However, the use of such chemical markers of pollution is limited by three principal factors:

1. It is difficult to account for the spatial variability of this type of pollutant in certain sites, called open sites, as opposed to closed sites that are quite homogeneous. Raoux and Garrigues (1993) have designed a model describing the sedimentation mechanisms of PAH in the coastal environment. This model differentiates the 'closed systems (or sites)' from 'open systems (or sites)'. In a closed system, the deposit of suspended particles, which can be called direct sedimentation, is followed by an accumulation of this fine material in the sediments. In an open system, characterized by an uneven relief, there is, in addition to a direct sedimentation, a transfer of fine material from shallower sediments to deeper sediments. This causes a concentration of fine particles in the deeper and more enclosed areas, so that only the coarser sediments remain to interpret the field studies. In a site corresponding to a territory covered by a sedentary fish, the phenomena of transfer of fine particles, vectors of contaminants in the water column, can lead to an

accumulation of these contaminants in certain areas. These phenómena make it difficult to effectively assess the contamination of the site and necessitate the optimization of samples in order to minimize the variability of measurements.

2. In addition to the pollutants being studied, the sediments collect a multitude of other substances such as metals and organometallic or organohalogenic compounds. These compounds may have synergistic or antagonistic effects on the biological parameters studied. The mechanisms of combined effects of pollutants are complex and poorly understood.

3. The quality of sediments (coarse sand or fine silt) and the concentration of organic matter in the water (in dissolved or colloidal form) are also important factors of variability in the bioavailability of contaminants with respect to the sentinel organisms used and especially so for the filtering molluscs. Experiments carried out on organisms exposed to compounds of the PAH type have shown that the presence of dissolved organic matter (DOM) in a larger quantity reduces the absorption of PAH by animals. Narbonne et al. (1992) showed in a microcosm study on mussels that the regular return to suspension of sediments contaminated with [^{3}H] BaP does not favour the lasting absorption of [^{3}H] BaP by animals. In this experiment, the brief increase in the radioactivity measured in whole mussel after the return of the sediment to suspension suggests a passage of the fraction of [^{3}H] BaP associated with particles through the digestive tract without penetration in the organism. Those particles are thus directly excreted.

As in the case of chemical markers, the results have shown how difficult but necessary it is to control at least the variability of biochemical parameters measured, for activities of metabolization as well as cholinesterase activities. Despite a range of values, correlations can be established with the presence of pollutants in the medium. However, a greater refinement must be achieved if one wishes to obtain a more precise estimation of degrees of pollution and of the threshold at which an environment can be considered polluted. These fluctuations arise from the genetic heterogeneity of individuals, but also from various biotic, abiotic, and nutritional factors, such as characteristics of samples and measurement techniques.

4. TRANSPLANTATION OF ORGANISMS INTO THE ARCACHON BASIN

Without quite dropping sampling campaigns such as those carried out thus far, it seemed advantageous to be able to work on animals transplanted into the sites under study.

For this purpose, we conducted a preliminary experiment in transplantation in the Arcachon basin (Gironde, France). Mussels taken from a non-polluted site of the basin (Petit Nice) were transplanted around the Arcachon port and left in place for 10 days. The digestive glands of molluscs were then removed, in order to measure five biomarkers of pollution on cellular fractions of that tissue and to compare these values to those obtained on animals taken from the collection site. Biomarkers of phase I enzymatic activities (activation of BaP in mutagenic metabolites, BaPH), phase II enzymatic activities (glutathion S-transferases (GST)), oxidative stress (TBARs), and genotoxicity (DNA adducts) were measured in parallel with the PAH concentration in sediments (Table 5).

The sediments around the port of Arcachon were much more contaminated by PAH than those at the site from which the mussels were collected. The transplanted animals showed no sign of loss of girth or weight or reduction in concentration of proteins in the subcellular fraction. The mortality of the animals after transplantation was lower than 5%. The parameters measured showed that, overall, the bivalves reacted strongly to this pollution with, especially, a significant increase in GST activity and above all a growth in TBARs in the cellular fractions of the animals transplanted, which indicates a considerable oxidative stress. Simultaneously, an increase was reported in the potential for BaP activation in mutagenic products by the cellular fractions and the appearance of DNA adducts. On the other hand, the BaPH activity was not modified. This activity catalyses the metabolization of BaP into

Table 5. Measurement of certain biomarkers in cellular fractions of the digestive gland of mussel (*Mytilus galloprovincialis*) transplanted for 10 days around the port of Arcachon

	Petit Nice mussels	Transplanted mussels
PAH ng/g sediment dry	60.3 ± 6.0	5604 ± 560***
DNA adducts no. adducts/10^8 nucleotides	not detected	1.34 ± 0.89
BaP activation Ames test/TA98 revertants	74 ± 12	118 ± 8**
GST activity μmol/min/g tissue	6.4 ± 0.3	7.4 ± 0.2***
[TBARs] cytosol nmoles/g tissue	342.7 ± 62.4	657.7 ± 81.3***
[TBARs] microsomes nmoles/g tissue	18.8 ± 1.1	28.8 ± 0.8***
BaPH activity pmoles $_3$OHBO/min/g tissue	10.2 ± 3.0	10.8 ± 2.8

$n = 4 \pm SD$; significantly different from the control: *$p < 0.05$, **$p < 0.01$, ***$p < 0.001$. $n = 10$ for the measurement of adducts; no sample from Petit Nice showed adducts, as opposed to 5 for the transplanted mussels.

phenols of BaP. The absence of BaPH induction and the results of the measurement of BaP activation indicate a quantitative and/or qualitative change in the BaP metabolism, which would lead to the formation of metabolites other than phenols of BaP (diols and quinones of BaP).

These results confirm, if confirmation were necessary, the advantage of a multi-marker approach for this type of study.

5. CONCLUSIONS

Chemistry and biology are fundamentally inseparable in the understanding of processes of dispersal and transport of contaminants for the purpose of studying their toxicity and applying this knowledge to the improvement of biomarkers of pollution.

The application of biomarkers as tools of monitoring pollution has evolved. In the present state of our understanding, this approach must be coupled with a systematic detection of major contaminants such as the hydrocarbons (10–15 significant PAHs), the organochlorines (20–22 major PCB, chlorine pesticides), heavy metals, organometallic compounds (especially the organostannic compounds), and pesticides. In the near future, we can envisage an approach chiefly based on biomarkers as they become better controlled and their use becomes somewhat 'validated'. Thus, the use of biomarkers, which is confined for now to the domain of research, will enter a phase of routine use in large-scale monitoring programmes, owing especially to the improvement of simplified and automated measurement techniques. Chemical measurements, often very expensive, play a less systematic role, while biomarkers indicate the presence of an unidentified pollution. Such indications will become more precise as biomarkers are specific for a type or group of contaminants.

The results that we have obtained during our different study campaigns have shown how difficult, but nevertheless essential, it is to better control the variability of chemical or biochemical parameters measured. Despite the range of values obtained, correlations have been established with the presence of pollutants in the environment. These results, very encouraging and already containing significant information, must motivate us to aim for greater precision in our measurements. The effort will enable us to better define estimations of degrees of pollution and the threshold at which an environment can be considered polluted.

Finally, the use of biomarkers has advantages and significance only within a multi-marker approach. Only such an approach can provide valuable expertise in the context of mediums having multiple sources of pollution. The choice of biomarkers must take into account major mechanisms of response to the presence of a toxic compound. This includes biomarkers linked to the biodegradation of a toxin, its genotoxicity, its neurotoxicity, and its immunotoxicity, to cite only the most important aspects.

At present, GICBEM is extending the understanding acquired in the marine environment to freshwater and terrestrial environments, using, respectively, the Asian clam *Corbicula fluminea* (a bivalve mollusc) and the earthworm as sentinel organisms.

ACKNOWLEDGEMENTS

First of all, we wish to thank the persons who participated in this work: Mrs. Pierrette Cassand (toxicology laboratory, University of Bordeaux 1), who conducted the mutagenesis tests; Prof. Arnoux and Mr. J.L. Monod of the faculty of pharmacy of Marseilles, who did the PCB measurements for the GICBEM mission in 1989; Prof. A. Leskowicz and Mr. Y. Gross, who gave us the benefit of their expertise on measurement of DNA adducts; and Mr. E. D'jomo, who completed the work on the kinetics of PAH absorption.

We thank the following organizations, groups, and societies that supported our research: the European Union; the IFREMER, the Ministry of Environment; l'Agence de l'eau Adour/Garonne; the Aquitaine Region; the CEA/CESTA (Bordeaux); the Elf Aquitaine; and the ADEME.

REFERENCES

Bayne B.L. (1989). Measuring the biological effects of pollution: the Mussel Watch approach. *Wat. Sci. Tech.*, 21: 1089–1100.

Boryslawskyj M., Garrod A.C. and Pearson J.T. (1988). Elevation of glutathione S-transferase activity as a stress response to organochlorine compounds in the freshwater mussel, *Sphaerium corneum*. *Mar. Environ. Res.*, 24:101–104.

Buege J.A. and Aust S.D. (1978). Microsomal lipid peroxidation. *Meth. Enzymol.*, 50:302–310.

Burke M.D. and Mayer R.T. (1974). Ethoxyresorufin: direct fluorimetric assay of a microsomal O-dealkylation which is preferentially inducible by 3-methylcholanthrene. *Drug Metab. Dispos.*, 2:583–588.

Chevreuil M and Granier L. (1992). Les PCB: des polluants difficiles à éliminer. *La Recherche*, 23:484–486.

Claisse D. (1989). Chemical contamination of French coasts. The results of a ten years mussel watch. *Mar. Pollut. Bull.*, 20:523–528.

Collier T.K., Connor S.D., Eberhart B.T.L. et al. (1992). Using cytochrome P450 to monitor the aquatic environment: initial results from regional and national surveys. *Mar. Environ. Res.*, 34:195–199.

Dehnen W., Tomingas R. and Roos J. (1973). A modified method for the assay of benzo(a)pyrene hydroxylase. *Anal. Biochem.*, 53:373–383.

Ellman G.L., Courtney K.D., Andres V. and Featherstone R.M. (1961). A new and rapid colorimetric determination of acetylcholinesterase activity. *Biochem. Pharmacol.,* 7:88–95.

Fischer W., Bauchot M.L. and Schneider M. (1987). *Fiches FAO d'Identification des Espèces pour les Besoins de la Pêche (Révision 1). Méditerranée et Mer Noire. Zone de Pêche 37. Vol. I. Végétaux et Invertébrés.* FAO, Rome.

Garrigues Ph., Raoux C., Lemaire P. et al. (1990). *In situ* correlations between polycyclic aromatic hydrocarbons (PAH) and PAH metabolizing system activities in mussels and fish in the Mediterranean Sea: preliminary results. *Int. J. Environ. Anal. Chem.,* 38:379–387.

Garrigues Ph., Narbonne J.F., Lafaurie M. et al. (1993). Banking of environmental samples for short-term biochemical and chemical monitoring of organic contamination in coastal marine environments—the GICBEM experience (1986–1990). *Sci. Total Environ.,* 139/140:225–236.

Goksøyr A. (1991). A semi-quantitative cytochrome P-450 IAI-ELISA: a simple method for studying the monooxygenase induction response in environmental monitoring and ecotoxicological testing of fish. *Sci. Total Environ.,* 101:253–261.

Labrot F., Ribera D., Saint Denis M. and Narbonne J.F. (1996). *In vitro* and *in vivo* studies of potential biomarkers of lead and uranium contamination: lipid peroxidation, acetylcholinesterase, catalase and glutathione peroxidase activities in three nonmammalian species. *Biomarkers,* 1:21–28.

Lafaurie M., Mathieu A., Salaun J.P. and Narbonne J.F. (1992). Biochemical markers in pollution assessment: field studies along the north of the Mediterranean Sea. In: *The Report of the FAO/IOC/UNEP Training Workshop on the Techniques for Monitoring Biological Effects of Pollutants in Marine Organisms,* Nice, France, 14–25 September 1992. *Mediterranean Action Plan Technical Series,* pp. 49–70.

Livingstone D.R. and Farrar S.V. (1984). Tissue and subcellular distribution of enzyme activities of mixed-function-oxygenase and benzo(a)pyrene metabolism in the common mussel *Mytilus edulis. Sci. Total Environ.,* 39:209–325.

Livingstone D.R., Martinez P.G., Michel X.R. et al. (1990). Oxyradical production as a pollution mediated mechanism of toxicity in the common mussel, *Mytilus edulis* L. and other molluscs. *Funct. Ecol.,* 4:415–424.

Michel X.R., Salaun J.P. Galgani F and Narbonne J.F. (1994). Benzo(a)pyrene hydroxylase activity in the marine mussel, *Mytilus galloprovincialis:* a potent marker of contamination by polycyclic aromatic hydrocarbons. *Mar. Environ. Res.,* 38:257–273.

Michel X.R., Suteau P., Robertson L.W. and Narbonne J.F. (1993). Effects of benzo(a)pyrene, 3,3',4,4'-tetrachlorobiphenyl and 2,2',4,4',5,5'-hexachlorobiphenyl on the xenobiotic-metabolizing enzymes in the mussel (*Mytilus galloprovincialis*). *Aquat. Toxicol.,* 27:335–344.

Monod J.L. and Arnoux A. (1978). Étude des composés organochlorés (PCB-DDT) dans l'environnement marin de l'île des Embiez (Var, France). *4ème Journées Etud. Pollutions, Antalya,* CIESM, pp. 147–148.

Narbonne J.F., Garrigues Ph, Ribera D. et al. (1991). Mixed-function oxygenase enzymes as tools for pollution monitoring: field studies on the French coast of the Mediterranean Sea. *Comp. Biochem. Physiol.,* 100C:37–42.

Narbonne J.F., Ribera D., Garrigues Ph. et al. (1992). Different pathways for the uptake of benzo(a)pyrene adsorbed to sediment by the mussel *Mytilus galloprovincialis. Bull. Environ. Contam. Toxicol.,* 49:150–156.

Narbonne J.F. and Michel X.R. (1997). Systèmes de biotransformation chez les mollusques aquatiques. In: *Biomarqueurs en Écotoxicologie: Aspects fondamentaux.* Lagadic L., Caquet Th., Amiard J.C. and Ramade F. (eds.). Masson, Paris, pp. 11–32.

Raoux C.Y. and Garrigues Ph. (1993). Modelling of the mechanisms of PAH contamination of marine coastal sediments from the Mediterranean Sea. In: *Synthesis, properties, analytical measurements, occurrence and biological effects.* Garrigues Ph. and Lamotte M. (eds.). Pol. Arom. Comp. (sup. vol. 3), Gordon & Breach Science Publishers, pp. 443–450.

Raoux C., Lemaire P., Mathieu A. et al. (1989). Bioprotective system activities in *Serranus scriba* from the Mediterranean Sea: relation between pollutant content and enzymatic activities. *Oceanis*, 15:623–627.

Ribera, D., Narbonne J.F., Suteau Ph. et al. (1989). Activities of the PAH metabolizing system in the mussel as a biochemical indicator for pollution: French coasts of the Mediteranean sea. *Oceanis*, 15:443–449.

Ribera D., Narbonne J.F., Michel X.R. et al. (1991). Response of antioxidants and lipid peroxidation in mussels to oxidative damage exposure. *Comp. Biochem. Physiol.*, 100C:177–181.

Romeo M., Mathieu A., Gnassia-Barelli A. et al., (1994). Heavy metal content and bio-transformation enzymes in two fish species from the NW Mediterranean. *Mar. Ecol. Prog. Ser.*, 107:15–22.

Soclo H.H., Garrigues P. and Ewald M. (1986). Analyse quantitative des hydrocarbures aromatiques polycycliques dans les sédiments marines récents par chromatographie en phase liquide et détection spectrofluorométrique. *Analysis*, 14:344–350.

Stegeman J.J. and Lech J.J. (1991). Cytochrome P-450 monooxygenase systems in aquatic species—carcinogen metabolism and biomarkers for carcinogen and pollutant exposure. *Environ. Health Perspect.*, 90:101–109.

Van Veld P.A., Westbrook D.J., Woodin B.R. et al. (1990). Induced cytochrome P-450 in intestine and liver of spot *Leiostomus xanthurus* from a polycyclic aromatic hydrocarbon contaminated environment. *Aquat. Toxicol.*, 17:119–132.

Wenning R.J. and Di Giulio R.T. (1988). The effects of paraquat on microsomal oxygen reduction and antioxidant defenses in ribbed mussels (*Geukensia demissa*) and wedge clam (*Rangia cuneata*). *Mar. Environ. Res.*, 24: 301–305.

3

Application of EROD in Marine Fish in a Multidisciplinary Surveillance Programme in the North Sea

Th. Burgeot and F. Galgani

INTRODUCTION

The use of biomarkers in environmental monitoring has developed considerably in the past few years because of two major advances (Koeman et al., 1993). The first corresponds to field experience in the use of biochemical techniques and advances in molecular biology, which enable us to understand mechanisms involved in interactions between chemicals and organisms. The second is the refinement and improvement of new techniques allowing the detection and identification of biological effects of environmental pollutants.

However, detection and quantification of biological effects of pollutants in the context of environmental monitoring remains a complex task because it implies the integration of a large number of variables (such as non-specificity of responses, synergy and antagonism between pollutants, adaptability of species, latency of some responses). The use of biomarkers as tools for prediction of ecotoxicological risks must therefore be integrated in an overall approach of environmental monitoring combining several disciplines (including biology, biogeochemistry, biostatistics, and physical geography).

The multidisciplinary programme organized in the North Sea was an unprecedented experiment in Europe concerning the use of biomarkers, with the objective of monitoring biological effects on a site characterized by intense industrial activity. It was the first time an

enzymatic biomarker (EROD activity) was used in the context of biological effects monitoring as part of an international programme combining varied disciplines such as physical oceanography, chemistry, and
marine biology. The use of a biomarker in an international monitoring
context constitutes an interesting and useful, even indispensable, exercise for testing what was still a very recent tool.

The biological part of this programme, which is explained in greater
detail in this chapter, was based on four major points: the frequency of
various diseases in fish, bioassays on embryos of bivalves, studies of the
benthic community, and measurement of EROD activity in fish. The
study of the benthic community on such a large scale proved to be
poorly effective but remained useful in the case of highly contaminated
sites. However, biomarkers also have limits (e.g., specificity to some
groups of contaminants, seasonal variations, poorly understood mechanisms of interactions, absence of observation of dose-response relations
in the field) and cannot be used alone (Livingstone, 1993). The bioassays
on oyster (*Crassostrea gigas*) embryos seemed to be well adapted to analysis of water and sediment quality in the contaminated zones, but the
absence of consensus on the choice of an analytical methodology made
it difficult to interpret them (ICES, 1990). The survey of diseases in fish
was also criticized, because of the inherent variability of a study conducted in such a vast zone (ICES, 1989). The data on pathologies presented by fish will, however, be discussed in this chapter, as a useful
element in the interpretation of measurement of EROD activities.

Such use of a biomarker was a novel experiment. Although it was
acknowledged during working group meetings, in 1986 at Oslo and in
1990 at Bremerhaven, that biological effects of pollutants can only be
diagnosed by a multi-marker approach (North Sea Task Force, 1993),
the scientists involved chose EROD because it is a biomarker with which
they have the largest experience with regard to application in marine
environments. The choice of EROD was made on the basis of an
intercalibration exercise on a pollution gradient in the Elbe estuary
(Stebbing and Dethlefsen, 1992) and of research carried out on marine
organisms in laboratories and in the natural environment over about
thirty years (Payne, 1976; Stegeman, 1981; Stegeman et al., 1986; Addison,
1988). Various methods of measurement on the liver of dab (*Limanda
limanda*) were compared. Campaigns were organized between 1990 and
1992 in the North Sea for collection of this sentinel species. As the French
representative, IFREMER participated in several sampling campaigns
and in various exercises to compare methods. The numerous environmental factors affecting EROD activity and the complexity of mixtures
of chemical pollutants present in the studied stations were identified as
the major parameters that could hamper interpretation of results. Other
campaigns were organized, over a period of two years, to collect dabs

along the coast of Belgium. The objective was to identify sources of intra-annual variations in EROD activity in polluted and unpolluted sites (Cooreman et al., 1993). The physical, chemical, and biological information obtained enabled researchers to perform a preliminary assessment of the health of the North Sea. In addition to information collected on physical geography, comparison of chemical data seemed very useful in interpreting variations in a biomarker of exposure such as EROD. This study showed the advantages and limits of the use of this biomarker in a marine zone of 750,000 km² area.

In the first part of this chapter we summarize the objectives of the monitoring programme in the North Sea and provide a geographic description of the site in order to justify the need for and advantage of multidisciplinary information in diagnosing biological effects of pollutants. The specific difficulties related to the use of a biomarker of exposure such as EROD in the context of such a programme will then be discussed, based on the results obtained on contamination of biotopes and organisms. Finally, we present the work undertaken by IFREMER on the French coasts. A significant part of this chapter is inspired by a report produced in 1993 on the state of health of the North Sea (North Sea Task Force, 1993).

1. THE NORTH SEA MONITORING PROGRAMME

In 1987, a ministerial declaration at the Second International Conference for the Protection of the North Sea defined gaps in scientific knowledge about the North Sea environment. This knowledge was considered a prerequisite to strategic decisions relating to environmental protection as well as evaluation of the efficacy of measures already taken (North Sea Task Force, 1993).

The ministers concerned felt that the best way to tackle the problem posed by the gaps in knowledge about the quality of the North Sea environment was to create a task force that would undertake research leading, within a reasonable period of time, to a reliable and complete identification of the structure of circulation, sources, and dispersion pathways of pollutants, ecological conditions, and effects of human activities in the North Sea.

To carry out this mission, the North Sea Task Force was created under the common responsibility of the Conseil International pour l'Exploration de la Mer (CIEM) and the Commissions of Oslo and Paris (OSPARCOM). Representatives of the European Commission, acting under the joint patronage of the OSPARCOM and CIEM, as well as eight coastal States of the North Sea (around the English Channel and the Skagerrak: Belgium, Denmark, France, Germany, the Netherlands,

Norway, Sweden, and the United Kingdom) were designated members of this group. The North Sea Task Force started its work at the end of 1988. Its chief mission was to publish an assessment of the condition of the North Sea for 1993.

1.1 Setting up the Monitoring Programme

A number of important stages had to be considered in setting up a programme to monitor the quality of a marine environment (Wolfe, 1992), the objectives being to define the nature of pollutants, their spatio-temporal distribution, their bioavailability, and their effects on the ecosystem. For a monitoring trial of long-term biological effects within a general programme involving the North Sea, three major phases were recommended (Fig. 1):

1. An initial phase in which the objectives, study sites, and parameters of the study are defined.

2. A training stage including the definition of analytical methods, sampling frequency, and sentinel species, field validation and comparison of the chosen methods, and integration of biomarkers in multidisciplinary monitoring programmes.

3. A final phase that can progressively lead to a routine surveillance system. For this purpose, the components of phase 2 must be refined, the analysis and sampling protocols must be standardized, and the greatest effort must be taken to standardize interpretation of results, which depends chiefly on experience acquired in the field and on accumulation of data.

In the face of various constraints, and especially the cost of a long-term monitoring programme, it was necessary to attempt a preliminary assessment of detection of biological effects on a pilot site on which chemical pollutants had been found. This exercise was done on a pollution gradient in German Bight (Stebbing and Dethlefsen, 1992). Then, it was necessary to define the constraints of monitoring and optimize sampling from sites on which significant biological effects due to chemical pollutants had been observed (Wolfe et al., 1987).

1.2 Definition of Study Sites

The North Sea, including the English Channel and Skagerrak, was considered a unique geographic zone, made up of characteristic ecological regions. The differences observed among these regions are due to variations in latitude and depth of water, as well as their geographic location in relation to sources of pollution and the extent of human activity. It was therefore considered appropriate to divide the North Sea into

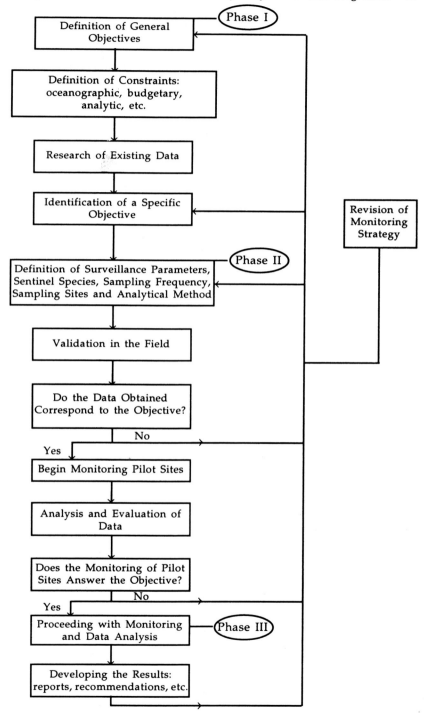

Fig. 1. Scheme for setting up a monitoring programme (modified from McDonald and Smart, 1993)

several sub-regions (Fig. 2) and to compile an evaluation report on each of them. On the basis of the sub-regional evaluation reports, a preliminary analysis of the health of the North Sea was performed. It constitutes a practical overall evaluation and a consensus of opinions on the present state of the North Sea, in physical, chemical, and biological terms.

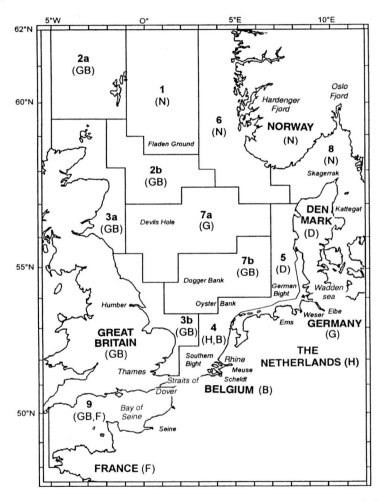

Fig. 2. Sub-regions of the North Sea monitoring programme (modified from North Sea Task Force, 1993)

2. GENERAL DESCRIPTION OF THE NORTH SEA

The North Sea is located on the continental plateau of northeastern Europe. It opens on the Atlantic Sea to the north, the Baltic Sea to the east, and the English Channel to the south. It extends over a surface of 750,000 km^2 and has a volume of 94,000 km^3. It is characterized by shallow beds (30 to 200 m), except for the steep drops bordering the Norwegian coasts (700 m). Sand covers most of the beds, while gravel zones are chiefly found in the English Channel and along the coast. Mud deposits are found in the estuaries and the deepest areas: Devils Hole, Dogger Bank, Oyster Bank, Fladen Ground, Skagerrak, and the Norwegian Deep (Fig. 2). The water temperature is affected by currents from the Atlantic Ocean, winds of highly variable direction and speed, and significant precipitation (425 mm/year).The most significant variations in annual temperature are found in the German Bight (0–21°C). Differences in temperature and salinity among the various sub-regions of the North Sea are greater than those found in the same latitudes in the North Atlantic. The circulation of water masses in relation to the topographic nature of the beds is also affected by the currents from the Atlantic Ocean (Fig. 3) and the English Channel.

The hydrographic basin of the North Sea has a surface of 850,000 km^2. Rivers such as the Elbe, Weser, Rhine, Meuse, Scheldt, Seine, Thames, and Humber, the basins of which are among the most industrialized in Europe, discharge 300 km^3 of fresh water into it each year. This hydrographic basin, also characterized by a dense population and intensive agriculture, constitutes a significant source of man-made contamination. In the estuaries the biomass is abundant, with significant nurseries for fishes, and fishing activities are especially developed. These are also areas in which numerous contaminants have accumulated.

Finally, since the late 1960s, oil and gas drilling constitutes the most important economic activity of the North Sea. Around 300 off-shore oil platforms produce 92.5×10^9 m^3 of gas and 183×10^6 tonnes of petrol a year, which are transported across a network of 10,000 km of pipelines (Fig. 4).

3. POLYCYCLIC AROMATIC HYDROCARBONS (PAH) AND POLYCHLOROBIPHENYLS (PCB) IN THE NORTH SEA

Contaminants are introduced into the North Sea from various sources. In addition to natural and accidental contaminants, there are those due to oil drilling and transport by atmosphere and rivers as well as

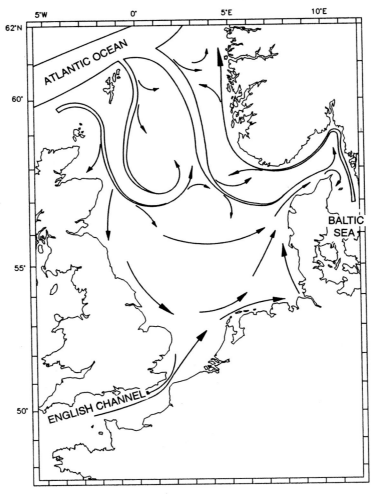

Fig. 3. General circulation of water masses in the North Sea (modified from Turrell et al., 1992)

discharge and incineration of industrial and domestic wastes. The equivalent of 130 kg/year of PCB is thus deposited from the atmosphere in sub-region 6 of the North Sea, and around 10,000 tonnes of wastes were incinerated there from 1981 to 1988.

Even though there is a considerable amount of chemical data on organic contaminants in the North Sea, a detailed interpretation on such a vast zone remains difficult. Organic pollutants have been studied in water, sediments, and organisms. Such a study, the first over such a vast area, necessarily involves several participants. In some cases, lack of information on quality control procedures relating to analytic protocols has hampered the comparison of results. Nevertheless, the data enabled

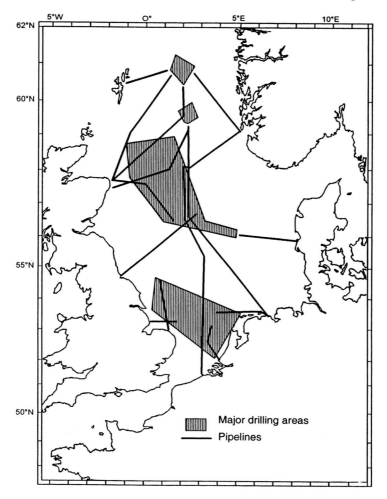

Fig. 4. Areas of gas and oil drilling and major pipelines in 1991 (modified from North Sea Task Force, 1993)

researchers to draw significant conclusions on differences in spatial distribution of organic contaminants such as PCB and PAH.

3.1 Measurement of PCB and PAH in Water

Few data on traces of organic contaminants in water have proved to be useful. These lipophilic compounds are not very soluble in water and are more generally associated with the suspended particle phase. The amount of PCB detected in the dissolved phase corresponds to the number of congeners on which research is recommended by the

International Union of Pure and Applied Chemistry (IUPAC) (i.e., those that must be studied on a priority basis). These are chlorobiphenyls (CB) 28, 52, 101, 118, 138, 153, and 180. Concentrations measured in surface waters on the coasts of France and Belgium (16 pg/l) and generally along the German coasts (29 pg/l) are higher than those observed in the North Atlantic, in the Bay of Biscay (3.2–12 pg/l). Transects conducted in the English Channel and in the central part of the western North Sea showed concentrations lower (6.3–11 pg/l) than those measured in Scotland (> 33 pg/l). However, the highest concentrations were found in the German Bight (> 178 pg/l), with a very clear gradient in the northwest direction. Significant concentrations (83 pg/l) were also observed on a transect crossing the oil drilling zones in the centre of the North Sea.

The PAH concentration in water was determined in the English Channel in the Straits of Dover. Samples were taken 1 m below the water surface and 3 m above the bed. The results obtained suggested hydrocarbon pollution in relation to maritime traffic. Measurements in the German Bight, on the other hand, showed profiles of pollution characterized by incomplete incineration of domestic wastes.

3.2 Measurement of PCB and PAH in Sediments

Only the data obtained for six out of seven congeners of PCB recommended by the IUPAC were retained (CB 28, 52, 101, 138, 153, and 180) because of lack of coordination of analysis protocols. A statistical study showed good correlation of CB 153 with all the other congeners, except for CB 52. CB 153 was therefore chosen as the CB representative of the group of PCBs and CB 52 was considered separately.

Concentrations of CB 153 varied from 0.01 to 11 µg/kg dry weight of sediment, with a mean of 0.57 µg/kg. The highest values were observed in the coastal region in the southeast of the North Sea, in the Scheldt estuary, on the Belgian coasts, and on the German coasts, including the estuaries of Ems, Weser, and Elbe. High concentrations were also measured in areas of deep sediments of Skagerrak, in the Norwegian Deep as well as north of Scotland.

PCB show a spatial distribution similar to that observed for metals in sediments. PCB is peculiar in that the highest concentrations occur in areas in which sediments are deposited, especially in areas characterized by fine sediments and high concentration of organic matter (Devils Hole, Oyster Bank, German Bight, and Norwegian Deep).

A statistical study of PAH concentrations in sediments led to the selection of some compounds considered representative of PAH: benzo(a)pyrene (BaP), a compound of high molecular weight produced mainly by combustion, and phenanthrene and naphthalene, compounds of low molecular weight, produced by combustion or originating from oil drilling activities. Alkyls C1 derived from naphthalene were also studied as geochemical markers of contamination originating from oil drilling. Except for the Scheldt estuary, the samples were taken from the central and northern parts of the North Sea, and in the Skagerrak and Kattegat.

The BaP concentrations ranged from 0.0006 to 0.24 µg/kg dry weight of sediment. The highest concentrations were measured in the Skagerrak and Kattegat as well as in the fjords of Oslo and Hardenger. Standardization of results in relation to organic carbon from sediments showed that contamination in sub-region 7a, in the centre of the North Sea, was higher than in Skagerrak and Kattegat. The highest concentrations of phenanthrene were also observed in Skagerrak, the Oslo fjord, and Kattegat. Concentrations ranged from 0.001 to 0.15 mg/kg dry weight of sediment. Naphthalene showed a distribution similar to those of phenanthrene and BaP, before and after standardization of results in relation to total organic content in sediments, with concentrations between 0.001 and 0.055 mg/kg dry weight of sediment. The distribution profile of alkyl derivatives of naphthalene was also identical to that of BaP.

In conclusion, PCB contamination seems particularly significant in the large estuaries that are under a high industrial pressure. The PAH profiles are a result of pollution related to oil drilling activities, particularly intensive in the central and northern areas of the North Sea (Fig. 4). Moreover, the measurement of total hydrocarbons enabled us to observe that off-shore oil platforms are a major source of pollutants. Very high quantities of hydrocarbons were measured near off-shore oil platforms and in various oil wells. Concentrations 100 to 1000 times as high as those measured in unpolluted zones were seen in a radius of 0.5 to 1 km around the off-shore oil platforms. This radius may extend up to 5 km, or even 7 km for certain rigs in Norway.

3.3 PCB and PAH Distribution in Marine Organisms

Temporal variations in concentrations of contaminants were analysed in the tissues of fish and invertebrates. The species studied were dab (*Limanda limanda*), cod (*Gadus morhua*), whiting (*Merlangius merlangus*), flounder (*Platichthys flesus*), plaice (*Pleuronectes platessa*) (Fig. 5), and mussel (*Mytilus edulis*).

A statistical model was developed to optimize estimation of variations in contaminant content, taking into account biological factors (e.g., sex, age, size, period of reproduction) that could affect the accumulation and elimination of contaminants. The results, expressed in mg/kg fresh weight, allowed us to indicate some significant trends.

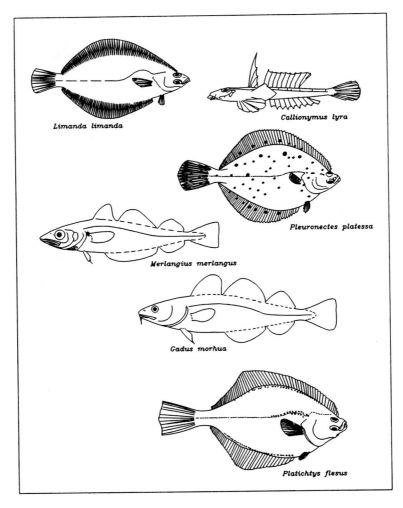

Fig. 5. Fish species studied in the English Channel and North Sea: dab (*Limanda limanda*), plaice (*Pleuronectes platessa*), dragonet (*Callionymus lyra*), cod (*Gadus morhua*), whiting (*Merlangius merlangus*), and flounder (*Platichthys flesus*)

Among the group of organisms studied, only some organic compounds could be used. PAH were not retained, but PCB allowed us to obtain useful information. Unfortunately, for a single species, the analyses done on PCB did not cover the entire North Sea. Moreover, some countries involved did not follow up with a congener-by-congener analysis. Nevertheless, the results selected from seven congeners analysed in cod liver (CB 28, 52, 101, 118, 138, 153, and 180) enabled us to indicate

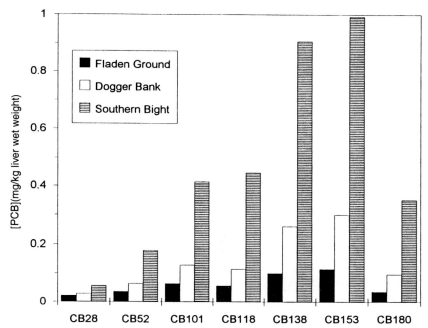

Fig. 6. Concentrations of seven PCB congeners analysed in cod liver in the North Sea (1979–1987). The results show an increasing contamination gradient from north to south, from Fladen Ground to the Southern Bight, passing through Dogger Bank (De Boer, 1988).

the presence of an increasing contamination gradient from north to south (Fig. 6). Contamination of cod living in the southern zone, generally in the coasts of the Netherlands, is linked to transport of domestic and industrial pollutants. This result conforms to the presence of PCB in sediments of this region.

Total PCB concentrations measured in dab, cod, plaice, and flounder varied from 0.01 to 1.3 mg/kg fresh weight. In mussel, they varied from 1.5 to 78 μg/kg fresh weight (Fig. 7).

The highest concentrations were found in the Scheldt and Ems estuaries as well as the Wadden Sea, which receives flows from the Ems, Weser, and Elbe. The total PCB measured in mussel also showed the existence of high levels of contamination in the Thames estuary and the Bay of Seine and in several sites along the Belgian coast, as well as along the French coasts of the English Channel (Duinker et al., 1988).

Finally, the study of variations in contamination levels from 1979 to 1991 on the German coasts enabled us to draw a more encouraging trend with a reduction in rates of PCB contamination in cod.

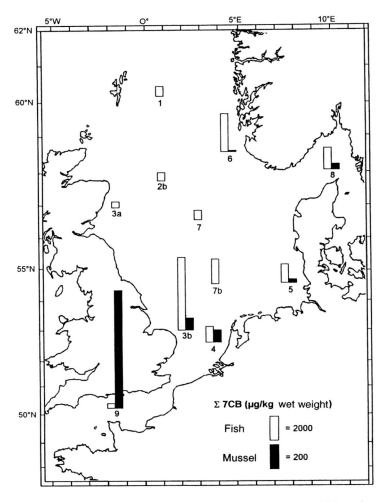

Fig. 7. Concentrations of the seven PCB congeners measured in livers of fish and in mussels taken from sub-regions of the North Sea (modified from North Sea Task Force, 1993)

4. EXPOSURE OF ORGANISMS TO POLLUTANTS: MEASUREMENT OF EROD ACTIVITY

In the North Sea, the highest concentrations of contaminants seem relatively independent of geographic characteristics of the site, since they are observed near off-shore oil platforms, on the coasts, and in estuary zones. Organisms living in these areas are therefore the most exposed to the effects of contaminants.

4.1 Factors Affecting Variations in EROD Activity

Measurement of EROD activity, enzymatic activity associated with certain cytochrome P450, is one of the biomarkers that give an early warning of the effect of certain organic pollutants such as PCB and PAH.

Some pollutants such as PCB fix themselves on lipid-rich tissues in which they accumulate (Varanasi et al., 1992). PCB and other pollutants are thus frequently detected in the fat and liver of fish. Lipid contents in dab liver decrease in winter and at the beginning of spring, probably because of changes in fat metabolism due partly to a variation in feed regime. These variations in lipid content, linked to the life cycle of organisms, particularly during the reproductive period, affect the accumulation and elimination of PCB and other lipophilic contaminants. A study of intra-annual variations of EROD activity in dab, done over two years on the Belgian coasts, showed that EROD activity and lipid contents have an inverse relationship (Figs. 8A and B; Cooreman et al.,

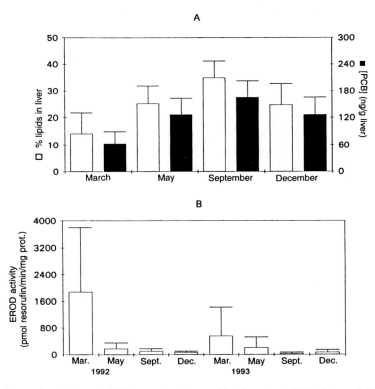

Fig. 8. A. Seasonal variations in PCB concentrations and lipid contents (%) in livers of dab from the Belgian coast. B. Seasonal variations in EROD activity in dab in 1992 and 1993 (Cooreman et al., 1993).

1993). The relation between PCB concentrations and lipid percentage in liver remains relatively constant during the year. Increases in EROD activity will be directly linked to the utilization of lipid reserves contaminated by pollutants such as PCB. An increase in EROD activity in March was therefore followed by a decrease in May and September, while lipid contents increased over the same period. These results show that a good knowledge of biological rhythms in the species studied is needed for interpretation of measurements of EROD activity.

Apart from the temperature (EROD values for dab living in water at 8°C are higher than those obtained for animals taken from water at 14°C; Cooreman et al., 1993), the maturation stage and age of male or female organisms, the geographic location, and the duration of exposure to contaminants are factors that can cause variation in EROD activity.

Observations pertaining to sources of variation in EROD activity show the complexity involved in interpreting a biomarker of exposure to pollutants. Except in the case of specific contaminants such as those related to oil drilling, the relation between EROD activity and the effect of contaminants present in the environment is not always direct.

4.2 Measurement of EROD Activity and Survey of Diseases in Fish

The timely evaluation of effects of low levels of chemical pollutants on the health of organisms is one of the concerns of the Paris Commission, which aims to fight chemical pollution. Such an evaluation is complex because the effects on organisms result from a combination of interacting factors. The disturbances linked to the presence of these pollutants can, in the very long term, influence the immune defences and reproductive capacity of the animals studied. Interpretation of biological effects measured with biomarkers is complicated by the fact that biomarkers integrate, at least partly, the natural and/or man-made sources of variation. However, biomarkers generally offer a rapid response from the organism exposed to an environmental stress.

Variations in EROD activity can be considered early biological events that may manifest themselves before the appearance of infectious diseases in fish. Such diseases have been studied in relation to the contamination of individuals for dab and flounder. The pathologies most often observed are the presence of papilloma (which can be linked to viruses), lymphocystis (viral infections), and various ulcerations caused by bacteria (Oslo Commission, 1991). The occurrence of cutaneous tumours can be associated with the presence of certain pollutants because such tumours have been observed in the region of titanium dioxide discharges in German Bight (sub-region 5), where contaminants transported by the

Elbe were also found. The frequency of liver tumours in flounder, studied over a period of 5 years, was related to the presence of PAH in the sediments (Malins et al., 1987). It is nevertheless advisable to remain cautious in drawing conclusions, because the prevalence of nodules in dab liver was not shown to be directly related to pollutants present in the medium (Vethaak and Van Der Meer, 1991), and no relation has been shown in flounder with PCB contamination (Vethaak, 1991). On the other hand, it has been shown that fluctuations in salinity in the Wadden Sea favour the development of skin diseases in flounder (Vethaak, 1991).

In consequence, pollution cannot be considered the sole cause of the occurrence of tumours and infections in North Sea fish, but only one of several factors that can favour their occurrence. Still, a determination of pathologies is an important part of setting up a monitoring programme. However, in the case of the North Sea monitoring programme, the variability of sampling on such an extensive area and the absence of a long-term study prevents us from identifying precise trends (ICES, 1989).

4.3 Mapping EROD Activity in the North Sea

EROD activity was measured on dab liver in the North Sea sub-regions. A comparison of results showed that measurements can vary by up to 40% (Stagg and Addisson, 1994). However, despite the difficulty in coordinating analytical protocols, differences in exposure of organisms were observed according to the zones studied (Fig. 9). The highest variations in EROD activity were found in the central zone of the North Sea, on the Dogger Bank site. This zone is characterized by a turbulent current, the presence of silty sediments, and above all a significant concentration of off-shore oil platforms, which are major sources of hydrocarbons (PAH in particular). High EROD activity has also been observed in the petroleum complex located in sub-region 3b, south of Oyster Bank. EROD activities measured in sub-region 8 showed contamination of deep sediments by PAH and PCB. Intermediate values were obtained in sub-region 5, on the German and Danish coasts. Chemical measurements of contaminants showed that this zone, including German Bight, is particularly exposed to flows of man-made contaminants (PCB and PAH). Nodules were present in 28% of the dab liver samples taken from this region. Variations in EROD activity and malformation of fish embryos were also observed in dab and flounder. The presence of titanium dioxide and various effluents in the Rhine estuary and in German Bight can be considered responsible for the epidermal pathology observed (Oslo Commission, 1991).

The lowest values were measured on the coasts of the English Channel in sub-region 9, and in sub-region 4 on the Belgian and Dutch coasts.

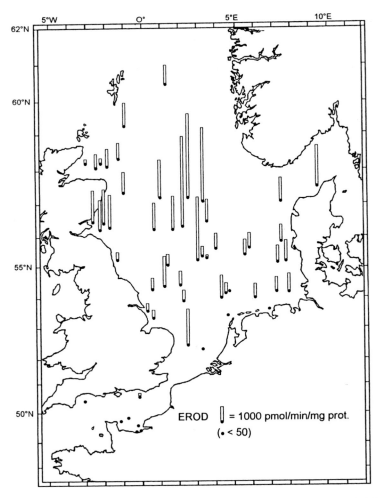

Fig. 9. Measurement of EROD activity in dab liver in the North Sea in 1991 (after data from ICES, 1992)

Chemical measurements showed that oil drilling is a source of hydrocarbon contamination. The level of EROD activity is, in this case, a useful marker of exposure to indicate the presence of hydrocarbons. For zones of diffuse contamination such as the estuaries and coastal areas, the results are more difficult to interpret. Amounts of hydrocarbons introduced in the estuaries of the North Sea are clearly higher than those from oil drilling activities. However, in the coastal zones, it was sometimes impossible to prove a direct relationship between the presence of PCB and PAH, the occurrence of diseases, and variations in EROD activity.

Moreover, the fact that the samples were taken from the various sub-regions during different seasons limits the possibilities of comparison of data.

4.4 Monitoring Biological Effects on the French Coasts of the Bay of Seine

IFREMER participated in collection of data on biological effects in the English Channel (sub-region 9). Measurements of EROD activity in dab liver were provided as part of the North Sea programme to the ICES, which was responsible for compiling the data. These same data were also integrated into a French programme, the Réseau National d'Observation (RNO). The experience acquired during the programme has enabled the development of a section pertaining to biological effects of contaminants on two pilot sites of the French coasts since 1991: Fos Marseilles, on the Mediterranean Sea coast, and the Bay of Seine, in the English Channel. The Bay of Seine site was included in the North Sea programme and in the RNO programme because it is the most contaminated site on the French coasts along the Atlantic front (Cossa et al., 1994).

The study conducted in the Bay of Seine, on levels of EROD activity in dab, enabled us to indicate the difficulties of collecting this sentinel species during all the seasons. The Bay of Seine is a spawning area for dab, which considerably affects the mobility of this species inside the Bay. In summer, it comes close to the coasts and with the arrival of the first autumn frosts it moves away (Quero, 1984). In a monitoring programme, sampling is directly dependent on the abundance of individuals in the various zones explored. The distribution of dab in March allows sampling throughout the bay, while in August the collection of dab is limited to the coastal area (Fig. 10). A second sentinel species was thus proposed, because it is particularly well represented on the Atlantic coast: the dragonet (*Callionymus lyra*; Fig. 5). Studies conducted on this species in the Bay of Seine since 1990 indicate that it meets most of the accepted criteria for use of a species in a monitoring programme (Galgani et al., 1991, 1992; Burgeot et al., 1993, 1994). No sentinel species can meet all of those criteria. The choice of a species results most often from a compromise between feasibility and representativity. The chief criteria that have been met in the case of dragonet are representativity, low mobility, a mode of living in contact with sediment, and easy collection.

The dragonet lives in the littoral zone on beds up to 430 m deep (Fricke, 1986). This benthic fish prefers beds of sand, gravel, and fine sediments. Its geographic distribution extends from south of Iceland to south of Mauritania. It is also found in the Mediterranean Sea, from

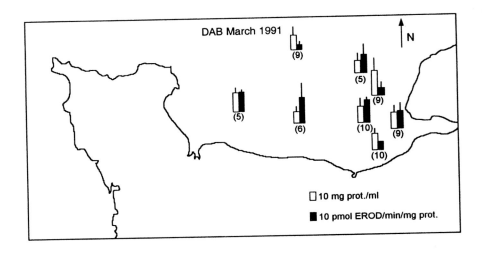

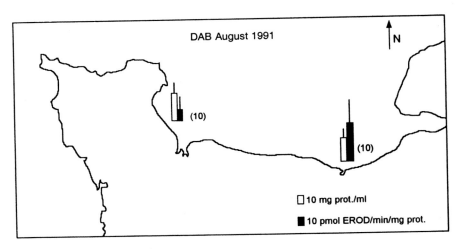

Fig. 10. Measurement of EROD activity in dab liver in the Bay of Seine (after Burgeot et al., 1994)

Gibraltar to the Black Sea. Its distribution in the Bay of Seine allows a more regular sampling following the seasons. The reproductive period lasts from January to August, depending on the zones, with a peak in activity from February to April. In the North Sea, the spawning season is from April to August, while on the coasts of Britanny intense activity is observed from February onwards (Quiniou, 1978). The dragonet is a carnivorous fish that feeds mainly on polychaetes and crustaceans, but

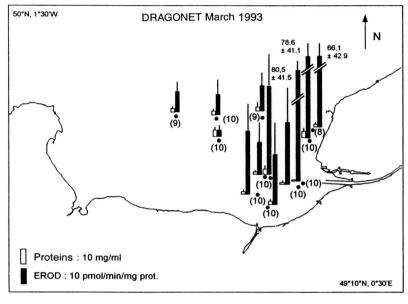

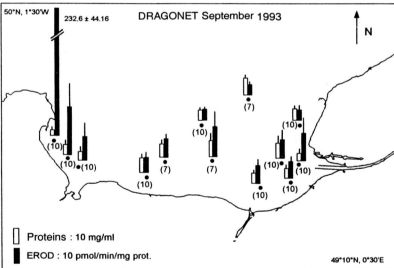

Fig. 11. Measurement of EROD activity in liver of dragonet caught in the Bay of Seine in 1993 (Burgeot, 1994)

it also consumes a large amount of molluscs and echinoderms. A clear reduction in its feeding frequency was observed in the coldest months (Quiniou, 1978; King et al., 1994). Information on the biology of the dragonet can prove very useful in interpretation of measurements of enzymatic activity. The rhythm of reproduction and the feeding regime are examples of factors shown to influence EROD activity in dab (Cooreman et al., 1993). The utilization of lipid reserves during winter and at the beginning of spring thus seems to favour elimination of bioaccumulated PCB, which therefore may also cause variations in EROD activity. This phenomenon, presently being studied in dragonet, may prove very interesting because it concerns a fish in which lipid contents (87%) are much higher than those observed in dab (57%) caught in the Bay of Seine in September (Burgeot, 1994).

Measurements of EROD activity in dragonet liver enabled us to demonstrate that in March 1993 the activities were high in the zone of influence of the Seine but in September the highest activities were observed in the Bay of Veys (Fig. 11).

5. CONCLUSION

The evaluation of the condition of the North Sea presented at the Conference Internationale sur la Protection de la mer du Nord in 1993 is a reference for the application of biomarkers in a large-scale monitoring programme. Although several problems were identified, such as lack of coordination in sampling, rhythm of sampling, and analytical methods, the programme is unequalled in the field of international monitoring and represents a scientific basis that allows us to take corrective measures for anthropogenic pollution in the North Sea. The difficulties of coordinating an international programme involving eight countries over such a large area resulted in data that were sometimes unusable. Lessons can be drawn from this for future programmes. Rigorous quality control of the rhythm of sampling, choice of species, and standard analytical protocols is a key element to the success of such a programme.

The results relating to the biomarker EROD show that, despite all the difficulties of interpretation (mobility of fish, complexity of pollutant mixtures, combination of induction and inhibition of enzymatic activity according to the chemical molecules present, natural variations linked to reproduction, sex, and size of individuals), the measurement of this parameter enabled us to obtain a preliminary cartography of interaction between contaminants and organisms. The interpretation of biomarker results remains now the most delicate phase of monitoring programmes. Laboratory and field studies are still needed for us to understand the phenomena of bioavailability and metabolization of pollutants. Routine surveillance cannot be set up unless the standardization of interpretation of EROD measurements is perfected. These problems of interpretation have also proved that it is difficult to identify a unique biomarker perfectly adapted to the surveillance of such a vast geographical area. Only a multi-marker approach developed in a programme including histopathological analyses and chemical measurements on tissues and sediments will enable us to establish a diagnostic of the effects of pollutants.

The North Sea data nevertheless made it possible to reveal hydrocarbon contamination on oil drilling sites. However, this contamination remains very low compared to the pollution carried by rivers in the coastal areas, the diffuse nature of which will require suitable measurement techniques (Stagg, 1991). In the case of environmental monitoring, the difficulty lies in the need for techniques that are precise and resilient enough to sustain the rigour demanded of an international programme and to face the operational constraints caused by the involvement of such a large number of countries.

ACKNOWLEDGEMENTS

The authors acknowledge the Commissions of Oslo and Paris (New Short, 48 Carey Street, London) for their authorization to publish some extracts of the 'North Sea Quality Status Report, 1993' and the International Council for the Exploration of the Sea (Palaegade 2-4-DK-1261, Copenhagen) for providing EROD data.

REFERENCES

Addison R.F. (1988). Biochemical effects of a pollution gradient—summary. *Mar. Ecol. Prog. Ser.*, 46:45–47.

Burgeot T., Bocquené G., Truquet P. et al. (1993). The dragonet *Callionymus lyra*, a target species used for evaluation of biological effects of chemical contaminants on the French coasts. *Mar. Ecol. Prog. Ser.*, 97 : 309–316.

Burgeot T., Bocquené G., Pingray G. et al. (1994). Monitoring biological effects of contamination in marine fish along French coasts by measurement of ethoxyresorufin-O-deethylase activity. *Ecotoxicol. Environ. Saf.*, 29 : 131–147.

Burgeot T. (1994). L'ethoxyrésorufine-O-déethylase, les adduits à l'ADN et les micronuclei dans les organismes marins. Application à la surveillance des effets biologiques sur les côtes françaises. Doctoral thesis, University of Nantes. 267 pp.

Cooreman C., Roose P. and Vyncke W. (1993). EROD monitoring in dab from the Belgian continental shelf: seasonal variation and relation to organochlorines. Scientific Symposium on the 1993 North Sea QSR, 11pp. ICES Palaegade 2-4-DK-1261, Copenhagen.

Cossa D., Meybeck M., Idlafkih Z. and Bombled B. (1994). Étude pilote des apports en contaminants par la Seine. Rapport IFREMER, Agence de l'Eau Seine-Normandie et Ministère de l'Environnement. *R. Int. DEL. 94.13*. Nantes.

De Boer (1988). Trends in chlorobiphenyl contents in livers of Atlantic cod (*Gadus morhua*) from the North Sea, 1979–1987. *Chemosphere*, 17:1811–1919.

Duinker J.C., Knap A.H., Binkley K.C. et al. (1988). Method to represent the qualitative and quantitative characteristics of PCB mixtures: Marine mammal tissues and commercial mixtures as examples. *Mar. Poll. Bull.*, 19:74-79.

Fricke R. (1986). Callionymidae. In: Fishes of the North-eastern Atlantic and Mediterranean, vol. 3. Whitehead P.J.P., Branchot M.L., Hureau J.C., Nielsen J. and Tortonese E. (eds). Unesco, Paris, pp. 1086-1093.

Galgani F., Bocquené G., Luçon M. et al. (1991). EROD measurements in fish from the northwest part of France. *Mar. Poll. Bull.*, 22:494–500.

Galgani F., Bocquené G., Truquet P. et al. (1992). Monitoring of pollutant biochemical effects on marine organisms of the French coasts. *Oceanol. Acta*, 15:355–364.

ICES (1989). Methodology of fish disease surveys. Report of an ICES Sea-going Workshop held on U/F Argos, 16–23 April 1988. *Coop. Res. Rep.* 166.

ICES (1990). Report of the Working Group on Biological Effects of Contaminants, Nantes, 24–27 April 1990. CM 1990. ICES Palaegade 2-4-DK-1261, Copenhagen.

ICES (1992). Report of the Working Group on the Biological Effects of Contaminants. CM 1992, Env 3, Ref E. ICES Palaegade 2-4-DK-1261, Copenhagen.

King P.A., Five J.M. and McGrath D. (1994). Reproduction, growth and feeding of the dragonet, *Callionymus lyra* (Teleostei: Callionymidae), in Galway bay, Ireland. *J. Mar. Biol. Ass. U.K.*, 74:513–526.

Koeman J.H., Kolher-Gunter A., Kurelec B. et al. (1993). Applications and objectives of biomarker research. In: Biomarkers. Research and Application in the Assessment of Environmental Health. D.B. Peakall and L.R. Shugart (eds). NATO Advanced Science Institutes Series Vol. H 68. Springer Verlag, Berlin, Heidelberg, pp. 1–13.

Livingstone D.R. (1993). Biotechnology and pollution monitoring: use of molecular biomarkers in the aquatic environment. *J. Chem. Tech. Biotechnol.*, 57:195–211.

Malins D.C. (1987). Sediment associated contaminants and liver diseases in bottom-dwelling fish. *Hydrobiologia*, 149:67–74.

McDonald L.H. and Smart A. (1993). Beyond the guidelines: practical lessons in monitoring. *Environ. Monit. Assess.*, 26: 203–218.

North Sea Task Force (1993). North Sea quality status report 1993, Bilan de santé de la mer du Nord 1993. Oslo and Paris Commissions, London. Olsen & Olsen, Fredensborg.

Oslo Commission (1991). Results of monitoring activities in the former dumpsite for waste from the TiO$_2$ production. SACSA 18/INFO.17, Germany.

Payne J.F. (1976). Field evaluation of benzo(a)pyrene hydroxylase induction as a monitor for marine pollution. *Science*, 191:945–946.

Quero J.C. (1984). Les poissons de mer des pêches françaises. Éditions J. Gaucher, Paris.

Quiniou L. (1978). Les poissons démersaux de la baie de Douarnenez: alimentation et écologie. Univ. thesis, Université de Bretagne Occidentale, Brest.

RNO (1990). Réseau National d'Observation de la qualité du milieu marin: Surveillance du milieu marin. Travaux du RNO, 1989–90. IFREMER and Ministere de l'Environnement.

Stagg R.M. (1991). North Sea Task Force biological effects monitoring program. *Wat. Sci. Tech.*, 24:87–98.

Stagg R.M. and Addison R.F. (1994). An inter-laboratory comparison of measurements of ethoxyresorufin-O-deethylase activity in dab (*Limanda limanda*) liver. *Mar. Environ. Res.*, 40:93–108.

Stebbing A.R.D. and Dethlefsen V. (1992). Introduction to the Bremerhaven Workshop on biological effects of contaminants. *Mar. Ecol. Prog. Ser.*, 91:1–8.

Stegeman J.J. (1981). Polynuclear aromatic hydrocarbons and their metabolism in the marine environment. In: Polycyclic Hydrocarbons and Cancer, vol. 3. Gelboin H.V. and Ts'o P.O.P. (eds.). Academic Press, New York.

Stegeman J.J., Kloepper-Sams P.J. and Farrigton J.W. (1986). Monooxygenase induction and chlorophenyls in the deep sea fish *Coryphaenoides armatus. Science*, 231:1287–1289.

Turrel W.R., Henderson E.W., Slesser G. et al. (1992). Seasonal changes in the circulation of the northern North Sea. *Continental Shelf Res.*, 12:257–286.

Vethaak D. (1991). On the occurrence of liver tumours in flatfish in Dutch waters. *Tidal Waters report GWAO*-91-005.

Vethaak D. and Van Der Meer J. (1991). Fish disease monitoring in the Dutch part of the North sea in relation to the dumping of waste from titanium dioxide production. *Chem. Ecol.,* 5:149–170.

Wolfe D.A. (1992). Selection of bioindicators of pollution for marine monitoring programs. *Chem. Ecol.* 6:149–167.

Wolfe D.A., Champ M.A., Flemer D.A. and Mearns A.J. (1987). Long-term biological data sets: their role in research, monitoring and management of estuarine and coastal marine systems. *Estuaries,* 10:91–193.

4

Use of EROD Enzymatic Activity in Freshwater Fish

P. Flammarion, J. Garric and G. Monod

INTRODUCTION

The affirmation of the need for global management of aquatic environments in France and in Europe and preservation of their biological integrity, as defined in the French Water Act of January 3, 1992, as well as the proposition of the European directive on ecological quality of waters (94/C 222/06), confirm the need for sensitive and relevant tools to evaluate the health of aquatic ecosystems.

In this context, the use of biochemical markers measured at the individual level complements more traditional chemical analysis and biocoenotic measurements, for a better understanding of modes of action of chemicals on organisms as well as for management of aquatic ecosystems.

Some biomarkers are characterized by rapid and relatively specific responses. This is the case with cytochrome P4501A (Monod, 1997), the synthesis of which is induced, in a reversible manner (Melancon et al., 1987), in organisms exposed to certain families of contaminants, particularly polycyclic aromatic hydrocarbons (PAH) or polychlorobiphenyls (PCB), which are widespread in aquatic environments.

This biomarker can thus be used as a tool to assess environmental quality, detect polluted areas, identify the sources of pollution, and judge the efficacy of restoration. Its use was first proposed by Payne in 1976. Since then, many authors have described significant inductions of cytochrome P4501A in freshwater fish in relation to pollution of various origins (Kezic et al., 1983; Payne et al., 1987; Haux and Förlin, 1988; Lindstrom-Seppa and Oikari, 1988; Monod et al., 1988; Vindimian and Garric, 1989).

In the field, various catalytic activities are used to measure induction of cytochrome P4501A: aryl hydrocarbon hydroxylase (AHH), ethoxycoumarin O-deethylase (ECOD), ethoxyresorufin O-deethylase (EROD), pentoxyresorufin O-depenthylase (PROD). The immunochemical detection of the cytochrome P4501A is also recommended to refine the interpretation of induction levels measured through catalytic activities (Stegeman et al., 1987; Goksøyr and Förlin, 1992).

Among the measurements of monooxygenase activities, EROD activity appears to be the most sensitive. For example, induction of ECOD activity, in fish caught in the River Rhone and subjected to PCB contamination, was found to be lower than that of AHH and EROD activities (Monod et al., 1988). The fact that ECOD measurement would be less sensitive than EROD measurement has been confirmed by other studies (Larsson et al., 1988; Ahokas et al., 1994). Similarly, PROD activity, unlike AHH and EROD activities, could not distinguish sites polluted by paper industry effluents (Lindstrom-Seppa and Oikari, 1990). Finally, although AHH and EROD activities are effectively well correlated, EROD activity remains generally more sensitive than AHH activity and is moreover easier to measure (Stegeman et al., 1987; Monod et al., 1988).

The first part of this chapter consists of a synthetic presentation of representative research on the *in situ* use of this biomarker. A second part is devoted to the analysis of a case indicating the influence of factors other than chemical contamination of the environment. The third part concerns the data obtained *in situ* that give indication on the relations that can exist between the response of the biomarker 'induction of EROD activity' and responses provided by other biological variables. The last part is devoted to a current assessment of the use of this biomarker.

1. EROD INDUCTION *IN SITU* AND ITS RELATION TO CHEMICAL CONTAMINATION

The advantage of using this biomarker as a pollution indicator was confirmed by many field studies, some examples of which, conducted in fresh water, are shown in Table 1.

The pollutants referred to are those found in paper industry effluents (chlorine bleached pulp mill effluents containing organochlorine compounds) as well as PCB, PAH, and dioxins.

Field sites in which contamination was less defined have also been studied, e.g., downstream of industrial chemical pollution (Vindimian et al., 1991; Ueng et al., 1992) or in relation to non-point

Table 1. Induction of EROD activity measured in some freshwater species exposed to chemical contaminations *in situ*

Sites	Contamination identified	Sampling period	Species	EROD[a] Reference sites	EROD[a] Contaminated sites	Induction	Authors
Lake Coleman (Australia)	Treated effluent (wood pulp) AOX (water), EOX (sediment), PCDD/TCDD (muscle)	August	*Cyprinus carpio* (Cyprinidae) males and females	4.6×10^3 to 6×10^3	17×10^3 to 31×10^3	yes ($\times 3$ to 6)	Ahokas et al., 1994
Lake Kallavesi (Finland)	Treated effluent of unbleached wood pulp	January-February, April, June, September-October	*Perca fluviatilis* (Percidae) males and females	27 (winter) to 16 (summer)	36 (winter) to 45 (summer)	Yes ($\times 1.9$ to 2.9)	Huuskonen et al., 1995
Lake Saima (Finland)	Treated effluent of wood pulp bleached with chlorine	September, October (post-spawning)	*Perca fluviatilis* (Percidae), *Rutilus rutilus* (Cyprinidae), *Abramis brama* (Cyprinidae) females	3-16 0.3-0.8 7.7	12-43 3-24 37-50	yes	Lindstrom-Seppa and Oikari, 1990
Lakes Gaasperplas and Nieuwe Meer (Holland)	PAH, PCB, OC in sediments and fish	December	*Rutilus rutilus* (Cyprinidae) females	1.3	1.6	no	Van der Oost et al., 1994
River Rhone (France)	PCB in fish	March, June, November	*Chondrostoma nasus* (Cyprinidae) males and females	6-60 (males) 3-60 (females)	20-100 (males) 10-110 (females)	yes ($\times 2$ to 5)	Masfaraud et al., 1990
River Rhone (France)	PCB in fish	Spring	*Chondrostoma nasus* (Cyprinidae), *Rutilus rutilus* (Cyprinidae), *Thymallus thymallus* (Salmonidae)	4-13 7 76-136	36 (± 1) 52 (± 6) 480 (± 73)	yes ($\times 6$ to 9)	Monod et al., 1988
River Durance (France)	Chemical effluent (organochlorine contaminants, including γ-HCH, DDT PCB in bryophytes)	October, February, May	*Chondrostoma nasus* (Cyprinidae), *Leuciscus cephalus* (Cyprinidae), *Barbus barbus* (Cyprinidae) males and females	30-48 12-29 11-27		yes ($\times 6$ to 20)	Vindimian et al., 1991

[a] pmol/min/mg of protein.

source agriculture contamination (Vindimian et al., 1993). In these studies, the authors were also able to indicate a positive relation between the chemical contamination and the response of EROD activity.

Table 1 shows EROD activities measured in fish from sites identified by the authors as reference sites (i.e., non-polluted stations). These activities show wide variations due to measurements made in different reference stations (for the same author) (Monod et al., 1988; Lindstrom-Seppa and Oikari, 1990; Vindimian et al., 1991; Ahokas et al., 1994), different sampling periods (Masfaraud et al., 1990), or different measurement methods (see below).

Several authors have indicated close correlation between the response of EROD activity and the concentrations of inducers (PAH, PCB, dioxins) measured in the environment (water, sediment) or in the organism (Table 2).

In these examples, only Van der Oost et al. (1994), in contrast to results they obtained in 1991 on the same sites, did not indicate significant induction of EROD activity in relation to a contamination gradient by PCB, PAH, and organochlorine molecules. In this case, the contamination level of fish might have been too low to lead to a significant induction of EROD activity. In fact, the authors noted that PCB concentrations measured in the fish were below 100 µg/kg (fresh weight), i.e., were 15 to 50 times lower than those measured by Monod et al. (1988), who observed a significant induction of monooxygenase activity in fish from the River Rhone.

2. OTHER FACTORS MODULATING EROD INDUCTION IN FISH

Apart from pollution, environmental studies must take into account various biotic factors (species, sex) and abiotic factors (temperature, season), which do modulate cytochrome P4501A and/or associated enzymatic activities (Table 3). Besides, the influence of measurement conditions has to be taken into account.

We will not go into detail about the influence of these parameters on induction of EROD activity (Andersson and Förlin, 1992; Monod, 1997).

2.1 Measurement Conditions

Analysis of EROD activity does not present major difficulties. However, given the well-known fragility of enzymatic systems, several factors can influence the result, from the moment fish are captured to the time of the measurement of enzyme activity.

Table 2. Concentration-response relations observed *in situ* (established from data published in the articles cited)

Inductor	Compartment	Species	Concentration	Unit	EROD (pmol/min/mg)	Correlation EROD-dose(r^2)	Authors
PCB	muscle	*Chondrostoma nasus* *Rutilus rutilus*	0.9-4.5 1.6-5.3	mg/kg fresh	4-36 7-52	0.88	Monod et al. 1988
AOX TCDD	water muscle	*Cyprinus carpio* *Cyprinus carpio*	300-400 1-4	µg/l pg/g sec	5,000-35,000 5,000-35,000	0.93 0.88	Ahokas et al., 1994
PAH	sediment	*Pleuronectes americanus*	7-200	µg/g sec	300-2,000	0.98	Vignier et al., 1994
PCB, PAH/ OC	sediment	*Anguilla anguilla*	0.1-0.9 25-128 0.05-0.2	µg/g organic carbon,	128-238	—	Van der Oost et al., 1991
PCB, PAH/ OC	muscle	*Anguilla anguilla*	2-13 2.4-1.8 1.4-8	µg/g lipids	128-238	positive correlation	
PCB, PAH/ OC	muscle	*Rutilus rutilus*	1.4-5.7 3.3-1.9 0.7-3.8	µg/g lipids	0.6-3.5	positive correlation	
PCB, PAH/ OC	sediment	*Rutilus rutilus* 0.08-0.4	0.1-0.8 15-80 carbon,	µg/g organic	1.6-2.0	no significant EROD induction	Van der Oost et al., 1994
PCB, PAH/ OC	muscle	*Rutilus rutilus*	2.5-7.0 4.0-3.0 0.5-1.3	µg/g lipids	1.6-2.0	no significant EROD induction	

Table 3. Factors modulating P450 system in fish (after Goksøyr and Förlin, 1992)

species	sex	feed regime
stock	temperature	inducer agents
age	sexual maturation	antagonistic agents

The liver sample must be taken as quickly as possible after the fish are caught. Stress conditions (anoxia) may lead to a significant loss of enzymatic activity (Monod et al., unpublished data).

In certain conditions, enzymatic levels can be measured at the site of capture (see chapter 3). But if the samples (generally liver) are brought to the laboratory, they must be well preserved. Most researchers use liquid nitrogen. Various works have shown that conditions in which liver is preserved can affect the EROD activity. Monod and Vindimian (1991) observed that homogenization of liver in the presence of glycerol before immersion in liquid nitrogen enabled an EROD activity that was twice as high. This result may be explained by the capacity of glycerol to protect the membranous structures (endoplasmic reticulum) in which cytochrome P4501A is included.

Although the methods of preparation of subcellular fractions (post-mitochondrial or microsomal fraction) are well standardized, the measurement conditions vary from one laboratory to another, which makes it difficult to compare EROD activities from different studies. An inter-laboratory comparison of EROD activities in fish exposed or not exposed to paper mill effluents (Munkittrick et al., 1993) showed that the measurement conditions (pH, temperature, and substrate concentration), which varied considerably among the participating laboratories, and the measurement of protein content in the fractions considerably influenced the measured EROD activities, but also that the levels of induction were approximately the same from one laboratory to another.

Immunological measurements need fewer precautions. Förlin and Celander (1993) observed an identical level of induction using an enzymatic measurement made on microsomes from livers directly immersed in liquid nitrogen after sampling, or using an immunochemical measurement made on microsomes from livers frozen in dry ice, while these latter microsomes did not show significant EROD activity using enzymatic measurement.

Thus, when catalytic activity is well preserved, there is a good statistical correlation between the measurements of the activity and immunochemical detection (Collier et al., 1992; Van der Weiden et al., 1993). However, an induction can still be measured by immunochemical analysis even if there is a loss of activity (inhibition and/or alteration of the enzyme).

2.2 Species

Basal levels of EROD activity (in non-polluted situations) as well as induced levels vary according to the species (Monod et al., 1988; Vindimian et al., 1991; Collier et al., 1995). This species-linked variability is well established in controlled laboratory situations (Monod, 1997). In natural environments, the habitat (pelagic or benthic) can be an important factor to consider. Vindimian et al. (1991) noted a much greater induction in barbel (*Barbus barbus*) than in nase (*Chondrostoma nasus*) and chub (*Leuciscus cephalus*) caught in the same stations, downstream of an industrial effluent. Such a result is no doubt linked to the benthic nature of barbel, which is more exposed to contamination through contact with a sediment.

2.3 Season and the Sex of Individuals

Male fish generally show higher EROD activities than females, by a factor of 1 to 2 according to the season and the species (Vindimian et al., 1991). This sexual difference appears clearly more significant during the period of sexual maturation (Masfaraud et al., 1990; Fig. 1).

The practical implications may be considerable: A sampling leading to an unequal ratio of sexes upstream and downstream may significantly alter the description of the pollution situation. Figure 1 reproduces various scenarios, some of which lead to an erroneous conclusion. EROD induction in fish captured downstream of the point of pollution is shown for each measurement campaign if the individuals are of the same sex upstream and downstream of the pollution (scenarios A and B). Scenario C leads to an overestimation of EROD induction in March and November (periods of sexual maturation in the species studied) because the individuals captured downstream of the pollution were male and had a higher basal EROD activity than female individuals. This difference in basal activity level thus adds to the phenomenon of induction that was observed in male individuals (see scenario A). Scenario D is the most unfavourable as it leads to non-detection of induction in March and November because the individuals captured downstream of the pollution were females and had a lower basal EROD activity than male individuals. This differential 'cancels out' the induction in females (see scenario B). Comparison of different stations must therefore be made on individuals of the same sex.

During a single sampling period, some authors did not observe differences linked to sex (Andersson et al., 1988; Gagnon et al., 1995). These contradictory results could be due to the influence of reproductive hormones on monooxygenase activities (Jimenez and Stegeman, 1990; Andersson and Förlin, 1992). Since the period of the year during which

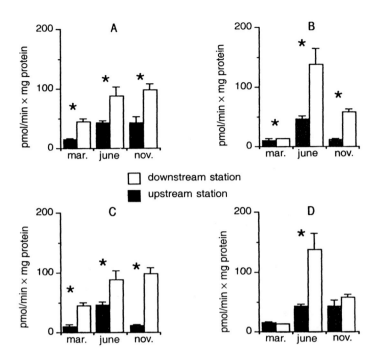

Fig. 1. Hepatic EROD activity in nase (*Chondrostoma nasus*) captured upstream (reference station) and downstream of a PCB elimination plant during 1988 (after Masfaraud et al., 1990):
A. The individuals compared were male.
B. The individuals compared were female.
C. The individuals compared were male downstream and female upstream.
D. The individuals compared were female downstream and male upstream.
*statistically significant difference between upstream and downstream.

sexual maturation occurs varies from one species to another, sexual dimorphism of EROD activity seems also to vary according to the species during those periods. Moreover, various studies have shown that juveniles of both sexes have higher activities than mature individuals (Forlin and Lindman, 1981; Singh et al., 1985). Thus, in order to avoid sex-linked variations in activity levels, some authors prefer to measure monooxygenase activity on juveniles (Collier et al., 1995). This is generally very difficult to do in freshwater ecosystems, in which the juvenile stages are often more difficult to sample than mature stages. It is easier to measure EROD activity from adults of a single sex, preferably males (which have lesser seasonal variations than females).

2.4 Water Temperature

In fish, temperature is known to determine the period of sexual maturation. In different field sites with varying water temperature levels, therefore, there may be differences in the sexual maturity of the females, leading to differences in EROD activity (see above). The direct influence of water temperature might then be hidden by the sex effect.

Nevertheless, the influence of temperature on the kinetics of EROD induction has been demonstrated in controlled laboratory conditions (Andersson and Förlin, 1992; Jimenez et al., 1990). Induction occurs more quickly and reaches higher levels in trout acclimated at 17°C than in those acclimated at 5°C (Andersson and Koivusaari, 1985). In the field, however, water temperature does not appear to modify induction (Andersson et al., 1988) or explain differences in induction (Vindimian et al., 1991). These results confirm the influence of season, in conjunction with sexual maturation rather than with temperature.

2.5 Diet

It is very difficult to accurately estimate the nature of diet of wild fish during the period preceding capture, whereas the direct influence of diet quality on monooxygenase activity and its induction has been proved in the laboratory (Ankley and Blazer, 1988). Diet can also play an indirect role, as a factor of individual variability. It is a major route of contamination of fish by persistent and potentially bioaccumulable micropollutants, among them known inducers of cytochrome P4501A (e.g., dioxin, PCB).

2.6 Inhibition and Interactions

Some laboratory studies reported inhibition of monooxygenase activities in the presence of high doses of hepatotoxic inducer compounds or organic or metallic inhibitors (Monod, 1997). Several authors have also suggested that these phenomena occur *in situ*. EROD activity measured in individuals caught in stations close to effluent discharges may therefore be lower than those measured further downstream (Vindimian and Garric, 1989; Lindström-Seppa and Oikari, 1990; Adams et al., 1992). In a later study, Huuskonen and Lindström-Seppa (1995), comparing induction of protein P4501A and EROD activity measured in perch exposed to unbleached pulp mill effluents, discussed the potential presence of inhibitor compounds in these effluents, which would explain the differences in induction according to the measurement method (immunochemical or enzymatic).

Some apparent inhibitions may be due to toxic phenomena affecting the tissue on which the analysis is done (most often liver), particularly when the level of pollution is high (Biagianti-Risbourg, 1997). This was suggested by Vindimian et al. (1991), who showed that the lowest rate of induction, observed in barbel, was found in individuals captured in the station nearest the discharge, and therefore most exposed to the pollutants.

2.7 Strategies Currently Planned

The possible influence of several factors on the induction level in fresh-water fish is illustrated in the previous section. Although it is easy to come to a conclusion in clear-cut situations, given the high level of EROD induction (Collier et al., 1995), in less obvious cases the variability due to these factors, if not quantified, can considerably modify the environmental assessment. It is thus important to choose a reliable reference station.

Table 4 shows the activities used as reference in various works. The reliability of the reference obviously affects the interpretation of results in terms of induction.

An alternative to measurements made on wild fish subjected to factors of variability (other than chemical pollution) is to use caged fish. This solution has been applied successfully (Grizzle et al., 1988; Haasch et al., 1993; De Flora et al., 1993; Soimasuo et al., 1995). First of all, it enables a comparison of groups of fish that are considered historically and biologically 'homogeneous', and it resolve problems linked to migration of fish, particularly at the time of spawning, and, to some extent, the problem of a reliable reference.

However, this approach has limits that cannot be ignored, such as stress caused by confinement and difficulties of access to food (Soimasuo et al., 1995). Moreover, the feasibility of such a strategy remains to be validated in the perspective of regular study of a biomarker such as EROD in a monitoring network, including a large number of highly diverse stations.

3. RESEARCH ON RELATIONS BETWEEN EROD INDUCTION AND EFFECTS ON OTHER BIOLOGICAL VARIABLES

Studies suggest that cytochrome P4501A induction could be, under certain conditions, an early warning of more important biological alterations (Monod, 1997). This possibility was foreseen in the context of several large-scale programmes aiming to evaluate *in situ* the impact of pulp mill effluents. This type of effluent was extensively studied after the discovery of contamination by dioxins. In addition to EROD activity, a large number of biological variables were considered (Table 5).

Table 4. EROD activities for some cyprinid species

Species	Season	Sex	EROD[a]	Authors
Barbus barbus	mixed	F	12 (11)	Vindimian et al., 1991
		M	21 (25)	
Barbus barbus	mixed	F	4 (3)	Richert, 1994
		M	7 (9)	
Chondrostoma nasus	spring		4-13 (1)	Monod et al., 1988
Chondrostoma nasus	mixed	F	16 (15)	Vindimian et al., 1991
		M	53 (45)	
Chondrostoma nasus	mixed	F	6 (5)	Richert, 1994
		M	8 (5)	
Cyprinus carpio			20 (10)	Van der Weiden et al., 1989
Cyprinus carpio			10 (2)	Ueng et al., 1992
Cyprinus carpio	summer		5 (1)	Ahokas et al., 1994
Gobio gobio	mixed	F	15 (10)	Richert, 1994
		M	58 (20)	
Leuciscus cephalus	mixed	F	19 (24)	Vindimian et al., 1991
		M	22 (17)	
Leuciscus cephalus	mixed	F	12 (31)	Richert, 1994
		M	13 (24)	
Rutilus rutilus	spring		7 (1)	Monod et al., 1988
Rutilus rutilus	winter	F	1.6 (1.6)	Van der Oost et al., 1994

[a] EROD activity, mean (standard deviation) in pmole/min/mg of protein.

The advantages of EROD activity as a biomarker of chemical contamination of the environment were clearly confirmed. A clear relationship seems also to be established between this biomarker and the alteration of biological variables with more obvious toxicological significance. Most of the studies indicate that EROD induction was correlated with disturbance in reproductive parameters, particularly variations in levels of blood steroids, and sometimes with an alteration in fecundity. All the observed effects may be linked to disturbance in energy metabolism due to the exposure to effluents. These studies have nevertheless measured mainly individual level parameters. Only the studies of Adams et al. (1992) and Kloepper-Sams et al. (1994) integrated population level variables. However, while Adams et al. (1992) observed concomitantly an EROD induction, a modification of the age distribution in *Lepomis auritus,* and an alteration in the index of biological integrity (IBI) relative to fish populations, Kloepper-Sams et al. (1994) did not observe a correlation between such an induction and an effect on the age distribution or on the index of species abundance.

Table 5. Research on induction of hepatic EROD activity in relation to chemical contamination of the environment (pulp mill effluents) and alteration of other biological variables in freshwater fish

Species and biological variables measured	Relation of EROD to chemical pollution	Relation of EROD to other biological variables	Authors
Lepomis auritus sexual steroids, DNA integrity, HSi,[a] morphometry, growth, fecundity, age structure, IBI	+ (downstream of effluents)	+ +	Adams et al., 1992
Perca fluviatilis sexual steroids, hepatic glycogen, muscular and hepatic ascorbic acid, osmotic balance, hematology, GSi,[b] HSi	+ (downstream of effluents)	+	Andersson et al., 1988
Catostomus commersoni sexual steroids, GSi, morphometry, growth, fecundity, age of first maturity	+ (phenols, guaiacols,...)	+	Gagnon et al., 1995
Catostomus catostomus and *Prosopium williamsoni* hematology, sexual steroids, HSi, GSi, histology, morphometry, growth, age of first maturity, fecundity, age structure, abundance	+ (dioxins, furans)	−	Kloepper-Sams et al., 1994
Catostomus commersoni sexual steroids, spermogram, sperm fertility, GSi, fecundity, morphometry, growth, age of first maturity	+ (downstream of effluents)	+	Munkittrick et al., 1994
Catostomus commersoni sexual steroids, HSi, GSi, morphometry, growth	+ (dioxins, furans)	+	Munkittrick et al., 1994

[a] Hepatosomatic index
[b] Gonadosomatic index

4. SUMMARY AND PERSPECTIVES

The P4501A biomarker can be used in two complementary approaches to study aquatic contamination. First, to detect the exposure to chemical contamination by a certain type of pollutant (polycyclic and plane chemical structure) and second, to monitor the health of organisms exposed to anthropogenic influences. The first approach tends to use this biological response as indicator of environmental contamination. The second aims to interpret it as an early indication of biological alteration in the exposed organisms.

Mayer et al. (1992) gave a few criteria for the development of a biomarker and its use in the field. It must be a simple measurement, with a response depending simply on the dose and duration of exposure. The measurement must have good sensitivity and a variability linked to non-toxic factors (organisms, environment, method) that are well understood and within acceptable limits, and finally it must have some biological significance (size or condition of the population).

Cytochrome P4501A induction meets most of these criteria. The available data show that this biomarker has, for more than a decade, been widely used in the study of environmental quality and its use is enhanced by advanced knowledge (Monod, 1997).

Moreover, the measurement methodology is relatively easy. It must nevertheless be emphasized that enzymatic analysis is costly because of the quality of material required (centrifuge, spectrophotometer, spectrofluorimeter), as well as the time and technical know-how needed to accomplish it. Moreover, the dosage of this biomarker is still not standardized, even though that would be easy enough with the knowledge presently available.

Immunodosage (ELISA) can also be standardized, as long as commercial sources of antibodies of a consistent quality can be arranged for. Antibodies presently used have been produced by some laboratories in limited quantities and for their own research needs. The quality of polyclonal antibodies depends closely, first, on the source and level of purification of the cytochrome P4501A used as immunogen, and, second, on the original nature (not exactly reproducible) of immune reaction when the polyclonal antibody was raised. Myers et al. (1993) have demonstrated the immunogenic character of a synthetic peptide that is identical to that of cytochrome P4501A of trout, and propose to use it as immunizing reactant, allowing thus the production of a more consistent quality of antibody. The ideal would nevertheless be the production of monoclonal antibodies. Private organizations, moreover, already market monoclonal antibodies directed against cytochrome P4501A of some fish species. It remains to be demonstrated whether a limited number of antibodies will allow detection of cytochrome P4501A in all the species that can be studied. In the contrary case, such a strategy must undoubtedly be evaluated in economic terms at the outset.

Furthermore, complementary efforts are required to better define the significance of natural variability in induction of EROD activity, particularly in some species that can be easily and widely sampled in fresh water. Analysis of this natural variability must be associated with a definition of basal activities in the studied species. In the context of use of this biomarker for environmental monitoring, the hierarchy of these factors must be considered to determine from the measured response how much is due exclusively to contamination, what are the factors of variability that are to be taken into account on priority, and what are their respective contributions to that response. If results of EROD activity measurements are to be correctly understood, particularly in complex situations (e.g., multi-source pollution, considerable dilution), pertinent information must be obtained on the response of this biomarker according to the gradient and typology of pollutant transport.

The biomarker EROD was used as a fitted tool in numerous field studies. Yet, in a large-scale biomonitoring perspective, it is crucial to the validity of the biological assessment to define reference conditions and to establish a classification that is effective in detecting changes due to pollution by adequately separating natural variation from variation caused by anthropogenic impacts (Mayer et al., 1992; Nixon et al., 1996). Most studies consider upstream-downstream comparisons. However, when suitable upstream values are not available, it might be of great interest to assess EROD levels in a contaminated area in relation to background EROD levels in non-contaminated areas that may be accepted as reference values of fish EROD activity. Yet, several natural factors modulate the EROD levels, including sex, reproductive status, season, and species (Goksøyr and Förlin, 1992). As a consequence, it is necessary to determine baseline values in sentinel fish species as well as to precisely quantify the biotic and abiotic sources of variability. However, Lindström-Seppa and Oikari (1990) showed the difficulty of choosing several appropriate control areas.

From our collection of cyprinids EROD data from various sites of the Rhone watershed, we conclude that no statistical difference in the EROD activity was observed between three river areas, all of which were *a priori* considered to uncontaminated by effluents likely to induce the P4501A system (Flammarion and Garric, 1997). Hence, we concluded that EROD levels measured in these sites (Ain, Ardèche and Drôme) may be considered as useful and relevant control values to be compared with levels in other areas where no suitable upstream values can be obtained. However, establishing relevant reference EROD values needs further interannual validation in these sites as well as in other mildly polluted sites from other watersheds.

The alternative proposed through the caging method offers undeniable advantages but must be rigorously evaluated (see above). The duration of exposure being in any case relatively limited, this method does not allow access to information (other biological parameters) requiring a relatively long temporal integration, corresponding to long-term effects.

The constraints and sources of variation linked to measurement of EROD or cytochrome P4501A from fish captured from or caged on the study sites led some authors to consider a different strategy. This strategy could be elaborated only after the possibility of measuring EROD and its induction in cell cultures was demonstrated and the concept of dioxin equivalent was enunciated (Monod, 1997). Zacharewski et al. (1989) calculated the dioxin equivalents present in fish from the Great Lakes by measuring the rate of induction of EROD activity in H4II4E cell line cells exposed to organic extracts of these fish. Following the same procedure, Van den Heuvel et al. (1994) evaluated the level in dioxin equivalents in fish exposed to pulp mill effluents. In this study, the bioassay response was correlated to hepatic EROD level of fish. A later study showed that use of cell line from trout (RTL-W1) showed greater sensitivity (Clemons et al., 1994). The use of techniques of molecular biology allowed the creation of transgenic cells from which the presence of inducer of cytochrome P4501A in the culture medium is shown very easily by bioluminescence (Anderson et al., 1995). The use of such a strategy can enable assessment of the presence of dioxin equivalents in other environmental compartments (water, sediments).

The criteria listed by Mayer et al. (1992) (see above) address the problem of the significance of the response given by a biomarker with respect to effects at higher levels of biological organization. Given that cytochrome P4501A induction is not an (eco)toxicological effect in itself, two aspects must nevertheless be addressed, one pertaining to the state of health of wild organisms in which an induction is observed, the other pertaining to the precocity of the response of this biomarker in relation to other variables that indicate the state of health of the environment.

Some studies showed pathological traits at the individual level (histopathology, alteration of reproductive and growth parameters) and sometimes at the population level in fish in which cytochrome P4501A induction can be related to chemical contamination of their environment (see Table 5). A good agreement between induction of EROD activity in wild fish and an impact on the invertebrate communities was proved on several sites, in relation to the presence of a toxic chemical contamination (Richert, 1994). However, these observations are always done concomitantly and can never really prove that the response of the biomarker precedes the occurrence of other disorders. A hasty interpretation of the expression 'early warning system' (Payne et al., 1987) can lead us to postulate that P4501A induction is a signal indicating that a pathological process (in the general sense) with significant ecotoxicological consequences is under way, as if there is a cause-and-effect link between the occurrence of the signal and the process. In fact, as far as we now know, the most reasonable position would be to consider EROD induction as a biomarker of the presence of molecules representing an ecotoxicological hazard (not forgetting that P4501A induction is due to compounds such as PAH, PCB, and dioxins). The conditions of relationships between P4501A induction and significant disturbances in the biology of organisms are therefore still wide open to research (Monod, 1997).

A relevant use of this biomarker relies therefore on its practicability in the context of *in situ* studies and on a thorough understanding of 'risk' for hydrosystems. That understanding depends on characterization of the relationship between induction responses, sensitivity to inducer pollutants, and the possible consequences for contaminated individuals and populations. It seems essential to refine the typology of responses according to the characteristics of the environments studied, in the ecological as well as toxicological context.

Nevertheless, for the present, cytochrome P4501A induction is a biomarker that, associated with other indicators (e.g., toxicity tests, ecological indicators), can contribute to a strategy of chemical pollution management (such as comparison of effluent treatment processes or manufacturing processes) and the identification of areas to be restored.

REFERENCES

Adams S.M., Crumby W.D., Greeley M.S. et al. (1992). Relationships between physiological and fish population responses in a contaminated stream. *Environ. Toxicol. Chem.*, 11:1549–1557.

Ahokas J.T., Holdway D.A., Brennan S.E. et al. (1994). MFO activity in carp (*Cyprinus carpio*) exposed to treated pulp and paper mill effluent in Lake Coleman, Victoria, Australia, in relation to AOX, EOX and muscle PCDD/PCDF. *Environ. Toxicol. Chem.*, 13:41–50.

Anderson J.W., Rossi S.S., Tukey R.H. et al. (1995). A biomarker, P450 RGS, for assessing the induction potential of environmental samples. *Environ. Toxicol. Chem.*, 14:1159–1169.

Andersson T. and Förlin L. (1992). Regulation of the cytochrome P450 enzyme system in fish. *Aquat. Toxicol.*, 24:1–20.

Andersson T. and Koivusaari U. (1985). Influence of environmental temperature on the induction of xenobiotic metabolism by β-naphthoflavone in rainbow trout *Salmo gairdneri*. *Toxicol. Appl. Pharmacol.*, 80:43–50.

Andersson T., Förlin L., Hardig J. and Larsson A. (1988). Physiological disturbances in fish living in coastal water polluted with bleached kraft pulp mill effluents. *Can. J. Fish. Aquat. Sci.*, 45:1525–1536.

Ankley G.A. and Blazer V.S. (1988). Effect of diet on PCB-induced change in xenobiotic metabolism in the liver of channel catfish. *Can. J. Fish. Aquat. Sci.*, 45:132–137.

Biagianti-Risbourg S. (1997). Les perturbations (ultra) structurales du foie des poissons utilisées comme biomarqueurs de la qualité sanitaire des milieux aquatiques. In: Biomarqueurs en écotoxicologie: aspects fondamentaux. Lagadic L., Caquet Th., Amiard J.C. and Ramade F. (eds.). Masson, Paris, pp. 355–391.

Clemons J.H., Van Den Heuvel M.R., Stegeman J.J. et al. (1994). Comparison of toxic equivalent factors for selected dioxin and furan congeners derived using fish and mammalian liver cell lines. *Can. J. Fish. Aquat. Sci.*, 51:1577–1584.

Collier T.K., Anulacion B.F., Stein J.E. et al. (1995). A field evaluation of cytochrome P4501A as a biomarker of contaminant exposure in three species of flatfish. *Environ. Toxicol. Chem.*, 14:143–152.

Collier T.K., Connor S.D., Eberhart B.T.L. et al. (1992). Using cytochrome P450 to monitor the aquatic environment: initial results from regional and national surveys. *Mar. Environ. Res.*, 34:195–199.

De Flora S., Vigano L., D'Agostini F. et al. (1993). Multiple genotoxicity biomarkers in fish exposed *in situ* to polluted river water. *Mutation Res.*, 319:167–177.

Flammarion P. and Garric J. (1997). Cyprinids EROD activities in low contaminated rivers: a relevant statistical approach to estimate reference levels for EROD biomarker? *Chemosphere*, 35:2375–2388.

Forlin L. and Celander M. (1993). Induction of cytochrome P5401A in teleosts: environmental monitoring in Swedish fresh, brackish and marine waters. *Aquat. Toxicol.*, 26:41–56.

Forlin L. and Lindman U. (1981). Effects of Clophen A50 and 3-methylcholanthrene on the hepatic mixed function oxydase system in female rainbow trout, *Salmo gairdneri*. *Comp. Biochem. Physiol.*, 70C:297–300.

Gagnon M.M., Bussieres D., Dodson J.J. and Hodson P.V. (1995). White sucker (*Catostomus commersoni*) growth and sexual maturation in pulp mill-contaminated and reference rivers. *Environ. Toxicol. Chem.*, 14:317–327.

Goksøyr A. and Förlin L. (1992). The cytochrome P-450 system in fish, aquatic toxicology and environmental monitoring. *Aquat. Toxicol.*, 22:287–312.

Grizzle J.M., Horowitz S. and Strength D.R. (1988). Caged fish as monitors of pollution: effects of chlorinated effluent from a wastewater treatment plant. *Water Resour. Bull.*, 24:951–959.

Haasch M.L., Prince R., Weijksnora P.J. et al. (1993). Caged and wild fish: induction of hepatic cytochrome P-450 (CYP1A1) as an environmental biomonitor. *Environ. Toxicol. Chem.*, 12:885–895.

Haux C. and Förlin L. (1988). Biochemical methods for detecting effects of contaminants on fish. *Ambio*, 17:376–380.

Huuskonen S. and Lindström-Seppa P. (1995). Hepatic cytochrome P4501A and other biotransformation activities in perch (*Perca fluviatilis*): the effects of unbleached pulp mill effluents. *Aquat. Toxicol.*, 31:27–41.

Jimenez B.D. and Stegeman J.J. (1990). Detoxication enzymes as indicator of environmental stress on fish. *Amer. Fish. Soc. Symp.*, 8:67–79.

Jimenez B.D., Oikari A., Adams S.M. et al. (1990). Hepatic enzymes as biomarkers: interpreting the effects of environmental, physiological and toxicological variables. In: Biomarkers of Environmental Contamination. McCarthy J.F. and Shugart L.R. (eds.). Lewis Publishers, Boca Raton, pp. 123–142.

Kezic N., Britvic S., Protic M. et al. (1983). Activity of benzo(a)pyrene monooxygenase in fish from the Sava river, Yugoslavia: correlation with pollution. *Sci. Total Environ.*, 27:59–69.

Kloepper-Sams P.J., Swanson S.M., Marchant T. et al. (1994). Exposure of fish to biologically treated bleached-kraft effluent. 1. Biochemical, physiological and pathological assessment of rocky mountain whitefish (*Prosopium williamsoni*) and longnose sucker (*Catostomus catostomus*). *Environ. Toxicol. Chem.*, 13:1469–1482.

Larsson A., Anderson T., Förlin L. and Hardig J. (1988). Physiological disturbances in fish exposed to bleached kraft mill effluents. *Water Sci. Technol.*, 20:67–76.

Lindström-Seppa P. and Oikari A. (1988). Hepatic biotransformations in fishes exposed to pulp mill effluents. *Water Sci. Technol.*, 20:167–170.

Lindström-Seppa P. and Oikari A. (1990). Biotransformation activities of feral fish in waters receiving bleached pulp mill effluents. *Environ. Toxicol. Chem.*, 9:1415–1424.

Masfaraud J.F,. Monod G. and Devaux A. (1990). Use of the fish cytochrome P-450 dependent 7-ethoxyresorufin O-deethylase activity as a biochemical indicator of water pollution. Study of the liver and the kidney of male and female nase (*Chondrostoma nasus*) from the river Rhône. *Sci. Total Environ.*, 97/98:729–738.

Mayer F.L., Versteeg D.J., MacKee M.J. et al. (1992). Physiological and nonspecific biomarker. In: Biomarkers: Biochemical, Physiological, and Histological Markers of Anthropogenic Stress. Huggett R.J., Kimerle R.A., Mehrle P.M. and Bergman H.L. (eds.). Lewis Publishers, Chelsea, pp. 5–86.

Melancon J.M., Steven E.Y. and Lech J.J. (1987). Induction of hepatic microsomal monooxygenase activity in fish by exposure to river water. *Environ. Toxicol. Chem.*, 6:127–135.

Monod G. (1997). L'induction du cytochrome P4501A1 chez les poissons. In: Biomarqueurs en écotoxicologie: Aspects fondamentaux. Lagadic L., Caquet Th., Amiard J.C. and Ramade F. (eds). Masson, Paris, pp. 33–54.

Monod G. and Vindimian E. (1991). Effect of storage conditions and subcellular fractionation of fish liver on cytochrome P-450-dependent enzymatic activities used for the monitoring of water pollution. *Water Res.*, 25:173–177.

Monod G., Devaux A. and Rivière J.L. (1988). Effects of chemical pollution on the activities of hepatic xenobiotic metabolizing enzymes in fish from the river Rhône. *Sci. Total Environ.*, 73:189–201.

Munkittrick K.R., Van Den Heuvel M.R., Metner D.A. et al. (1993). Interlaboratory comparison and optimization of hepatic microsomal ethoxyresorufin O-deethylase activity in white sucker (*Catostomus commersoni*) exposed to bleached kraft pulp mill effluent. *Environ. Toxicol.*, 12:1273–1282.

Myer C.R., Sutherland L.A., Haasch M.L. and Lech J.J. (1993). Antibodies to a synthetic peptide that react specifically with rainbow trout hepatic cytochrome P450 1A1. *Environ. Toxicol. Chem.*, 12:1619–1626.

Nixon S.C., Mainstone C.P., Iversen T.M. et al. (1996) The harmonised monitoring and classification of ecological quality of surface waters in the European Union. Final report. Water Research Ref CO 4150. 289 p.

Payne J.F., Fancey L.L., Rahimtula A.D. and Porter E.L. (1987). Review and perspective on the use of mixed-function oxygenase enzymes in biological monitoring. *Comp. Biochem. Physiol.*, 86C:233–245.

Richert C. (1994). Comparaison de marqueurs biochimiques chez les poissons et d'indicateurs écologiques pour le diagnostic *in situ* de la pollution toxique dans les cours d'eau. Doctoral thesis, Université Claude Bernard-Lyon I.

Singh H., Pavgi-Singh S., Kezic N. and Kurelec B. (1985). Xenobiotic and endobiotic induction of mixed function monooxygenase in carp *Cyprinus carpio*. *Sci. Total Environ.*, 44:123–133.

Soimasuo R., Jokinen I., Kukkonen J. et al. (1995). Biomarker responses along a pollution gradient: effects of pulp and paper mill effluents on caged whitefish. *Aquat. Toxicol.*, 31:329–346.

Stegeman J.J., Teng F.Y. and Snowberger E.A. (1987). Induced cytochrome P450 in winter flounder (*Pseudopleuronectes americanus*) from coastal Massachusetts evaluated by catalytic assay and monoclonal antibody probes. *Can. J. Fish. Aquat. Sci.*, 44:1270–1277.

Ueng T.H., Ueng Y.F. and Park S.S. (1992). Comparative induction of cytochrome P-450-dependent monooxygenases in the livers and gill of tilapia and carp. *Aquat. Toxicol.*, 23:49–64.

Van Den Heuvel M.R., Munkittrick K.R., Van Der Krak G.J. et al. (1994). Survey of receiving-water environmental impact associated with discharges from pulp mills. 4. Bioassay-derived 2,3,7,8-tetrachlorodibenzo-p-dioxin toxic equivalent concentration in biochemical indicators of impact. *Environ. Toxicol. Chem.*, 13:1117–1126.

Van Der Oost R., Heida H., Opperhuizen A. and Vermeulen N.P.E. (1991). Interrelationships between bioaccumulation of organic trace pollutants (PCBs, organochlorine pesticides and PAHs), and MFO-induction in fish. *Comp. Biochem. Physiol.*, 100C:43–47.

Van Der Oost R., Van Gastel L., Worst D. et al. (1994). Biochemical marker in feral roach (*Rutilus rutilus*) in relation to the bioaccumulation of organic trace pollutants. *Chemosphere*, 29:801-817.

Van Der Weiden M.E.J., Celander M., Seinen W. et al. (1993). Induction of cytochrome P450 1A in fish treated with 2,3,7,8-tetrachlorodibenzo-p-dioxin or chemically contaminated sediment. *Environ. Toxicol. Chem.*, 12:989–999.

Van Der Weiden M.E.J., Craane L.H.J., Evers E.H.G. et al. (1989). Bioavailability of PCDDs and PCDFs from bottom sediments and some associated biological effects in the carp. *Chemosphere*, 19:1009–1016.

Vignier V., Vandermeulen J.H., Singh J. and Mossman D. (1994). Interannual mixed function oxidase (MFO) activity in winter flounder (*Pleuronectes americanus*) from a coal tar contaminated estuary. *Can.J. Fish. Aquat. Sci.*, 51:1368–1375.

Vindimian E. and Garric J. (1989), Freshwater fish cytochrome P450 dependent enzymatic activities: a chemical pollution indicator. *Ecotoxicol. Environ. Saf.*, 18:277–285.

Vindimian E., Namour P., Migeon B. and Garric J. (1991). *In situ* pollution induced cytochrome P450 activity of freshwater fish barbel (*Barbus barbus*), chub (*Leuciscus cephalus*) and nase (*Chondrostoma nasus*). *Aquat. Toxicol.*, 21:255–266.

Vindimian E., Namour P., Munoz J.F. et al. (1993). Ethoxyresorufin-O-deethylase induction in fish from a watershed exposed to a non point source pollution of agricultural origin. *Water Res.*, 27:449–455.

Zacharewski T., Safe L., Safe S. et al. (1989). Comparative analysis of polychlorinated dibenzo-p-dioxin and dibenzofuran congeners in Great Lakes fish extracts by gas chromatography-mass spectrometry and *in vitro* enzyme induction activities. *Environ. Sci. Technol.*, 23:730–735.

5

Use of Metallothioneins as Biomarkers of Exposure to Metals

R.P. Cosson and J.-C. Amiard

INTRODUCTION

The physico-chemical forms in which metals are stored in living tissues affect their reactivity in relation to their intracellular environment. 'Fossilized' as inert compounds in the tissues, metals, even toxic metals (such as cadmium and mercury), represent no further hazard except for an eventual predator, provided that the predator possesses the enzymatic capacity to degrade the complex mineral formed. On the other hand, an association with a proteinic ligand is never indefinitely stable. This type of ligand can exchange metals that it fixes with other ligands, according to the reactions involved in the affinities of metals present for sites of fixation of the ligands themselves. This 'exchangeable' character of such fixed metals and their possible release in the cell during catabolism of the ligand means that their detoxification is not definitive. They could thus interfere with the normal metabolic function of tissues or organs in which they are stored.

It is a well-established fact that overall levels of trace elements in an organism cannot provide an acceptable response for the evaluation of the biocoenotic impact of their excessive presence in the environment, because they do not provide means to assess the possible toxicity of the accumulated element. It was therefore necessary to refine our understanding of their accumulation in organisms. A preliminary step was the identification of target organs, which vary according to the species, and which are used to quantify metals. Three types of organs were selected:

- organs located at the interface between the external environment and the organism (gills or digestive tract);
- organs in which metabolic functions can be altered by the presence of toxic metals or excessive levels of metals in their cells (liver or digestive gland);
- organs involved in storage or excretion of metals (kidney).

From such studies, a possible disturbance of physiological functions of organisms could be estimated, in relation to the organ or tissue in which metals tend to accumulate. The next stage took into account their distribution between the insoluble compounds and soluble compounds that can fix them within these organs and tissues. The metallothioneins (MTs), which are proteins that can be induced by metals and have a strong affinity for them, therefore appeared to be preferred intracellular ligands (Amiard and Cosson, 1997). Several studies indicated their participation in processes of detoxification and acquisition of some tolerance and resistance to metals (Kagi and Kojima, 1987). The spillover theory, proposed in the late 1970s, developed the idea that as long as a contaminated organism is capable of synthesizing enough MTs to prevent metals from fixing to other metalloproteins, no toxic effect manifests itself (Brown et al., 1977; Brown and Parsons, 1978). These considerations led the scientific community to consider the use of MTs as biomarkers of exposure to metals, since they appeared to be key molecules in regulation of intracellular content of metals and strongly implicated in the process of detoxification.

For MTs to be used as biomarkers of exposure to metals, they must have the following characteristics (Stegeman et al., 1992):

- The technique used for their quantification must simultaneously exhibit sensitivity, reliability, precision, good reproducibility, and simplicity of implementation (especially if it is to be used in the field).
- Their reference levels must be known, to allow us to establish a clear distinction between their natural variations and variations in response to stress.
- All the intrinsic and extrinsic factors modifying their level in the test organism must be perfectly known so that variations due to these factors are not attributed to the presence of a contaminant.
- It must be established whether changes in their levels are due to temporary modification of the physiology of an organism, or to a more profound modification, such as a genetic adaptation, in response to the presence of a contaminant.
- The occurrence of modifications in their basal level must be linked to disturbances in the health of the organisms.

1. GENERAL SURVEY OF ANALYTIC TECHNIQUES FOR DETECTION AND QUANTIFICATION OF METALLOTHIONEINS

The use of a potential biomarker cannot be examined independently of the evaluation of techniques that enable its quantification. We will therefore review the different techniques enabling quantification of MTs in organisms sampled *in situ*. The concentration of MTs in the soluble fraction of cells, homogenates of tissues, and biological fluids is lower than the concentration of other proteins. The quantification requires therefore a particular treatment designed to separate them from other proteins. An outline proposed by Summer and Klein (1993) summarizes the stages of their separation, detection, and quantification (Fig. 1).

1.1. Chromatographic Techniques

Chromatographic techniques permit the separation of MTs from other soluble compounds present in tissue homogenates. Given the absence of specific activity of MTs, the criteria used to characterize isolated fractions rest on the physico-chemical properties of these proteins (Amiard and Cosson, 1997) or on the detection of metals associated with them.

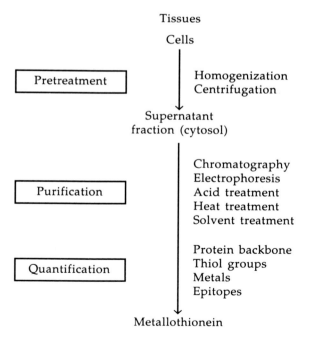

Fig. 1. Protocol of metallothionein extraction (according to Summer and Klein, 1993)

On gel permeation columns, the MTs elute like compounds of molecular weight 12,000 to 15,000 Da. To ensure that they are actually MTs, it is advisable to verify that these compounds persist in the denaturated homogenates at high heat and that they have a very low absorbency at a wavelength of 280 nm, because MTs do not consist of (or have very little of) aromatic amino acids. On the other hand, these compounds must show significant absorbence at a wavelength characteristic of metal-cystein liaisons (Zn-SH = 230 nm, Cd-SH = 250 nm, Cu-SH = 270 nm, Hg-SH = 310 nm).

If the preparation of samples and their conditions of elution on the column have respected the stability of metal-cystein liaisons (sensitive to oxidation), the peaks containing the MTs must show consistent levels of metals. It is possible from the measurement of metal concentrations in the fractions containing MTs to deduce their concentration, based on stoichiometric ratios of metal to apoprotein. The natural MTs are never saturated by a single metal. Different metals (silver, cadmium, copper, mercury, zinc) can be linked to proteins, which poses analytical and methodological problems. It is technically possible to measure simultaneously and precisely several metals present in low concentrations in a sample, but it requires a high performance and costly apparatus. Moreover, if the stoichiometry of MTs containing zinc or cadmium is well known (7 metallic atoms per molecule of apoprotein), it is not the same for those that contain copper (in principle 12 atoms per apoprotein). The uncertainty is even higher for mercury and silver (ratios not yet entirely clarified with regard to natural MTs, but possibly reaching 18 atoms per apoprotein; Lu et al., 1990; Zelazowski and Stillman, 1992). It therefore seems difficult to recommend such an approach to quantify the concentration of MTs. However, this analytic protocol may provide very useful information on the metallic composition of MTs when metals are analysed continuously in fractions separated by high performance liquid chromatography (HPLC; Suzuki, 1980). The applications of HPLC to separation and detection of MTs have been considerable. Presently, the use of anion exchange columns or reversed phase columns allows separation and quantification of different isoforms of MTs, either indirectly by measurement of associated metals (Suzuki et al., 1983), or directly by measurement of the surface of the corresponding peaks (Richards and Steele, 1987). Despite constant improvements, these techniques remain difficult to implement because they necessitate sophisticated analytic instruments and are relatively time-consuming, which is a serious handicap in analysis of a series of samples.

1.2 Metal Saturation Assay

The different protocols used are based on the difference in the affinity of metals for apoprotein, which decreases in the following order: Hg >

Ag > Cu > Cd > Zn. Once all the sites of MTs are saturated by the metal used (cadmium, mercury, or silver), this metal is quantified (via atomic absorption spectrophotometry or by measurement of the radioactivity of a marked isotope) and the MT content is calculated in reference to the stoichiometric ratios mentioned earlier.

The method proposed by Onosaka and Cherian (1982) and improved by Eaton and Toal (1982) relies on saturation of binding sites of metals of MTs by radioactive cadmium (^{109}Cd). This method was very frequently used, because the number of Cd ions fixed by a molecule of MT (7) is well established. The process of substitution, on the other hand, does not work perfectly except for MTs that contain only zinc and/or a very small amount of copper. It works even less well for MTs containing mercury or silver.

Piotrowski et al. (1973) used mercury (^{203}Hg), but the low specificity of mercury fixation seemed to introduce an overestimation of MT content and the method seemed to lack sensitivity (Webb, 1979; Dieter et al., 1987). It appears that these disadvantages can be overcome by the addition of ovalbumin (Dutton et al., 1993) or rat hemoglobin (Couillard et al., 1993).

Scheuhammer and Cherian (1986) developed a method very similar to the protocol used for cadmium, but replacing it with silver. The silver fixed by MTs can be quantified by atomic absorption spectrophotometry or using a radioactive isotope (^{110m}Ag). The strong affinity of silver for MTs excludes all possibility of substitution by metals initially present on the MTs (except perhaps mercury).

1.3. Polarographic Method

The polarographic method relies on the measurement of change in intensity of a current caused by the reduction of hydrogen of sulfhydryl groups of MT (Olafson and Sim, 1979). The analysis is not influenced by the nature of metals fixed on the protein or by the species (or tissue) analysed (Hogstrand and Haux, 1992). The use of a specific electrolyte, elimination of soluble compounds that are heat-sensitive (or that can be denatured by ethanol), and reduction in the temperature of the analysis allow us to directly quantify MTs in solution without major interference (Thompson and Cosson, 1984). The specificity of this analytic technique was confirmed by chromatographic study (permeation and anion exchange) of polarographic active peaks (Olafson and Olsson, 1991).

1.4 Immunological Methods (RIA and ELISA)

The quantification of MTs by RIA (radioimmunoassay) or by ELISA methods (enzyme-linked immunosorbent assay) was made possible by the production of polyclonal antibodies (Tohyama and Shaikh, 1978;

Van der Malie and Garvey, 1979; Thomas et al., 1986; Roesijadi et al., 1988). These techniques are much more sensitive than the preceding ones, but their use cannot be easily extended because they remain subject to the possession of an antibody that can react with MTs of the species studied. The antibodies produced against the MTs of mammals may have good reactivity against MTs of other species of mammals, but they react little with MTs of non-mammal species (Duquesne, 1992). There are antibodies produced against MTs of different fishes (Kito et al., 1986; Chatterjee and Maiti, 1987; Hogstrand and Haux, 1990; Norey et al., 1990) that can be used in an interspecific study because of the strong homology of MTs of fish (George and Olsson, 1994) and anti-MT antibodies of mussel (Roesijadi et al., 1988). The precise quantification of MTs by these highly effective methods demands therefore a study of the appropriate antibody used, particularly with respect to cross-reactions with MTs of the species studied, as well as to different isoforms present.

1.5 Quantification of mRNA

Regulation of MT synthesis occurs at the transcriptional level, during the production of messenger RNA (mRNA) coding for the apoprotein (Amiard and Cosson, 1997). It is therefore possible to consider measuring an increase in the production of these mRNA in response to a metallic contamination of the environment, rather than to seek to prove an increase in the corresponding protein. For that, it is necessary to use a nuclear probe (simple marked strand) that can link itself (hybridization by complementarity) with the mRNA and form a double strand. This technique, clearly more sensitive than earlier ones, allows us to detect responses at the cellular level. The preparation of specific probes requires cloning of the gene or genes coding for MTs in the species studied. Several cDNA have been cloned in fish (Bonham et al., 1987; Chan et al., 1989; Leaver and George, 1989; Kille et al., 1991), in a sea urchin (Nemer et al., 1985), and in an oyster (Roesijadi et al., 1991). For the present, this method has only been used operationally in evaluation of the impact of pollutants in the Forth estuary in Scotland (George and Olsson, 1994), and it has proved to be well adapted to this type of study.

1.6 Remarks

Comparative essays of different methods for quantification of MTs have been published, as well as refinements in the use of these methods in field studies (Onosaka and Cherian, 1982; Nolan and Shaikh, 1986; Dieter et al., 1987; Hogstrand and Haux, 1989; Petering et al., 1990;

Stegeman et al., 1992; Summer and Klein, 1993; George and Olsson, 1994). The authors agree in attributing a greater sensitivity to immunological techniques and to measurement of mRNA, but the routine use of these methods seems improbable, given the specificity of the responses involved. Moreover, these methods do not give information such as the metal associated with the apoprotein, which may pose problems in interpreting a variation in biomarker level (see later). Moreover, it seems difficult to use only the quantification of mRNA, because this indicates only the quantity of messengers present, and not the quantity of protein, which depends on rates of synthesis and degradation. The method of silver saturation (the most effective of saturation assays) is slightly less sensitive than the electrochemical technique. It is also more difficult to use. Here again, neither of the two methods provides information on the metals associated with the MTs. Only quantitative analysis of metals in fractions separated by chromatography can provide such data. Moreover, chromatography enables identification of different isoforms of MTs, which may be important if it is confirmed that the induction of each of them depends specifically on a metal. These analytic techniques therefore seem complementary, and the choice of one or another must depend on the particular characteristics of the study to be conducted rather than on the criterion of performance. It may also be advantageous to combine two of these techniques.

2. GENERAL SURVEY OF SOURCES OF VARIABILITY IN METALLOTHIONEIN CONTENT

Independently of metals, induction of MT synthesis may be triggered by several other factors (e.g., hormones, stress; Amiard and Cosson, 1997). This is why MT contents can vary with the species, physiological condition, and toxicological history of the organism, or any combination of these factors.

Most of the information we have on the regulation of genes coding for MTs and on the potential inducers of their synthesis comes from experiments done on mammals (Dunn et al., 1987; Palmiter, 1987). An abundant literature is devoted to responses of different fish species to an entire series of stresses provoking this synthesis (George and Olsson, 1994). Even though several authors have isolated (or revealed) MTs in invertebrates, the participation of these proteins in processes of detoxification of metals in these organisms is still not clearly understood (Roesijadi, 1992, 1993).

Variations in MT levels in different biological fluids (blood, urine) in response to exposure to metallic pollutants can be quantified, but it is rare to use mammal as sentinel organism in estimating environmental contamination. Nevertheless, such studies have been made on humans (Falck et al., 1983; Roels et al., 1983; Clough et al., 1986), horses (Koizumi et al., 1989), and aquatic mammals (Olafson and Thomson, 1974; Lee et al., 1977; Ridlington et al., 1981; Kwohn et al., 1986; Wagemann and Hobden, 1986; Caurant, 1994). The environmental impact of metallic wastes has generally been studied on invertebrates or fish, because the analytic techniques used involve killing the specimens. The choice of sentinel organism will, however, be important, because induction of MT synthesis in response to the presence of metals in the environment is not equivalent in all organisms.

2.1 Variations Linked to Physiological Condition of Organisms

In order to be able to use variations in MT levels as indicators of environmental contamination, we must define the reference level of MTs to which the observed levels can be compared. Unfortunately, MT levels in tissues vary naturally, independently of the presence of metals, which makes it very difficult to establish reference levels. These natural variations are more or less significant according to the taxonomic groups considered. They are frequently found in invertebrates, in which the metabolism itself varies according to season, in relation to moulting, reproduction, nutrition, and other factors.

2.1.1 Variability in Fish

In fish, variations in MT levels were shown to depend on age, sex, and sexual maturity (Olsson et al., 1987; Overnell et al., 1988; George, 1989; Hamza-Chaffai et al., 1995). Duquesne (1992), working on three species of fish collected in the Pas-de-Calais, showed that there it is not a general phenomenon (Fig. 2). In dab (*Limanda limanda*), hepatic levels of copper and zinc (associated with soluble ligands) varied with the sex of

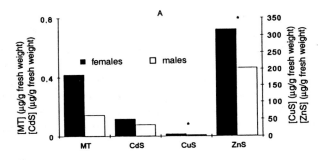

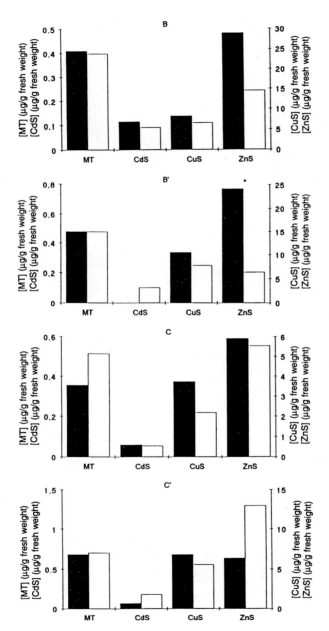

Fig. 2. Influence of sex on hepatic MT levels and levels of soluble metal in dab (A), lemon sole (B, B′), and cod (C, C′) (according to Duquesne, 1992): A and C, spring 1989; B′ and C′, winter 1989; B, random sample in the period. *t significant at 5%.

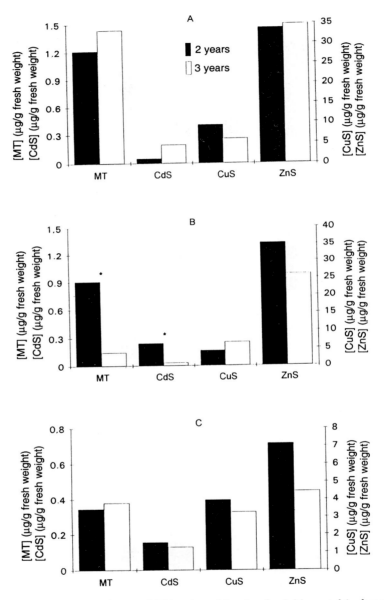

Fig. 3. Influence of age on hepatic MT levels and levels of soluble metal in female dab (A), lemon sole (B), and cod (C) (according to Duquesne, 1992): A, sampled in autumn 1989; B, spring 1989; C, winter 1989. *t significant at 5%.

the individuals captured, while MT levels were stable. In lemon sole (*Microstomus kitt*), only the soluble zinc level varied with the sex, while in cod (*Gadus morhua*) no significant difference in these levels could be established between the sexes. The same author has shown that hepatic levels of MT were also not systematically correlated with the age of the fish analysed (Fig. 3). Neither female dab nor female cod showed significant modification in hepatic MT level with age, whereas female lemon sole aged 3 years had MT levels significantly higher than females aged 2 years, in relation with a significant increase in cadmium level in their livers. It emerges from this that if a fish is chosen as sentinel organism, this variability in responses depending on sex and age must be taken into account in defining the reference levels to which the values measured *in situ* must be compared.

2.1.2 Variability in Invertebrates

As we have already mentioned, variability is much greater in invertebrates, because ligands other than MTs frequently affect the processes of sequestration of metals. These are:

- cytosolic proteins that fix metals (Dohi et al., 1983; Young and Roesijadi, 1983; Stone and Overnell, 1985; Webb et al., 1985; Brouwer et al., 1986; Dennai et al., 1986; Luten et al., 1986; Langston and Zhou, 1987; Lobel and Marshall, 1988; Nejmeddine et al., 1988; Langston et al., 1989; Roesijadi and Klerks, 1989);
- circulatory proteinic compounds or respiratory pigments containing copper, such as hemocyanins (George and Pirie, 1980; Pickwell and Steinert, 1984);
- mineral complexes forming granules and/or spherocrystals (Coombs and George, 1978; George and Pirie, 1979; Simkiss et al., 1982; George et al., 1984; Martoja and Martoja, 1984).

2.1.2.1 Molluscs

Generally, there is very little information on functional relationships between the different mechanisms of metal fixation in molluscs. In particular, we do not know whether they act alternatively, compete with each other, or are complementary. In mussel (*Mytilus edulis*), preferential accumulations of certain metals are observed in certain organs (copper in the digestive gland, cadmium and zinc in the kidney), where their half-life is variable ($t_{1/2}$ (Cu) = 6 d; $t_{1/2}$ (Cd) = 60 d; $t_{1/2}$ (Zn) = 300 d). These metals are very likely transported towards lysosomes associated with MTs, and their kinetics of accumulation or degradation are specific to the tissues concerned, in relation with the efficacy of the proteolysis, peroxidation, and lysosomal activity (George and Olsson, 1994). During the year, these molluscs show phases of weight loss (winter), accumulation and/or mobilization of reserves, and gonadic development, among others, which profoundly modify the metabolic activity of different organs depending on their involvement in the corresponding physiological processes. Variations in MT levels were studied in digestive gland of male and female clams (*Ruditapes decussatus*) for four months (Hamza-Chaffai et al., 1999). The fluctuations observed (Table 1) were greater in females than in males. This is related to the reproductive cycle (gamete emission and gonadic restoration). Male *R. decussatus* would therefore be more useful than females in a biosurveillance based on MTs. It is clear that the binding of metals by one intracellular ligand or another, in one organ or another, varies with the metabolism of the organism.

Table 1. Variations in MT levels in digestive gland of male and female clam collected over four consecutive months from a reference site. The levels are expressed in mg MT/ g fresh tissues (m, mean; SD, standard deviation).

		June	July	August	September
Females	m	2.65	5.99	4.20	2.71
	SD	0.83	1.22	1.69	1.05
Males	m	3.22	3.32	3.40	1.66
	SD	1.60	1.02	0.80	0.41

2.1.2.2 Crustaceans

The growth and moulting cycle of crustaceans profoundly complicate the metabolism of metals (Engel and Brouwer, 1984; Jenkins et al., 1984). Synthesis and degradation of hemocyanin along the moulting cycle is phased with fluctuations in levels of MTs rich in copper, themselves related to the relative abundance of MTs rich in zinc. The MTs function as a reservoir for the copper necessary to hemocyanin synthesis (Fig. 4). Regulation of this mechanism depends on steroid hormones that control moulting. Specific receptors of glucocorticoids have been indicated at the MT gene level in vertebrates (Hamer, 1986). The indication of regulator sites sensitive to steroids at the MT gene level in invertebrates would confirm this hypothesis (Roesijadi, 1993). Also, even though its use has been recommended (Hennig, 1986), the study of fluctuations in MT levels in crustaceans does not seem to be a reliable tool for estimating the degree of environmental pollution by metals.

2.2 Variations Linked to the Species Studied

The variations observed are of multiple origins. They derive mainly from the relative affinities of metals for different types of intracellular ligands present in the organs exposed to metals or in the organs ensuring their metabolization and/or storage. This intermetallic competition influences the binding of one metal or another by the MTs or by other

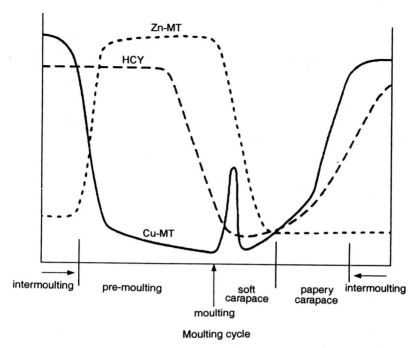

Fig. 4. Dynamics of metals, hemocyanin (HCY) and MTs (Zn-MT and Cu-MT) in the digestive gland of crab during moulting (according to Engel and Brouwer, 1993)

ligands. A more or less perceptible modification of MT levels will follow. The result is that the response to a given metal contamination varies with the zoological group, species, and organ involved.

2.2.1 Variability between Zoological Groups

2.2.1.1 Comparison between Amphibians and Fishes

The protective role of MTs against toxic metals (or excess metals) is not certain unless they bind them preferentially, and in a proportion sufficient to prevent them from denaturing other metalloproteins that play a more essential role. This process, which involves the synthesis of MTs, is observed only in some species. The protective role of MTs can be amplified by a prior exposure of organisms to a sublethal concentration of metals, which results in increased synthesis of MTs in response to a subsequent exposure to the same metal, at lethal or non-lethal concentrations. Woodall et al. (1988) compared the effect of pretreatment with zinc and cadmium of *Xenopus laevis* tadpoles and trout fry (*Salmo gairdneri*) on the tolerance of these two species to later exposure to either metal. Without pretreatment, zinc-tolerance was equivalent in the two species, while the trout fry proved to be 10 times as sensitive to cadmium as the tadpoles. For the tadpoles, the pretreatment with either metal led to increased tolerance to both, while for trout fry the results varied depending on the metal. Pretreatment with cadmium did not lead to increased tolerance to both metals, as did pretreatment with zinc. The authors interpreted these results in terms of differential redistribution of zinc and cadmium between the intracellular ligands of these metals in the two species. In the case of the tadpoles, pretreatment led to disturbance in the distribution of metals at the intracellular level, which would induce synthesis of MTs that could then have an impact on detoxification of excessive ions (whether Cd or Zn). In the case of trout fry, MT synthesis was induced only by pretreatment with zinc, which resulted in increased tolerance to both metals. Because of this, the tadpoles could be used as bioindicators of the presence of zinc or cadmium in water, via the monitoring of their MT levels, whereas the trout fry seem to respond only to the presence of zinc.

2.2.1.2 Comparison between Poikilotherms and Homeotherms

The use of MTs as biomarkers of exposure requires us to take into account the rapidity with which the response to metallic stress appears and the lapse of time during which this response remains detectable in tissues of the organisms studied. In mammals the response is detected some hours after the ingestion of contaminated feed, while in invertebrates a delay of several days has often been observed (Roesijadi, 1982; Young and Roesijadi, 1983). This delay is generally attributed to the influence of temperature on the rapidity of *de novo* synthesis of MTs, to the rate of penetration of metals in the organism, as well as to the speed with which they are transported towards the target organs. Few studies integrated time in their approach. Half-lives for Cd-MTs of 3–5 and 4–5 d have been observed respectively in the liver and kidneys of rat (Chen et al., 1975; Shaikh and Smith, 1976; Feldman et al., 1978; Held and Hoekstra, 1984). In plaice (*Pleuronectes platessa*) contaminated by zinc, Overnell et al. (1987) calculated half-life values of Zn-MTs varying from 22 to 32 d for liver and kidney. Bebianno and Langston (1993) showed that in mussels contaminated by cadmium, the half-life of MTs was 25 d, while that of cadmium itself was 300 d. There is therefore a significant difference between mammals and poikilothermic organisms, especially marine invertebrates. The determination of half-life of MTs in a sentinel organism seems therefore to be a prerequisite to its use in tracing environmental contamination.

2.2.2 Variability within a Single Zoological Group

2.2.2.1 Invertebrates

In crustaceans, drosophila and sea urchins, certain metals have been shown to specifically induce certain isoforms of MTs. Similarly, in the oyster *Crassostrea virginica*, there are non-acetylate and N-acetylated MTs. The acetylate form, which responds more quickly to exposure to cadmium (Roesijadi et al., 1991), is found in greater abundance in individuals taken from contaminated areas (Roesijadi, 1993). It seems desirable therefore, in certain cases, to isolate and quantify the different isoforms of MTs as well as the metals associated with them.

2.2.2.2 Vertebrates

George and Olsson (1994) listed and analysed the results of researches done for some years on MTs of teleost fishes. The structure of MTs of fishes is particularly well preserved, which makes the use of immunological techniques for their quantification possible (see above). However, there is no explanation for the presence of a single isoform in certain species (*Solea solea, Pleuronectes platessa, Scopthalmus maximus, Pseudopleuronectes americanus, Noemachelius barbatulus, Esox lucius*), while others have two (*Cyprinus carpio, Perca fluviatilis, Onchorynchus mykiss, Scorpaena guttata*). Furthermore, Harrison and Klaverkamp (1989) showed that the preferential sites of fixation of metals (tissues or organs) and their rate of fixation vary from one species to another, notably as a function of the mode of contamination (food, water). In consequence, the response to metallic pollutants, in terms of MT synthesis, is itself variable from one species to another, which leads to a specificity of sensitivity to toxicants (Brown et al., 1986).

2.2.3 Variability within a Single Species

2.2.3.1 Variability of Genetic Origin

It is generally acknowledged that a greater abundance of MTs in the tissues ensures better protection against the toxicity of metals. This abundance may result from a temporary physiological adaptation of the organism to the degradation of its environment, or it may indicate a genetic adaptation. In *Drosophila melanogaster*, two genes (Mtn and Mto) coding for two MTs were indicated (Lastowski-Perry et al., 1985; Mokdad et al., 1987). Even though they are considered homologues, these two genes code for proteins that have only 25% identity between their amino acids (Silar, 1990). Duplication of the gene Mtn was observed, associated with the acquisition of resistance to copper in natural populations (Maroni et al., 1987). On the other hand, duplication of the gene Mto was not observed in the natural environment. This gene would code for an MT more closely involved in homeostasis of copper and zinc, in relation with growth and cellular multiplication (Silar, 1990). Duplication of Mtn gene was observed only in industrialized regions, and it is interpreted as an adaptive response to environmental contamination (Roesijadi, 1993). Measurement of high MT levels in the individuals having a duplicated Mtn gene may allow us to conclude environmental contamination by metals. However, it would be false to deduce from this that these populations are in danger, because they have adapted to the chronic pollution of their environment. On the other hand, their resistance to metallic pollutants poses a threat for their predators, because of a possible trophic transfer of the contaminants (Stegeman et al., 1992; Dallinger, 1993).

2.2.3.2 Variability in Tissues

Induction of MT synthesis is triggered in response to several factors other than metals (Kagi, 1993), but these are the most powerful inducers. However, rates of synthesis may vary significantly according to the species, tissue (or organ), and inducer metal. The relative proportions of metals (silver, cadmium, copper, mercury, zinc) bound to the sites of MTs depend on their respective affinities for the -SH groups and their intracytoplasmic abundance. These metals are easily exchangeable between MT molecules or with other intracellular ligands. Displacement of a metal initially fixed by the MTs present in a tissue is not systematic and, if it happens, does not always lead to the *de novo* synthesis of MTs. In the gills of mussel, it is possible to induce MT synthesis by exposure to cadmium, copper, and mercury, but not by exposure to zinc (Roesijadi et al., 1988). In carp (*Cyprinus carpio*), MT synthesis in response to exposure to metals (silver, cadmium, mercury, zinc) varies according to the tissue studied (gills or liver), the metal used, and pretreatment with zinc of fish before exposure to other metals (Cosson, 1994a, 1994b).

At the branchial level, mercury is responsible for the greatest increase in MT concentration observed, whether or not there has been pretreatment with zinc, and this persists even after a week of decontamination (Fig. 5). Cadmium and silver produced similar increases, but the delay in response of gills is longer in the case of silver. Zinc only slightly modifies MT levels in gills, and only after a delay (Kito et al., 1982a, 1982b; Klaverkamp and Duncan, 1987).

In the liver, the greatest induction is due to mercury but, unlike what happens in gills, the decontamination phase leads to a decrease in MT content, more significant when the fishes have been pretreated with zinc (Fig. 6). As with gills, cadmium and silver produce an increase in MT level of liver, but here the delay in response is greater in the case of cadmium. However, while silver is associated with the cytosolic

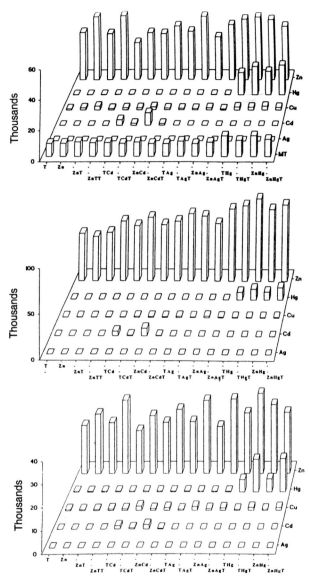

Fig. 5. Mean levels of metals in the supernatant (top) and pellet (middle) of homogenates of carp gills and in the pellet (bottom) obtained after heat-denaturation of the supernatant (ng/g); mean contents of MTs (10 × μg MT/g) in the heat-stable fraction of supernatant. T, control; Zn, 1 week pretreatment with zinc; ZnT, 1 week pretreatment with zinc followed by 1 week decontamination; ZnTT, 1 week pretreatment with zinc followed by 2 weeks decontamination; 1 week pretreatment with zinc followed by 1 week exposure to cadmium (ZnCd), silver (ZnAg), mercury (ZnHg), followed by 1 week decontamination (ZnCdT, ZnAgT, ZnHgT).

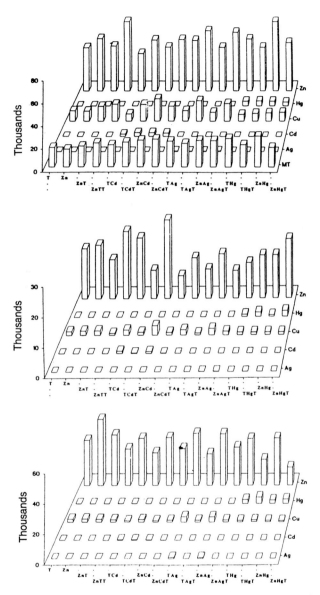

Fig. 6. Mean levels of metals in the supernatant (top) and pellet (middle) of homogenates of carp livers and in the pellet (bottom) obtained after heat-denaturation of the supernatant (ng/g); mean contents of MTs (10 × µg MT/g) in the heat-stable fraction of supernatant. T, control; Zn, 1 week pretreatment with zinc; ZnT, 1 week pretreatment with zinc followed by 1 week decontamination; ZnTT, 1 week pretreatment with zinc followed by 2 weeks decontamination; 1 week pretreatment with zinc followed by 1 week exposure to cadmium (ZnCd), silver (ZnAg), mercury (ZnHg), followed by 1 week decontamination (ZnCdT, ZnAgT, ZnHgT).

fraction containing MTs only after the decontamination phase, cadmium is closely associated with MTs from the contamination phase. As with gills, zinc modifies the MT levels in liver only after some delay, but the relative increase is twice as great.

These results may be interpreted taking into account ion flows (silver, cadmium, copper, mercury, zinc) between the different compartments (soluble and insoluble) and the different soluble ligands (heat-sensitive and heat-resistant). The studies show the absence of a systematic relationship between accumulation of metals and increase in MT levels in tissues. They emphasize the need to quantify the metals in the different cellular compartments so that corresponding fluctuations in MT levels can be interpreted.

3. EXAMPLES OF APPLICATIONS

The quantification of MTs was used by several authors to estimate the impact of metallic wastes in the environment. Most often, studies have addressed aquatic ecosystems in which pollutions of diverse origins have been found. These studies have used species belonging to various taxonomic groups:

- molluscs: Hennig, 1986; Imber et al., 1987;
- crustaceans: Hennig, 1986;
- fishes: Roch et al., 1982; Roch and McCarter, 1984; Benson and Birge, 1985; Olsson and Haux, 1986; Benson and Birge, 1987; Overnell et al., 1987; Hogstrand and Haux, 1990; Duquesne, 1992; Hylland et al., 1992; Stagg et al., 1992; Hamza-Chaffai et al., 1995;
- birds: Cosson, 1989;
- mammals: Olafson and Thompson, 1974; Lee et al., 1977; Kwohn et al., 1986; Tohyama et al., 1986; Wagemann and Hobden, 1986; Caurant, 1994.

In the rest of this chapter, we will describe some examples of applications, evaluating for each the limits of applications.

3.1 Use of Molluscs

Imber et al. (1987) used the quantification of MTs by polarography to evaluate the impact of zinc wastes produced by a paper factory, and of copper and cadmium discharged by a foundry on a population of oysters (*Crassostrea gigas*) in British Columbia (Canada). The levels of MTs, copper, cadmium, and zinc were measured in parallel on crude cytosols and on cytosols denatured with ethanol to precipitate proteins other than MTs. The results (Fig. 7) enabled indication of a significant difference between MT levels of oysters collected in the polluted sites (SP-CO) and those of oysters collected in the reference sites (NT-TP). This variation in MT level was significantly correlated to that of cytosolic copper levels. The MT quantified was a dimer of Cu-MT, synthesized in response to the presence of copper in the medium.

3.2 Use of Fishes

Hamza-Chaffai et al. (1995) determined the distribution of cadmium, copper, and zinc among the different fractions (insoluble and soluble) of gills and livers of three species of littoral fish of Tunisia (Gulf of Gabes). The MT contents were quantified in heat-stable soluble fractions from homogenates of the two organs. The authors demonstrated that MT levels were correlated to zinc levels in the liver, and to cadmium, copper, and zinc in the gills, and that these relations could be expressed in equations of the following type: $[MT] = a_0 + a_1[Cd] + a_2[Cu] + a_3[Zn]$. The relations they obtained are as follows:

Scorpaena porcus (liver)	$[MT] = 4.39-0.73 \ [Cd] + 2.32 \ [Zn]$	$r^2 = 0.81$
Scorpaena scrofa (liver)	$[MT] = 1.47-0.65 \ [Cu] + 1.35 \ [Zn]$	$r^2 = 0.91$
S. porcus + S. scrofa (gills)	$[MT] = 1.17-6.72 \ [Cd] + 0.82 \ [Cu] + 0.14 \ [Zn]$	$r^2 = 0.99$
Diplodus annularis (gills)	$[MT] = 0.59-21.04 \ [Cd] + 1.26 \ [Cu] + 0.21 \ [Zn]$	$r^2 = 0.99$

These results show that the correlations between MT levels and metal levels are complex and that they vary from one organ to another and with the species studied. Moreover, it has been confirmed during the course of this study that the maximum MT levels were found in mature females, which led the authors to suggest that the use of mature females as bioindicators be avoided.

Duquesne (1992) attempted to validate the use of quantification of MTs to estimate contamination by zinc, cadmium, and copper in three littoral fish species of the Straits of Dover (*G. morhua, L. limanda, M. kitt*). Hepatic MT levels in the three species were determined by immunological techniques (RIA and ELISA). The metals were quantified in whole livers and in the cellular compartment containing soluble metalloproteins (including MTs), obtained by homogenization and centrifugation of the organ. A principal components analysis was performed taking into account a group of individuals in each of the three species, characterized by their levels of MTs, total metal, and soluble metal (Fig. 8). The levels of metals determined axis I of the analysis (contribution = 42%), while the variable MT was independent and determined axis II

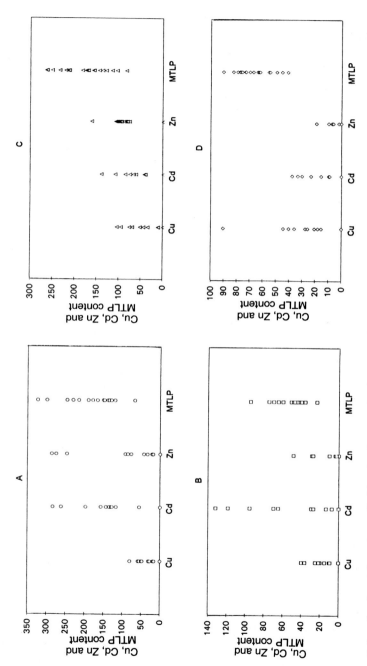

Fig. 7. Levels of metals and metallothioneins in cytosolic fractions of the digestive gland of oyster (*Crassostrea gigas*), taken from a clean site (A and B) and from a site contaminated by trace metals (C and D); A and C, crude homogenates; B and D, homogenates denatured with ethanol. The metal levels are expressed in relation to dry weight (according to Imber et al., 1987).

A:Cu (mg/kg).
B:Cd (mg/kg) × 10.
C:Zn (mg/kg) × 0.1.
D:metallothionein-like proteins (MTLP) (mg/kg) × 10.

of the analysis (contribution = 17%). According to observation of the distribution of individuals of the three species on the diagram (projection on the I–II plane), the lemon sole showed the highest metal levels, cod showed the lowest, and the group of points representing the dab was located in an intermediate position on axis I and was 'deformed' along axis II, indicating a tendency to high MT levels in this species. The species that showed highest MT (dab) was not the one that showed highest metal levels (lemon sole). The author concluded from this that determination of hepatic levels of MTs in fish did not suffice to estimate the degree of environmental contamination because of the non-agreement of levels of metals and MTs in the different species. This conclusion confirms those already mentioned pertaining to the strong specific variability of MT levels as a function of the physiological stage of fishes (e.g., sex, gonadal maturity, age).

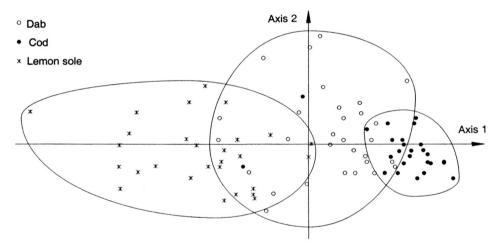

Fig. 8. Results of principal components analysis on metal and MT levels of groups of individuals in three species of littoral fishes caught in 1989. Around 36 individuals for each of the three groups (Duquesne, 1992).

3.3 Use of Birds

The existence of significant correlations between levels of metals (cadmium, copper, mercury, and zinc) and of MTs in the kidneys and livers of two bird species (the greater flamingo, *Phoenicopterus ruber*, and the little egret, *Egretta garzetta*) was established during a study on the possible metal contamination of the Camargue nature reserve (Cosson, 1989). For the flamingoes, significant correlations were established between levels of MTs (measured by polarography in heat-denatured

Table 2. Coefficients of correlation between levels of metals and MTs in livers and kidneys of flamingo and egret in Camargue (according to Cosson, 1989)

	Flamingo		Egret	
	Liver	Kidney	Liver	Kidney
Mercury	0.587	0.819**	-0.221	0.166
Cadmium	0.000	0.715**	0.247	0.587
Zinc	0.846**	0.823**	0.662*	0.283
Copper	0.873**	0.910**	0.457	0.197

*significant correlation, $p < 0.05$.
**significant correlation, $p < 0.01$.

cytosols) and copper and zinc levels in the liver. In the kidneys, the MT levels were correlated with levels of copper, zinc, cadmium, and mercury (Table 2). Only zinc levels in the liver were correlated with hepatic MT levels in egrets, no significant correlation having been proved in the case of renal MTs. A statistical analysis (step-by-step multiple regression) enabled an indication that, in flamingo, the levels of zinc and, secondarily, copper, were factors influencing mainly the production of MTs in the two organs studied. This result confirms that under natural conditions of exposure (as opposed to certain laboratory experiments conducted without regard for ecological or ethological realism, using, for example, excessive doses of intoxication, artificial means of intoxication such as multiple and various injections, disregard for physico-chemical forms of metals at the intracellular level, negation of processes of passage across membrane and transport in the organism), MTs participate in mechanisms of detoxification of metals only because of fortuitous interactions of foreign cations with the normal mechanism of homeostasis of zinc and copper (Cosson et al., 1991). Although it is possible to consider quantification of renal and/or hepatic MTs as a tool for evaluation of environmental contamination of organisms such as the greater flamingo, the same is not the case for egrets. This illustrates again the specificity of the metal-MT relation. Moreover, closer analysis of results shows that the variations in MT concentration are dependent on variations in the concentrations of essential metals (copper and zinc) rather than linked to bioaccumulation of toxic metals (cadmium and mercury) in the target organs.

3.4. Use of Mammals

Metallothioneins were isolated and quantified in many terrestrial mammals, and some studies pertain to aquatic mammals: the gray seal (*Halichoerus grypus*) and the sea lion (*Callorhinus ursinus*; Olafson and Thompson, 1974), the California sea lion (*Zalophus californianus*; Lee et al., 1977), the narval (*Monodon monoceros*; Wagemann and Hobden, 1986), the striped dolphin (*Stenella coeruleoalba*; Kwohn et al., 1986), and the harbour seal (*Phoca vitulina*; Tohyama et al., 1986). In relation to these species, the long-finned pilot whale (*Globicephala melas*) of the Faroe Islands are peculiar in that they show very high concentrations of cadmium in the liver and kidneys, and of mercury in the liver (Caurant, 1994). Caurant (1994) studied the relation between metal levels in organs (liver, kidneys) of two groups of pilot whale (groups III and XIII, comprising adults of both sexes, of which the females were gestating) and their respective MT levels (Table 3). The MT concentrations measured in the livers of individuals of group III were higher (5% of total proteins) than those determined in individuals of group XIII (1% of total proteins). The MT concentrations in liver of foetus were lower than

Table 3. Mean concentrations (standard deviations) of MTLP, in the liver and kidney of gestating females and foetuses of long-finned pilot whale of group III, and in the gestating females, foetuses, and other individuals of group XIII. The MTLP percentages are expressed in relation to total proteins (Caurant, 1994).

Group	Individuals	No. of individuals	Hepatic MTLP (mg/kg)	% Hepatic PTM	Renal MTLP (mg/kg)	% Renal MTLP
III	Gestating females	7	592.2 (200.0)	5.7 (2.1)	751.2 (213)	9.3 (1.5)
	Foetuses	5	268.7 (66.6)	5.7 (2.2)	121.3 (12.6)	1.5 (0.1)
XIII	Gestating females	7	166.9 (35.5)	1.0 (0.3)	—	—
	Foetuses	7	73.3 (31.0)	1.1 (0.5)	—	—
	Other individuals		135.3 (42.3)	1.3 (1.9)		

those measured in their mothers, but the MTs represented the same proportion of total hepatic proteins for the two types of organisms. This was not the case in the kidney, where the proportion of proteins represented by the MTs was much higher in the mothers (9.3%) than in the foetuses (1.4%). In the liver of the foetus, MT levels were significantly correlated to total levels of zinc ($r = 0.81$; $p < 0.01$). No correlation could be established in kidneys between MT levels and levels of metals (cadmium, copper, mercury, and zinc). In the liver, MT levels were significantly correlated with zinc levels for females of group III and for all the adults of group XIII (Fig. 9). Moreover, significant correlations between MT and cadmium levels could be shown in the two groups (Fig. 10).

The graphic representations of data (Figs. 9 and 10) show a significant discrepancy between the MT levels of the two groups, even though the metal levels were not significantly different. In the gestating females of group III, the existence of a correlation between levels of MTs and cadmium was established. The study of the intracellular distribution of metals between the insoluble and soluble ligands (heat-sensitive and heat-stable compounds) led the author to consider that the MTs do not play an essential role in the phenomena of accumulation and detoxification of metals in the long-finned pilot whale. For example, mercury is accumulated in high levels, but with a preferential

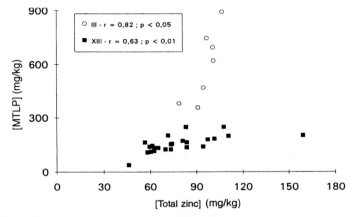

Fig. 9. Relation between concentrations of MTLP (in mg/kg) and total concentrations of zinc (mg/kg) in liver of gestating females of two groups of long-finned pilot whales. Coefficients of correlation were calculated for each group (Caurant, 1994).

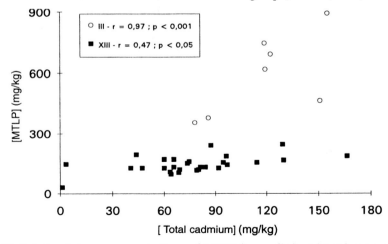

Fig. 10. Relation between concentrations of MTLP (in mg/kg) and total concentrations of cadmium (mg/kg) in liver of gestating females of two groups of long-finned pilot whale. Coefficients of correlation were calculated for each group (Caurant, 1994).

association with ligands other than the MTs (e.g., selenium, proteins of high molecular weight). Cadmium could be linked only transitorily to MTs and could be stored in the long term in an insoluble form. This study shows that, even in mammals, it is often difficult to associate bioaccumulation of metals with increase in MT levels.

4. CONCLUSION

As we have shown, the metallothioneins exhibit several of the properties necessary for their use as biomarkers. However, they form a rather complex family of proteins, which present enough specificities so that the properties indicated in certain species will not be directly attributed to others in a mistaken simplification or generalization. The synthesis proposed by Viarengo and Nott (1994) emphasizes the complexity of mechanisms for detoxification of metals in molluscs (Fig. 11). In this context, it seems difficult to use only the quantification of MTs to estimate the possible impact of a metal contamination at the level of essential metabolic functions. In studying MTs, only a fraction of the very complex whole is taken into account, which is not always essential as has been shown by various authors.

For metallothioneins to be usable as biomarkers of exposure to metals, we have seen that they must respond to certain criteria listed by Stegeman et al. (1992). There are presently several techniques enabling the precise, repeatable, and reliable quantification of MTs. However, each has advantages and disadvantages that are inherent to the property of MTs used in the analytic technique. Given the complexity of the metallothionein system in most organisms and the gaps that persist in the understanding that we have of them, especially concerning invertebrates, it seems necessary to use at least two complementary techniques to rigorously establish a diagnostic.

The distinction between the natural variations of MT levels and their variations in response to a metal stress is often not clearly made, above all in invertebrates. We have seen in particular that the reference levels vary because of changes in certain physiological parameters in relation to growth and reproduction.

The intrinsic and extrinsic factors that modify the MT levels are not all identified, for numerous species. It is necessary to keep in mind that participation of MTs in detoxification of metals essentially results from intermetallic competition for intracellular ligands of metals.

Although there is most often a correlation between MT levels in certain tissues or organs and the presence of metals in excess in the environment, it is nevertheless sometimes difficult to estimate the impact of metals on populations exclusively from the quantification of

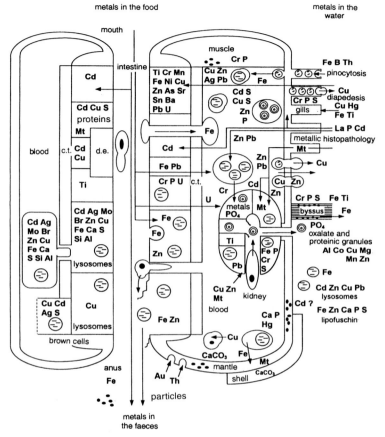

Fig. 11. Recapitulation of mechanisms for detoxification of metals in a mollusc (c.t., connective tissue; d.e., digestive epithelium) (Viarengo and Nott, 1994)

MTs. On the one hand, increased MT synthesis by certain organisms in response to metals could result from a phenotypical or genotypical adaptation to an unfavourable environment, and the corresponding population concerned may maintain itself rather than decline. On the other hand, increase in MT levels in tissues does not always accompany or precede the occurrence of symptoms of toxicity, but sometimes follows it (Lauren and McDonald, 1985, 1987). There may in fact be, depending on the species, redistribution of metals on the vital proteins.

One of the chief reservations about the use of MTs as biomarkers was expressed by Hamilton et al. (1987) after researches conducted on the stress provoked by cadmium in a salmonid (*Salvelinus fontinalis*). For these authors, it was not desirable to use MTs because there was no reliable correlation between their abundance and the accumulation of metal in the individuals and their mortality. Moreover, they did not prove proportionality between MT levels and environmental concentrations of toxicants.

Even though our understanding of MTs is greater than that of other biomarkers, the variability of responses of organisms in the presence of metals, in terms of increase in MT levels, is too significant for MTs to be used as a 'universal' biomarker. That is why we suggest, as others have before (George and Olsson, 1994), a limited use of metallothioneins as biomarker in fishes (taking into account the specificity of responses) and in vertebrates in general. All the same, invertebrates such as the mussel or oyster offer one great advantage for use of this biomarker, their sedentary nature, which justifies further research on these species.

REFERENCES

Amiard J.C. and Cosson R.P. (1997). Les métallothionéines. In: Biomarqueurs en écotoxicologie: Aspects fondamentaux. Lagadic L., Caquet Th., Amiard J.C. and Ramade F. (eds.). Masson, Paris, pp. 53–66.

Bebianno M.J. and Langston W.J. (1993). Turnover rate of metallothionein and cadmium in *Mytilus edulis*. Bio Metals, 6:239–244.

Benson W.H. and Birge W.J. (1985). Heavy metal tolerance and metallothionein induction in fathead minnows: results from field and laboratory investigations. *Environ. Toxicol. Chem.*, 4:209–217.

Benson W.H. and Birge W.J. (1987). Detection of cadmium-binding proteins in fish chronically exposed to heavy metals. *Environ. Toxicol. Chem.*, 6:623–626.

Bonham K., Zafarullah M. and Gedamu L. (1987). The rainbow trout metallothioneins: Molecular cloning and characterization of two distinct cDNA sequences. *DNA*, 6:519–528.

Brouwer M., Whaling P. and Engel D.W. (1986). Copper-metallothioneins in the american lobster, *Homarus americanus*: potential role as Cu (I) donors to apohemocyanin. *Environ. Health Persp.*, 65:93–100.

Brown D.A. and Parsons T.R. (1978). Relationship between cytoplasmic distribution of mercury and toxic effects to zooplankton and chum salmon (*Oncorhynchus keta*) exposed to mercury in a controlled ecosystem. *J. Fish. Res. Bd. Can.*, 35:880–884.

Brown D.A., Bawden C.A., Chatel K.W. and Parsons T.R. (1977). The wildlife community of Iona Island jetty. Vancouver, B.C. and heavy-metal pollution effects. *Environ. Conserv.*, 4:213–216.

Brown M.W., Thomas D.G., Shurben D. et al. (1986). A comparison of the differential accumulation of cadmium in the tissues of three species of freshwater fish. *Salmo gairdneri, Rutilus rutilus* and *Noemacheilus barbatulus*. Comp. Biochem. Physiol., 84C:213–217.

Caurant F. (1994). Bioaccumulation de quelques éléments traces (As, Cd, Cu, Hg, Se, Zn) chez le globicéphale noir (*Globicephala melas*, Delphinidé) pêché au large des îles Féroé. Doctoral Thesis, European Communities, University of Nantes.

Chan K., Davidson W.S., Hew C. and Fletcher G.L. (1989). Molecular cloning of metallothionein cDNA and analysis of metallothionein gene expression in winter flounder tissues. *Can. J. Zool.*, 67:2520–2527.

Chatterjee A. and Maiti J.B. (1987). Purification and immunological characterization of catfish (*Heteropneustes fossilis*) metallothionein. *Mol. Cell. Biochem.*, 78:55–64.

Chen R., Whanger P.D. and Weswig P.H. (1975). Biological function of metallothionein. I. Synthesis and degradation of rat liver metallothionein. *Biochem. Med.*, 12:95–105.

Clough S.R., Mitra R.S. and Kulkarni A.P. (1986). Qualitative and quantitative aspects of human fetal liver metallothioneins. *Biol. Neonate*, 49:241–254.

Coombs T.L. and George S.G. (1978). Mechanisms of immobilization and detoxication of metal in marine organisms. In: Physiology and Behaviour of Marine Organisms. McLusky D.S. and Berry A.J. (eds.). Pergamon Press, New York. pp. 179–187.

Cosson R.P. (1989). Relationship between heavy metal and metallothionein-like protein levels in the liver and kidney of two birds: the Greater Flamingo and the Little Egret. *Comp. Biochem. Physiol.*, 94C:243–248.

Cosson R.P. (1994a). Heavy metal intracellular balance and relationship with metallothionein induction in the gills of carp after contamination by Ag, Cd, and Hg following pretreatment with Zn or not. *Biol. Trace Elem. Res.*, 46:229–245.

Cosson R.P. (1994b). Heavy metal intracellular balance and relationship with metallothionein induction in the liver of carp after contaminatin by silver, cadmium and mercury following or not pretreatment by zinc. *BioMetals*, 7:9–19.

Cosson R.P., Amiard-Triquet C. and Amiard J.C. (1991). Metallothioneins and detoxification. Is the use of detoxication proteion for MTs a language abuse? *Water, Air, Soil Pollut.*, 57–58:555–567.

Couillard Y., Campbell P.G.C. and Tessier A. (1993). Response of metallothionein concentrations in a freshwater bivalve (*Anodonta grandis*) along an environmental cadmium gradient. *Limnol. Oceanogr.*, 38:299–313.

Dallinger R. (1993). Strategies of metal detoxification in terrestrial invertebrates. In: Ecotoxicology of Metals in Invertebrates. Dallinger R. and Rainbow P.R. (eds.). Lewis Publishers, Boca Raton, pp. 245–289.

Dennai N., Dhainaut-Courtois N., Bouquereau J.M. and Nejmeddine A. (1986). Effets du cadmium et du mercure sur un ver marin (*Nereis diversicolor* O.F. Muller). Mécanismes de détoxication. *C.R. Acad. Sci. Ser. III*, 302:489–494.

Dieter H.H., Muller L., Abel J. and Summer K.H. (1987). Metallothionein-determination in biological materials: interlaboratory comparison of 5 current methods. In: Metallothionein II. Kagi J.H.R. and Kojima Y. (eds.). Birkhauser Verlag, Basel, pp. 351–358.

Dohi Y., Ohba K. and Yoneyama Y. (1983). Purification and molecular properties of two cadmium-binding glycoproteins from the hepatopancreas of a whelk, *Buccinum tenuissimum*. *Biochim. Biophys. Acta*, 745: 50–60.

Dunn M.A., Blalock T.L. and Cousins R.J. (1987). Metallothionein. Minireview. *Proc. Soc. Exp. Biol. Med.*, 185:107–119.

Duquesne S. (1992). Bioaccumulation métallique et métallothionéines chez trois espèces de poissons du littoral Nord-Pas de Calais. Doctoral Thesis, University of Science and Technology of Lille.

Dutton M.D., Stephenson M. and Klaverkamp J.F. (1993). A modified mercury displacement assay for measuring metallothionein in fish. *Environ. Toxicol. Chem.*, 12:1193–1202.

Eaton D.L. and Toal B.F. (1982). Evaluation of the Cd/Hemoglobin affinity assay for the rapid determination of metallothionein in biological tissues. *Toxicol. Appl. Pharmacol.*, 66:134–142.

Engel D.W. and Brouwer M. (1984). Trace metal-binding proteins in marine molluscs and crustaceans. *Mar. Environ. Res.*, 13:177–194.

Engel D.W. and Brouwer M. (1993). Crustaceans as a model for metal metabolism: I. Effects of the molt cycle on blue crab metal metabolism and metallothionein. *Mar. Environ. Res.*, 35: 1–5.

Falck F.Y., Fine L.J., Smith R.G. et al. (1983). Metallothionein and occupational exposure to cadmium. *Br. J. Ind. Med.*, 40:305–313.

Feldman S.L., Squibb K.S. and Cousins R.J. (1978). Degradation of cadmium-thionein in rat liver and kidney. *J. Toxicol. Environ. Health*, 4:805–813.

George S.G. (1989). Cadmium effects on plaice liver xenobiotic and metal detoxication system: Dose-response. *Aquat. Toxicol.*, 15:303–310.

George S.G. and Olsson P.E. (1994). Metallothioneins as indicators of trace metal pollution. In: Biomonitoring of Coastal Waters and Estuaries. Kramer K.J.M. (ed.). CRC, Boca Raton, pp. 151–178.

George S.G. and Pirie B.J. (1979). The occurrence of cadmium in sub-cellular particles in the kidney of the marine mussel, *Mytilus edulis*, exposed to cadmium. *Biochim. Biophys. Acta*, 580:234–244.

George S.G. and Pirie B.J. (1980). Metabolism of zinc in the mussel, *Mytilus edulis* (L.): a combined ultrastructural and biochemical study. *J. Mar. Biol. Ass. U.K.*, 60:575–590.

George S.G., Pirie B.J., Frazier J.M. and Thomson J.D. (1984). Interspecies differences in heavy metal detoxication in oysters. *Mar. Environ. Res.*, 14:462–464.

Hamer D. (1986). Metallothionein. *Annu. Rev. Biochem.*, 55:913–951.

Hamilton S.J., Mehrle P.M. and Jones J.R. (1987). Evaluation of metallothionein measurement as a biological indicator of stress from cadmium in brook trout. *Trans. Am. Fish. Soc.*, 116:551–560.

Hamza-Chaffai A., Cosson R.P., Amiard Triquet C. and El Abed A. (1995). Physicochemical forms of storage of metals (Cd, Cu and Zn) and metallothionein-like proteins in gills and liver of marine fish from the Tunisian coast: ecotoxicological consequences. *Comp. Biochem. Physiol.*, 111C:329–341.

Hamza-Chaffai A., Amiard J.C. and Cosson R.P. (1999). Relationship between metallothioneins and metals in a natural population of the clam *Ruditapes decussatus* from Sfax coast: a non-linear model using Box-Cox transformation. *Comp. Biochem. Physiol.*, 123C, 153–163.

Harrison S.E. and Klaverkamp J.F. (1989). Uptake, elimination and tissue distribution of dietary and aqueous cadmium by rainbow trout (*Salmo gairdneri* Richardson) and lake whitefish (*Coregonus clupeaformis* Mitchill). *Environ. Toxicol. Chem.*, 8: 87–97.

Held D.D. and Hoekstra W.G. (1984). The effects of zinc deficiency on turnover of cadmium-metallothionein in rat liver. *J. Nutr.*, 114:2274–2282.

Hennig H.F.K.O. (1986). Metal-binding proteins as metal pollution indicators. *Environ. Health Persp.*, 65:175–187.

Hogstrand C. and Haux C. (1989). Comparison of polarography and radioimmunoassay for the quantification of metallothionein in perch (*Perca fluviatilis*). *Mar. Environ. Res.*, 28:187–190.

Hogstrand C. and Haux C. (1990). A radioimmunoassay for metallothionein in fish. *Mar. Environ. Res.*, 28:191–194.

Hogstrand C. and Haux C. (1992). Metallothionein as an indicator of heavy metal exposure in two subtropical fish species. *J. Exp. Mar. Biol. Ecol.*, 138: 69–84.

Hylland K., Haux C. and Hogstrand C. (1992). Hepatic metallothionein and heavy metals in dab *Limanda limanda* from the German Bight. *Mar. Ecol. Prog. Ser.*, 91:89–96.

Imber B.E., Thompson J.A.J. and Ward S. (1987). Metal-binding protein in the Pacific oyster, *Crassostrea gigas*: assessment of the protein as a biochemical environmental indicator. *Bull. Environ. Contam. Toxicol.*, 38:707–714.

Jenkins K.D., Sanders B.M. and Costlow J.D. (1984). Regulation of copper accumulation and subcellular distribution in developing crab larva: the role of metallothionein and other intracellular Cu-binding ligands. *Mar. Environ. Res.*, 14:474–475.

Kagi J.H.R. (1993). Evolution, structure and chemical activity of class I metallothioneins: an overview. In: Metallothionein III. Suzuki K.T., Imura N. and Kimura M. (eds.). Birkhauser Verlag, Basel, pp. 29–55.

Kagi J.H.R. and Kojima Y. (1987). *Metallothionein II*. Birkhauser Verlag, Basel.

Kille P., Stephens P.E. and Kay J. (1991). Elucidation of cDNA sequences for metallothioneins from rainbow trout, stone loach and pike liver using the polymerase chain reaction. *Biochim. Biophys. Acta*, 1089:407–410.

Kito H.,Ose Y. and Sato T. (1986). Cadmium-binding protein (metallothionein) in carp. *Environ. Health. Persp.*, 65:117–124.

Kito H., Tazawa T., Ose Y. et al. (1982a). Formation of metallothionein in fish. *Comp. Biochem. Physiol.*, 73C:129–134.

Kito H., Tazawa T., Ose Y. et al.(1982b). Protection by metallothionein against cadmium toxicity. *Comp. Biochem. Physiol.*, 73C:135–139.

Klaverkamp J.F. and Duncan D.A. (1987). Acclimation to cadmium toxicity by white suckers: cadmium binding capacity and metal distribution in gill and liver cytosol. *Environ. Toxicol. Chem.*, 6:275–289.

Koizumi N., Inoue Y. and Ninomiya R. et al. (1989). Relationship of cadmium accumulation to zinc or copper concentration in horse liver and kidney. *Environ. Res.*, 49:104–114.

Kwohn Y.T., Yamazaki S. and Okubo A. et al. (1986). Isolation and characterization of metallothionein from kidney of striped dolphin, *Stenella coeruleoalba*. *Agric. Biol. Chem.*, 50:2881–2885.

Langston W.J. and Zhou M. (1987). Cadmium accumulation, distribution and elimination in the bivalve *Macoma balthica*: neither metallothionein nor metallothionein-like proteins are involved. *Mar. Environ. Res.*, 21: 225–237.

Langston W.J., Bebianno M.J. and Zhou M. (1989). A comparison of metal-binding proteins and cadmium metabolism in the marine molluscs *Littorina littorea* (Gastropoda), *Mytilus edulis* and *Macoma balthica* (Bivalvia). *Mar. Environ. Res.*, 28:195–200.

Lastowski-Perry D., Otto E. and Maroni G. (1985). Nucleotide sequence and expression of *Drosophila* metallothionein. *J. Biol. Chem.*, 260:1527–1530.

Lauren D.J. and McDonald D.G. (1985). Effects of copper on branchial ionoregulation in the rainbow trout, *Salmo gairdneri* Richardson. *J. Comp. Physiol.*,155:635–644.

Lauren D.J. and McDonald D.G. (1987). Acclimation to copper by rainbow trout, *Salmo gairdneri:* physiology. *Can. J. Fish. Aquat. Sci.*, 44:99–104.

Leaver M.J. and George S.G. (1989). Nucleotide and deduced acid sequence of a metallothionein cDNA clone from the marine fish, *Pleuronectes platessa*. EMBL accession no. 56743.

Lee S.S., Mate B.R. , Von Der Trenck K.T. et al. (1977). Metallothionein and the subcellular localization of mercury and cadmium in the California sea lion. *Comp. Biochem. Physiol.*, 57C:45–53.

Lobel P.B. and Marshal H.D. (1988). A unique low molecular weight zinc-binding ligand in the kidney cytosol of the mussel *Mytilus edulis,* and its relationships to the inherent variability of zinc accumulation in this organism. *Mar. Biol.*, 99:101–105.

Lu W., Kasrai M., Bancroft G.M. et al. (1990). Sulfur L-EdgeXANES study of zinc, cadmium and mercury containing metallothionein and model compounds. *Inorg. Chem.*, 29:2561–2563.

Luten J.B., Bouquet W., Burgraaf M.M. and Rus J. (1986). Accumulation, elimination and speciation of cadmium and zinc in mussel, *Mytilus edulis,* in the natural environment. *Bull. Environ. Contam. Toxicol.*, 37:579–586.

Maroni G., Wise J., Young J.E. and Otto E . (1987). Metallothionein gene duplications and metal tolerance in natural populations of *Drosophila melanogaster*. *Genetics*, 117:739–744.

Martoja M. and Martoja R. (1984). La bioaccumulation de métaux, processus physiologique normal et conséquence de la pollution. *Cour. CNRS*, 54:32–37.

Mokdad R., Debec A. and Wegnez M. (1987). Metallothionein genes in *Drosophila melanogaster* constitute a dual system. *Proc. Natl. Acad. Sci. USA*, 84:2658–2662.

Nejmeddine A., Dhainaut-Courtois N., Baert J.L. et al. (1988). Purification and characterization of a cadmium-binding protein from *Nereis diversicolor* Annelida, Polycheta. *Comp. Biochem. Physiol.*, 89C:321–326.

Nemer M., Wilkinson D., Travaglini E.C. et al. (1986). Sea urchin metallothionein sequence: Key to an evolutionary diversity. *Proc. Natl. Acad. Sci. USA*, 82:4992–4994.

Nolan C.V. and Shaikh Z.A. (1986). Determination of metallothionein in tissues by radioimmunoassay and by cadmium saturation method. *Anal. Biochem.*, 154:213–223.

Norey C.G., Less W.E., Darke B.M. et al. (1990). Immunological distinction between piscine and mammalian metallothioneins. *Comp. Biochem. Physiol.*, 95B:597–601.

Olafson R.W. and Olsson P.E. (1991). Electrochemical detection of metallothionein. *Meth. Enzymol.*, 205:205–213.

Olafson R.W. and Sim R.G. (1979). An electrochemical approach to quantitation and characterization of metallothioneins. *Anal. Biochem.*, 100:343–351.

Olafson R.W. and Thompson J.A.J. (1974). Isolation of heavy metal binding proteins from marine vertebrates. *Mar. Biol.*, 28:83–86.

Olsson P.E. and Haux C. (1986). Increased hepatic metallothionein content correlates to cadmium accumulation in environmentally exposed perch *(Perca fluviatilis)*. *Aquat. Toxicol.*, 9:231–242.

Olsson P.E., Haux C. and Förlin L. (1987). Variations in hepatic metallothionein, zinc and copper levels during an annual reproductive cycle in rainbow trout, *Salmo gairdneri*. *Fish Physiol. Biochem.*, 3:39–47.

Onosaka S. and Cherian M.G. (1982). Comparison of metallothionein determination by polarographic and cadmium-saturation methods. *Toxicol. Appl. Pharmacol.*, 63:270–274.

Overnell J., McIntosh R. and Fletcher T.C. (1987). The levels of liver metallothionein and zinc in plaice, *Pleuronectes platessa* L., during the breeding season, and the effect of oestradiol injection. *J. Fish. Biol.*, 30:539–546.

Overnell J., Fletcher T.C. and McIntosh R. (1988). Factors affecting hepatic metallothionein levels in marine flatfish. *Mar. Environ. Res.*, 24:155–158.

Palmiter R.D. (1987). Molecular biology of metallothionein gene expression. In: Metallothionein II. Kagi J.H.R. and Kojima Y. (eds.). Birkhauser Verlag, Basel, pp. 63–80.

Petering D.H., Goodrich M., Hodgman W. et al. (1990). Metal-binding proteins and peptides for the detection of heavy metals in aquatic organisms. In: Biomarkers of Environmental Contamination. McCarthy J.F. and Shugart L.R. (eds.). Lewis Publishers, Boca Raton, pp. 239–254.

Pickwell G.V. and Steinert S.A. (1984). Serum biochemical and cellular responses to experimental cupric ion challenge in mussels. *Mar. Environ. Res.*, 14:245–265.

Piotrowski J.K., Bolanowska W. and Sapota A. (1973). Evaluation of metallothionein content in animal tissues. *Acta Biochim. Polonica*, 20:207–215.

Richards M.P. and Steele N.C. (1987). Isolation and quantitation of metallothionein isoforms using reversed-phase high-performance liquid chromatography. *J. Chromatogr.*, 402:243–256.

Ridlington J.W., Chapman D.C., Goeger D.E. and Whanger P.D. (1981). Metallothionein and Cu-chelatin: characterization of metal-binding proteins from tissues of four marine animals. *Comp. Biochem. Physiol.*, 70B:93–104.

Roch M. and McCarter J.A. (1984). Hepatic metallothionein production and resistance to heavy metals by rainbow trout *(Salmo gairdneri)*—II. Held in a series of contaminated lakes. *Comp. Biochem. Physiol.*, 77C:77–82.

Roch M., McCarter J.A., Matheson A.T. et al. (1982). Hepatic metallothionein in rainbow trout *(Salmo gairdneri)* as an indicator of metal pollution in the Campbell River system. *Can. J. Fish. Aquat. Sci.*, 39:1596–1601.

Roels H., Lauwerys R., Buchet J.P. et al. (1983). Significance of urinary metallothionein in workers exposed to cadmium. *Int. Arch. Occup. Environ. Health*, 52:159–166.

Roesijadi G. (1982). Uptake and incorporation of mercury into mercury-binding proteins of gills of *Mytilus edulis* as a function of time. *Mar. Biol.*, 66:151–157.

Roesijadi G. (1992). Metallothioneins in metal regulation and toxicity in aquatic animals. *Aquat. Toxicol.*, 22:81–114.

Roesijadi G. (1993). Response of invertebrate metallothioneins and MT genes to metals and implications for environmental toxicology. In: Metallothionein III. Suzuki K.T., Imura N. and Kimura M. (eds.). Birkhauser Verlag, Basel, pp. 141–158.

Roesijadi G. and Klerks P.L. (1989). Kinetic analysis of cadmium binding to metallothionein and other intracellular ligands in oyster gills. *J. Exper. Zool.*, 251:1–12.

Roesijadi G., Unger M.E. and Morris J.E. (1988). Immunochemical quantification of metallothioneins in a marine mollusc. *Can. J. Fish. Aquat. Sci.* 45:1257–1263.

Roesijadi G., Vestling M.M., Murphy C.M. et al. (1991). Structure and time-dependent behaviour of acetylated and non-acetylated forms of a molluscan metallothionein. *Biochem. Biophys. Acta*, 1074 : 230–236.

Scheuhammer A.M. and Cherian M.G. (1986). Quantification of metallothionein by a silver-saturation method. *Toxicology*, 36:2–3.

Shaikh Z.A. and Smith J.C. (1976). The biosynthesis of metallothionein in rat liver and kidney after administration of cadmium. *Chem.-Biol. Interact.*, 15:327–336.

Silar P. (1990). Régulation des gènes métallothionéine de drosophile, Doctoral thesis. Université Paris-Sud.

Simkiss K., Taylor M. and Mason A.Z. (1982). Metal detoxification and bioaccumulation in molluscs. *Mar. Biol. Lett.*, 3:187–201.

Stagg R., Goksoyr A. and Rødger G. (1992). Changes in branchial Na⁺K⁺ATPase, metallothionein and P450 1A1 in dab *Limanda limanda* in the German Bight: indicators of sediment contamination? *Mar. Ecol. Prog. Ser.*, 91:105–115.

Stegeman J.J., Brouwer M., Di Giulio R. et al. (1992). Enzyme and protein synthesis as indicators of contaminant exposure and effect. In: Biomarkers: Biochemical, Physiological, and Histological Markers of Anthropogenic Stress. Huggett R.J., Kimerle R.A., Mehrle P.M. and Bergman H.L. (eds.). Lewis Publishers, Boca Raton, pp. 235–335.

Stone H. and Overnell J. (1985). Non-metallothionein cadmium binding proteins. *Comp. Biochem. Physiol.*, 80C:9–14.

Summer K.H. and Klein D. (1993). Quantification of metallothionein in biological materials. In: Metallothionein III. Suzuki K.T., Imura N. and Kimura M. (eds.). Birkhauser Verlag. Basel. pp. 75–86.

Suzuki K.T. (1980). Direct connection of high-speed liquid chromatograph (equipped with gel permeation column) to atomic absorption spectrophotometer for metalloprotein analysis: Metallothionein. *Anal. Biochem.*, 102:31–34.

Suzuki K.T., Sunaga H., Aoki Y. and Yamamura M. (1983). Gel permeation, ion-exchange and reversed-phase columns for separation of metallothioneins by high-performance liquid chromatography–atomic absorption spectrophotometry. *J. Chromatogr.*, 281:159–166.

Thomas D.G., Linton H.J. and Garvey J.S. (1986). Fluorometric ELISA for the detection and quantitation of metallothionein. *J. Immunol. Meth.*, 89:239–247.

Thompson J.A.J. and Cosson R.P. (1984). An improved electrochemical methods for the quantification of metallothioneins in marine organisms. *Mar. Environ. Res.*, 11: 137–152.

Tohyama C. and Shaikh Z.A. (1978). Cross-reactivity of metallothioneins from different origins with rabbit anti-rat hepatic metallothionein antibody. *Biochem. Biophys. Res. Comm.*, 84:907–913.

Tohyama C., Himeno S.I., Watanabe C. et al. (1986). The relationship of the increased level of metallothionein with heavy metal levels in the tissue of the harbour seal (*Phoca vitulina*). *Ecotoxicol. Environ. Safety*, 12:85–94.

Van Der Malie R.J. and Garvey J.S. (1979). Radioimmunoassay of metallothioneins. *J. Biol. Chem.*, 254:8416–8421.

Viarengo A. and Nott J.A. (1994). Mechanisms of heavy metal cation homeostasis in marine invertebrates. Mini Review. *Comp. Biochem. Physiol.*, 104:355–372.

Wagemann R. and Hobden B. (1986). Low molecular weight metallothioneins in tissues of the narval (*Monodon monoceros*). *Comp. Biochem. Physiol.*, 84C: 325–344.

Webb M. (1979). *The Chemistry, Biochemistry and Biology of Cadmium*. Elsevier/North Holland, Amsterdam.

Webb J., Macey D.J. and Talbot V. (1986). Identification of ferritin as a major high molecular weight zinc-binding protein in the tropical rock oyster, *Saccostrea cuccullata*. *Arch. Environ. Contam. Toxicol.*, 14:403–407.

Woodall C., MacLean N. and Crossley F. (1988). Responses of trout fry (*Salmo gairdneri*) and *Xenopus laevis* tadpoles to cadmium and zinc. *Comp. Biochem. Physiol.*, 89C:93–99.

Young J.S. and Roesijadi G. (1983). Respiratory adaptation to copper-induced injury and occurrence of a copper-binding protein in the Polychaete, *Eudistylia vancouveri*. *Mar. Pollut. Bull.*, 14:30–32.

Zelazowski A.J. and Stillman M.J. (1992). Silver binding to rabbit liver zinc metallothionein and zinc a and b fragments. Formation of silver-MT with Ag(I):protein ratios of 6, 12, and 18 observed using circular dichroism spectrometry. *Inorg. Chem.*, 31:3363–3370.

6

Molecular Biomarkers of Exposure of Marine Organisms to Organophosphorus Pesticides and Carbamates

F. Galgani and G. Bocquené

INTRODUCTION

Preserving the quality of the marine environment is a priority for Europe as a whole. Public awareness and the anxiety of marine professionals and public authorities have led several industrialized countries to establish networks for the monitoring of the quality of the environment. France played a major role in this field in creating, in 1972, the Réseau National d'Observation de la Qualité du Milieu Marin (RNO), for the study of concentrations of certain metallic and organic pollutants in sea water, living matter, and sediments. However, several families of potentially toxic xenobiotics escape this surveillance, either because their analysis poses serious technical difficulties, or because the occurrence of such pollutants in the marine environment is poorly understood, or because priorities in surveillance are still being developed. Pesticides are among these marine pollutants.

1. PESTICIDES IN THE MARINE ENVIRONMENT: ORGANOPHOSPHATES AND CARBAMATES

The large majority of pesticides are compounds used chiefly in agriculture in order to eliminate weeds (herbicides) and to control pest insects (insecticides) or the development of pathogenic fungi (fungicides). They are also used to clear vegetation (e.g., from railway tracks, highways, and airport runways) and for the protection of wood as well as textiles.

The use of chemical molecules is the essential reason for increase in agricultural production in most countries, and though the generic term 'pesticides' has often been perceived negatively by the public, we must keep in mind that, even now, a third of world agricultural production is destroyed during growth, harvest, and storage.

In the United States, annual production of pesticides is 500,000 tons, representing 900 different active ingredients included in the composition of 25,000 registered formulations (Fikes, 1990). In France, in 1985, total production was 64,160 tons, of which around 10% were insecticides (6260 tons), 34% were fungicides, and 56% were herbicides (Tronczynski, 1990). The quantities used increased constantly for several years: in France the production increased from 24,400 tons in 1974 to 92,500 tons in 1987. This production seems to have stabilized at 100,000 tons for several years reaching 111,000 tons in 1997 (IFEN, 1999). It must be noted that information pertaining to production, sale, and use of these products is difficult to obtain because it often reveals confidential industrial or commercial information, and the figures available are sometimes estimated from treated agricultural areas and recommended treatment doses. However, reliable estimates may be obtained by cross-checking the information.

Pesticides may be mineral compounds or molecules from organic synthesis. The compounds most often used belong to families of triazines, phenylureas, organophosphates (OPs), carbamates, pyrethroids, and, in smaller amounts, organocholorines (OCs) since they were banned.

In the framework of several studies on the effects of pesticides, especially in the context of biological monitoring of environment, large-scale effects have presently been measured only with cholinesterases, the inhibition of which is specific to neurotoxic effects linked chiefly to OP pesticides and carbamates (Bocquené et al., 1997). More general information on the chemistry and toxicology of OPs and carbamates can be found in the works of Fest and Schmit (1982) and Ballantyne and Marrs (1992).

1.1 The Organophosphorus Compounds

The first report on the synthesis of an OP, tetraethylpyrophosphate, dates from 1854 and was presented at the French Académie des Sciences by Philippe de Clermont. However, it was only in 1932 that the inhibiting properties of OPs were described. The discovery of properties of OPs is linked to research on neurotoxic gases undertaken in the

United States and Europe after the First World War. The insecticidal advantages of these compounds appeared only later, but soon justified their industrial development. It was the German firm I.G. Farben that developed the first OP insecticides, notably parathion. This was followed by the synthesis of other chemical agents such as diisopropylfluorophosphate (DFP), then the powerful agents G (sarin and soman; Koelle, 1994). From the early 1970s, the development of OPs as insecticides was favoured by the limits and even bans on the use of OC compounds such as DDT, the use of which was prohibited in France in 1972. The OPs spread rapidly because they were highly effective, especially against insects, and their action on the environment was considered relatively inoffensive in comparison to that of the OCs. It is important always to qualify this judgement, because, although these compounds are effectively much more unstable than their chlorine equivalents, the persistence of their action may be significant, particularly in soils and sediments. In the light of their high toxicity, these molecules can in no case be considered ecotoxicologically negligible.

1.1.1 Release, Transfer and Levels of OPs in the Environment

It is difficult to assess the real quantities of OPs used, and although certain figures per product or per type of treatment are known, an overall estimation of the consumption of active ingredients is tricky. Nevertheless, according to the figures given by the Chambre Syndicale de la Phytopharmacie, 772 tons of these insecticides were used in France in 1978. Ten years later, in 1988, the OPs represented, according to the Union des Industries pour la Protection des Plantes (UIPP), a market of 1,366 tons of active ingredients. For 1990, the consumption of active ingredients of OPs in France was estimated at 1,797 tons.

Among the most widely used OPs are azinphos-methyl chlorfenvinphos, chlorpyrifos, demeton, diazinon, dichlorvos, enitrothion, malathion, parathion, phosalone, temephos, and terbufos (Tronczynski, 1990; Barcelo et al., 1991; Lartiges and Garrigues, 1993). Some OP compounds are not used for their insecticidal properties but as defoliants. This is the case of tributyl phosphorotrithioate (DEF), which is nevertheless a powerful inhibitor of cholinesterases (Habig and Di Giulio, 1988).

The release of pesticides in the environment, especially in aquatic ecosystems, is diffuse and chronic, except in the case of accidents. Pesticide introduction in the environment derives essentially from agricultural or anti-mosquito treatments, which may be repeated during the year. On grape or maize, for example, some treatments are done weekly. It is still more difficult to distinguish in pesticide contamination the contribution, clearly not negligible, of domestic uses.

Direct releases are most often accidental. Examples of cases of direct release, with often spectacular ecological effects, are accidental spills of disulfoton and thiometon (OPs) in the Rhine in 1988 (Wanner et al., 1989), and the aerial anti-mosquito treatment of coastal fields, and direct treatment of aquaculture farms with molecules such as dichlorvos (OP), to eliminate parasitic copepods (Cusack and Johnson, 1990). The fire at the Sandoz warehouses in Switzerland in November 1986 led to a serious and massive contamination of the Rhine, causing a veritable ecological catastrophe for the aquatic fauna and the killing of 220,000 kg of eels over 400 km of river (Capel et al., 1988). The more general cases, however, are those in which contaminants are released during treatments of agricultural areas far from seas and oceans.

Although it is acknowledged that water courses are the chief means of contamination of freshwater reservoirs and the ocean, the dispersal and transfer of OPs by atmospheric routes is not negligible, given the high vapour pressure of these products and the spreading techniques used. Barcelo et al. (1991) estimated that 40% of the quantities of fenitrothion applied is transferred to the atmosphere. However, the mechanisms that govern the migration of these molecules are extremely complex because they are specific to each compound and therefore not well known or quantified.

The OPs have a highly variable solubility in water, ranging from total insolubility (temephos) to very high solubility (acephate = 790 g/l), but the large majority have a relatively low solubility in water, between 0.4 mg/l (chlorpyrifos) and 150 mg/l (malathion). In fact, the persistence of these molecules appears to be directly linked to their solubility, reflected in the n-octanol-water partition coefficient (K_{ow}). The insolubility of temephos and the low solubility of chlorpyrifos in water explain why these compounds are present in the marine environment mainly in the form of molecules adsorbed on the sediments and on suspended particles. These laboratory data obviously cannot be directly extrapolated to real environmental conditions, but they enable us to affirm that the persistence of OPs is not zero, although they are much more unstable than OC compounds.

Data directly taken from measurements in the environment are rarer and more difficult to interpret. Nevertheless, some studies reported concentrations in water, sediments, and living matter that are equal to or higher than thresholds of induction of measurable effects on the activity of cholinesterases. Tully and Morrissey (1989) measured extreme levels of dichlorvos of 0.13 µg/l at 25 m depth in the waters of the Beirtreach Bui Bay in Ireland, following treatment of salmon stocks in open medium with this OP. Readmann et al. (1992) observed concentrations of chlorpyrifos equal to 34.2 mg/kg of sediment in some coastal areas of Costa Rica. Similarly, significant traces of temephos were found by Lores et al. (1985) in marshes of Florida, concentrations up to 62 µg/l having been measured in the marshes 48 h after aerial dispersal. On the other hand, Wang et al. (1987) concluded the non-persistence of fenthion in estuarine waters subject to tides: the concentrations found in the sea ranged from 1.69 µg/l, 45 min. after application, to a non-detectable level after 24 h.

Generally, the OPs that have the lowest solubilities in water show a lipophilic behaviour that explains their bioaccumulation in the fat and hepatic tissues of aquatic animals. In eels (*Anguilla anguilla*) contaminated by 0.056 mg/l of diazinon, the bioconcentration factors (BCF) were 800 in the liver and 1,600 in the muscle (Sancho et al., 1993). Barcelo et al. (1991) measured concentrations of fenitrothion up to 60 µg/kg of flesh in mussel (*Mytilus galloprovincialis*) and 90 µg/kg in clam (*Tapes semidecussatus*) taken from the Ebre delta (Spain) several weeks after aerial treatment of paddy fields in the region. The same product was found in a stable concentration of 3 µg/kg of sediment 8 months after the application (Durand and Barcelo, 1992).

1.1.2 Degradation of OP Compounds

For the same reasons of molecular diversity that affect the transfer of OPs between the different compartments of the environment, the degradation of these products is difficult to understand by a common model and the information, when it is available, is specific to a given molecule. It is, however, currently acknowledged that OPs can follow the two conventional ways of degradation of organic structures, chemical (photolytic and hydrolytic) and biological (metabolic). Although the mechanisms are at present not well understood, the different metabolites of OPs have only been described for some products. Among the metabolic ways of degradation of OPs, oxidative metabolism, ensured essentially by the hepatic monooxygenases, has a double role. It enables activation of thio forms to oxon forms, with increased toxicity (Bocquené et al., 1997), and contributes moreover to the detoxification of these products, affecting thus their occurrence in the organism.

It is almost certain that the OPs are not as persistent as OCs, but their persistence in the environment is highly variable. The half-life depends on several environmental parameters, among which pH and temperature, as well as exposure to light, play a major role. The persistence is clearly different according to mode of use of the insecticide, whether soil treatment or treatment of above-ground parts of crops (Table 1).

Table 1. Half-life (Franck et al., 1991) and persistence of action (ACTA, 1993) of some OPs and carbamates (C) among those most widely used

Compound	Half-life (d)		Persistence of action	
	at 4°C	at 21°C	Leaves	Soil
Chlorfenvinphos (OP)	—	—	2–3 wk	2–4 mo
Terbufos (OP)	34	2.3	—	2 mo
Dimethoate (OP)	142	34	2–3 wk	— ·
Chlorpyrifos (OP)	24	4.8	—	3 mo
Diazinon (OP)	45	14	—	—
Carbaryl (C)	—	10	15–21 d	—
Aldicarb (C)	—	—	—	50 d
Carbofuran (C)	—	—	—	2 mo
Furathiocarb (C)	—	—	—	1.5–3 mo

1.2 Carbamates

Carbamates first appeared in the late 1940s. They were used first as insecticides, then as fungicides, and finally as herbicides. Comprising about 50 different molecules, carbamates are still manufactured and used as insecticides.

The carbamates have several characteristics in common with OPs (Bocquené et al., 1997), and the quantities used in France are of the same order of magnitude. The sources of dispersion and mechanisms of transfer are also similar.

1.2.1 Release, Transfer and Levels of Carbamates in the Environment

The quantities of carbamates applied in 1990 in metropolitan France were of the same order as those of OPs, with 1,692 tonnes of active ingredient. It must be emphasized, however, that carbamates have formed a noticeably larger part of the insecticide market since 1985, when consumption was recorded at 1,136 tonnes. Among the most widely used carbamate insecticides are aldicarb, carbaryl, carbofuran, carbosulfan, and furathiocarb.

The potential sources of dispersion are of the same nature as those of OPs and essentially of agricultural origin, but it is also extremely difficult to quantify releases linked to domestic uses. Data on the levels of these molecules in the different compartments of the environment are still rarer than for the OPs. Carbamates are transferred toward the aquatic environment mainly by runoff from application areas. Carbofuran was found at concentrations reaching 2.1 µg/l in 1987 in the Sacramento river in the United States (Harrington and Lew, 1989). In France, carbofuran was detected in ditches of the marshes of Charente-

Maritime (Munschy, 1995) and in certain rivers of Britanny (the Seiche, the Horn) at concentrations of 0.25 µg/l (Gillet, 1994).

Generally, carbamates have higher solubility in water than OPs, and the persistence of their action varies considerably from one product to another (Table 1).

1.2.2 Degradation of Carbamates

The mechanisms of degradation are similar to those described for the OPs. The carbamates are among the most rapidly degraded insecticides, although they are rather stable in conditions of neutral or acid pH. Some products of degradation seem nevertheless more persistent than the molecule from which they arise. This is the case especially with carbofuran (Gupta, 1994), which paradoxically has a duration of action in the soil up to 50 d (ACTA, 1993).

1.3 Evaluating the Consequences of Pesticide Presence in the Marine Environment

The occurrence of pesticides in the environment is generally linked to the form in which the products are sold (Getzin, 1985), the nature of the soil (Gerstl, 1990), cultivation practices (Logan et al., 1987), and meteorological conditions. The physico-chemical properties of different molecules (e.g., volatility, solubility, degradability, persistence) influence their transfer towards aquatic environments. It must be kept in mind that pesticides can be totally degraded or transformed during the course of their transport. Nevertheless, the degradation products, which often have a toxicity that is far from negligible, must be considered in the field of ecological studies.

The biogeochemistry of pesticides, particularly OPs and carbamates, in the marine environment is complex. The particulate or free form of compounds and their partition in the biotic compartments (animal and plant) and abiotic compartments (water, sediments, suspended matter) affect their availability to living organisms. Also, their intrinsic properties (e.g., partition coefficient, solubility), as well as the physico-chemical properties of the medium (e.g., pH, temperature, salinity, carbon load, presence of fulvic acids) affect the occurrence of these molecules. Their accumulation in sediments (Gould, 1972) and in food chains is frequent and leads to significant levels of contamination for the higher trophic levels (Ramade, 1992).

The release of these pollutants in the marine environment and the disturbances that may result made it necessary to monitor contamination levels and their evolution and the effects of these compounds on the health of marine organisms. Techniques to measure a physiological or biochemical effect in relation to the presence of a toxicant are essential tools for such a study.

The diversity of molecules present in the environment makes it difficult to quantify them by chemical methods. Even though they are specific and quantitative, chemical measurements alone cannot allow us to conclude concerning the biological significance of the pollution of a medium. Finally, the presence of products of degradation and metabolites that are much more difficult to analyse, sometimes even impossible, poses problems in evaluating the actual contamination of the environment. In this context, the use of measurements related to the response of biological systems called biomarkers is increasingly widespread. With regard to pesticides, however, and despite several studies on the effects of different types of molecules, the only example of measurement of effects on the marine organisms that meets constraints on the use of biomarkers is that of cholinesterases, particularly acetylcholinesterase, the activity of which is inhibited mainly by OPs and carbamates.

2. ACETYLCHOLINESTERASE

Acetylcholinesterase (AChE) is a cholinesterase that hydrolyses esters of choline, especially acetylcholine, a chemical mediator that ensures the transmission of a nerve impulse at synapses of neuromuscular and interneuronal junctions (Bocquené et al., 1997). According to the nature of the substrate that they hydrolyse, the AChEs (EC 3.1.1.7), which hydrolyse acetylcholine preferentially, are distinguished from other cholinesterases, mainly butyrylcholinesterase (BuChE), which have a stronger affinity for other esters of choline.

2.1 Methods of Measuring AChE Activity

Various methods have been proposed for measuring cholinesterases activity. The first protocols were proposed several decades ago and used the principle of measuring the acidification of a reactive medium linked to the release of ions during hydrolysis of acetylcholine. These methods were subsequently optimized by the use of 'pH-stats' that allowed the reactive medium to be kept at a constant pH. They have now been replaced by colorimetric methods, which are simpler, more sensitive, and more precise.

The identification of a reactive specific to thiol groups, dithio-bis-nitrobenzoic acid (DTNB), allowed the use of a protocol that used a sulphurate derivate of acetylcholine, acetylthiocholine, as substrate of the reaction, the hydrolysis of which releases sulfhydryl groups that can be quantified by spectrophotometry. This method, described by Ellman et al. (1961), is the most widely used at present, on different types of tissues and organisms. It can be adapted to automatic analysers and overcomes constraints of analysis of large quantities of samples.

The principle of the reaction is described below:

$$\text{acetylthiocholine} \xrightarrow{\text{acetylcholinesterase}} \text{thiocholine} + \text{acetic acid}$$

$$\text{thiocholine} + \text{DTNB} \longrightarrow \text{thionitrobenzoate} \ (E_{412 \text{ nm}} = 13{,}400)$$

Other methods were later defined for applications more specific than routine measurement. Radiometric, chromatographic, or immuno-chemical measurements (Fairbrother et al., 1989) especially enable more specific measurements of different forms or isoforms of the enzyme, but their routine use remains uncertain.

2.2 Sources of Variability in AChE Activity

2.2.1 Variability Linked to Experimental Conditions

The knowledge of sources of variability in measurement is an important factor to be taken into account for the optimization and standardization of protocols. Among the known sources of variations are factors such as modalities of sampling, preparation and preservation of tissues, preparation of reactive agents, physico-chemical conditions of measurements, and instrumentation.

It is important to avoid the use of anticholinesterasic agents during the preparation of samples. In particular, the use of antibiotics, amino acids, anticoagulants, colorants, and inorganic ions can induce modifications in the measurement (Long, 1963). Similarly, the treatment of an organism by an anticholinesterasic compound may cause accumulation of this compound in the tissues. In this case, the inhibiting action of products must be taken into account during the extraction of tissues.

The stability of samples varies according to storage conditions. The maintenance of tissues or tissue extracts at ambient temperature immediately causes loss of cholinesterase activity. Similarly, lyophilization of tissues is responsible for degradation of enzymes in a few days. It is therefore essential to keep the tissues frozen at −85°C for long-term preservation. Under such conditions, AChE is stable for several months. However, apart from the absence or presence of inhibitors, stability also depends on the nature of the tissues. In fishes and marine invertebrates, cholinesterasic activity associated with AChEs and BuChEs varies according to type of tissue (Table 2). A study of their distribution shows higher activity of AChE in the muscle and brain for fishes, and mainly in the muscle for invertebrates.

Table 2. Specific activity (mean ± SD) of acetylcholinesterase and butyrylcholinesterase in different tissues of several species of fish and marine invertebrates (according to Bocquené, et al., 1990, modified)

Species	Organ	AChE	BuChE
Plaice	Muscle	13,252 ± 746	2065 ± 358
(*Pleuronectes platessa*)	Liver	2766 ± 479	1502 ± 260
	Brain	5946 ± 1020	1807 ± 346
	Heart	1875 ± 210	733 ± 130
Common sole	Heart	254 ± 31	131.4 ± 62
(*Solea solea*)	Muscle	3315 ± 507	1102 ± 297
	Brain	5131 ± 548	117 ± 94
European seabass	Muscle	1285	685
(*Dicentrarchus labrax*)	Liver	452	185
	Brain	4407	217
Mussel	Whole animal	311 ± 29	44 ± 20
(*Mytilus edulis*)	Hepatopancreas	151 ± 14	49 ± 18
	Mantle	394 ± 51	98 ± 29
	Adductor muscle	440 ± 43	100 ± 43
	Gill	1155 ± 114	372 ± 98
Oyster	Whole animal	1925 ± 305	439 ± 212
(*Crassostrea gigas*)	Hepatopancreas	632 ± 66	78 ± 27
	Adductor muscle	757 ± 112	77 ± 45
	Gill	841 ± 123	—
	Mantle	914 ± 68	126 ± 93
Common prawn	Whole animal	1686 ± 50	687 ± 180
(*Palaemon serratus*)	Cephalothorax	609 ± 85	518 ± 100
	Muscle	8125 ± 380	1438 ± 314
	Eggs	342 ± 34	—
	Hepatopancreas	1394 ± 188	—

The results are expressed in specific units (variation in optical density (DO)/1000 per minute and per mg of protein extract).

Similarly, the duration and speed of homogenization, the choice of extraction buffer, and the centrifugation procedure must be fixed to obtain reproducible results. Moreover, the use of non-ionic detergents, particularly Triton X-100, in the extraction buffer, by breaking the hydrophobic bonds between the enzyme and the cellular membrane, enables solubilization of the membrane-bound cholinesterases in order to measure the specific activity.

The biochemical properties of cholinesterases must be taken into account during these measurements. The optimal conditions of pH, temperature, and ionic force must be calculated systematically in order to standardize the procedures. Table 3 summarizes the optimal conditions for measurement of AChE activity in different tissues of some marine species of the French coasts.

Finally, the instrumentation is also an important factor. The nature of the homogenizer, the type of spectrophotometer, the wavelength, and the duration of measurement can influence the results.

The development of marine environment monitoring led laboratories to develop routine measurement procedures. In particular the use of microplate readers (Ashour et al., 1987; Galgani and Bocquené, 1991), developed at first for immunochemical measurements, enabled researchers to complete relatively quickly the large number of analyses needed for studies on the oceanographic scale. These techniques were validated in the framework of international programmes and allowed direct measurements on oceanographic ships (Fig. 1). Thus, the optimal conditions for measurement of AChE activity were defined with a view to evaluating the effects of pesticides on the marine environment.

2.2.2. An Example of Protocol for Measuring AChE Activity Adapted for Marine Species

Following several studies done in the past few years (reviewed in Bocquené and Galgani, 1996), a protocol that can be used for marine species was defined in the context of international programmes. Typically, the fresh or frozen tissues are homogenized for 1 min. in phosphate buffer 20 mM, pH 7, with an Ultra-Turrax homogenizer (1–4 ml buffer for 1 g of tissues), then centrifuged at 10,000 g for 20 min. The supernatant can then be preserved for several months at −85°C or used immediately for analysis. A measurement of proteins is then done in order to be able to compare several samples. The AChE activity is then measured in phosphate buffer 20 mM containing DTNB at a concentration of 10 mM and substrate (acetylthiocholine) at a concentration of 2.5 mM. The variations in optical density of the sample at 412 nm are read for 1 to 5 min. with the help of a spectrophotometer.

Non-enzymatic hydrolysis of substrate can be measured with the help of a control reaction in the absence of extract containing the enzyme and subtracting the activity measured in the samples.

Table 3. Optimal conditions for measurement of acetylcholinesterase activity of various marine organisms (according to Bocquené, et al., 1990; Salles, 1991)

Species	Tissue	pH	Temp. (°C)
European seabass	Brain	9	34
(*Dicentrarchus labrax*)	Muscle	8	> 40
	Liver	8.5	—
Plaice	Brain	8.5	34
(*Pleuronectes platessa*)	Muscle	8.5	30
Mussel	Whole animal	8.5	42
(*Mytilus edulis*)			
Great Atlantic scallop	Gills	9	25
(*Pecten maximus*)			
Tiger shrimp	Muscle	9	22
(*Penaeus japonicus*)			

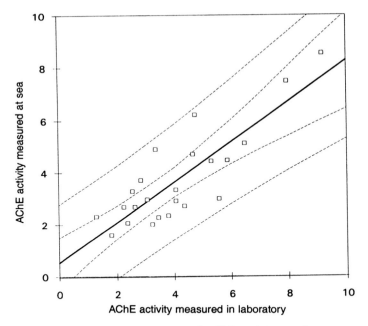

Fig. 1. Correlation between measurements of AChE activity made at sea and in the laboratory with a microplate reader (Galgani et al., 1992)

The variations in optical density of enzymatic activity are converted according to the following formula:

$$\text{Activity} = \frac{DO_{412\,nm} \times Vt \times 1000}{13{,}600 \times L \times Ve \times [P]}$$

where $DO_{412\,nm}$ is the variation in optical density per minute, Vt is the total reactive volume, 13,600 is the molecular coefficient of absorption of TNB, L is the optical path length in cm, Ve is the volume of the sample, and [P] is the concentration of proteins in the tissue extract.

The results are then expressed in micromoles of thiocholine released per minute and per mg of proteins, which corresponds to the specific cholinesterase activity of the extract.

2.2.3 Natural Variability of AChE

Since AChE is implicated in the transmission of nerve impulses, it is not surprising that this enzyme should be localized mainly in the nerve tissues and muscle tissues (Table 2). In these tissues, in spite of endogenous regulation, AChE activity is subject to variations linked to environmental factors.

Even though changes in cholinesterase activity have been observed during embryonic development in different species (Rattner and Fairbrother, 1991), few variations exist in adults, except those relating to blood cholinesterases. For marine fishes, no difference has been observed between organisms of variable size after the appearance of secondary sexual characteristics (Table 4). Similarly, the differences in activity between males and females do not seem significant except during periods of reproduction (Galgani et al., 1992), for which no information is available in marine organisms.

Among the environmental variables having a notable effect on the activity of the enzyme, temperature is the most important (Edwards and Fisher, 1991). Temperature variations in the medium have a direct effect on the level of the enzyme (Macek et al., 1969; Hogan, 1970). Also, variation in toxicity of pesticides in relation with the temperature has been demonstrated, caused by changes in reactivity of AChE (Chapman et al., 1982). These effects can explain, at least partly, the monthly variations observed in certain marine species (Fig. 2). Moreover, variations in enzyme activity were observed during the manipulation of organisms and during their adaptation to new experimental conditions. Finally, the interpretation of variations in AChE activity in marine organisms must take into account the possible alterations in cholinesterases that are linked to nutrition and physiopathology of the animals (Rattner and Fairbrother, 1991).

3. TOXICOLOGY OF ORGANOPHOSPHORUS PESTICIDES AND CARBAMATES PRESENT IN THE ENVIRONMENT

The characteristic symptoms of effects of pesticides that inhibit cholinesterases have been abundantly described (Kuhr and Dorough, 1976; Ballantyne and Marrs, 1992), especially for aquatic species (Zinkl et al., 1991). The diagnostic of an intoxication is characterized by infor-

Table 4. Variation of cholinesterase activity* (mean $\pm$ SD) of dab (*Limanda limanda*) in relation to size (according to Galgani et al., 1992)

Size (cm)	AChE	BuChE
< 12	3298 $\pm$ 1449	2001 $\pm$ 631
12–20	2654 $\pm$ 710	1487 $\pm$ 229
20–25	2689 $\pm$ 1700	1398 $\pm$ 1024

*Specific activity (variation in optical density (DO)/1000 per minute and per mg of protein extract).

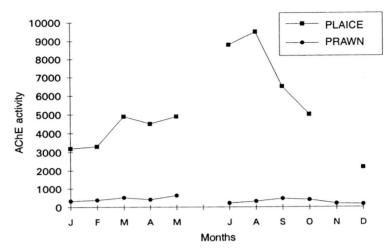

Fig. 2. Seasonal variations in AChE activity in muscle of plaice (*Pleuronectes platessa*) and prawn (*Palaemon serratus*) taken from the English Channel (unpublished data). Measurements are expressed in specific activity (variation in optical density (DO)/1000 per minute and per mg of protein extract).

mation on the history of organisms, clinical signs, and laboratory data. In general, the organisms affected by OPs and carbamates show signs of paralysis, hyperactivity, and loss of equilibrium. However, the mode of action differs with the nature of the compound. Most OPs are considered to have an irreversible action on cholinesterases, the duration of *de novo* synthesis of the enzyme being longer than the duration of dissociation of the OP-AChE complex. The carbamates, on the other hand, have an action that is reversible, sometimes within a very short period (Bocquené et al., 1997).

There is no doubt that OP pesticides and carbamates inhibit the AChE of marine organisms, as they do in terrestrial species, although the response depends on the species. The toxicity is greater for the oxidate (oxon) forms of OPs than for the thio forms, because of the biochemical properties of cholinesterases (Bocquené et al., 1997). Moreover, even in a single species, toxicity depends on the nature of the compound and also on biotic and abiotic factors such as persistence in the marine environment, metabolism, and physiological factors, particularly temperature (Zinkl et al., 1991). Apart from these general rules, certain characteristics apply more specifically to effects of these pesticides on marine species. Species belonging to the class Crustacea are more sensitive than species of other taxonomic classes. The explanation clearly lies in the manufacturers' and users' choice of molecules. Molecules are chosen mainly for their toxicity for Arthropoda of the class Insecta with a view to their use as insecticides, and the Crustaceae are Arthropoda that are taxonomically close to Insecta because they both belong to the sub-branch of Mandibulates or Antennates.

Basically, the toxicity of a product can be evaluated either from *in vitro* studies of toxicity, from biological tests on organisms maintained in controlled conditions, or by direct observations of animals in the natural environment.

3.1 *In Vitro* Toxicity Tests of OPs and Carbamates

In vitro measurements of toxicity of OP compounds and carbamates rely on tests of inhibition of cholinesterases extracted from tissues. The measurements are the inhibiting effect of the compounds tested (Fig. 3) or the kinetic parameters of cholinesterases (Table 5), which make the characterization of the affinity of a compound possible.

The major advantage of these measurements is that compounds can be compared in terms of toxicity linked to direct interaction between the inhibitor and the enzyme. They also enable us to find out the effects of molecules other than pesticides, the presence of which could lead to errors in interpretation of results showing variations in cholinesterase activity. The studies of Olson and Christensen (1980) showed an inhibition of cholinesterases of fish by many compounds other than OPs and carbamates. These risks of interference, however, remain secondary, given the maximum concentrations of these products in the marine environment (Bocquené et al., 1997).

The main limitation of this type of test is that physiological factors are not taken into account during the measurements. The interpretation of results makes the extrapolation of potential effects on the animal itself more tricky, and even more so for extrapolations on the natural environment. The evaluation of toxicity must therefore be complemented by *in vivo* measurements.

Table 5. Constant of apparent affinity of AChE of muscle in various marine species and values of the constant of inhibition of carbaryl (unpublished data)

Species	Km* (µmoles)	Ki** (µg/l)
Patella vulgata (limpet)	182	49
Carcinus maenas (green crab)	86	62
Cancer pagurus (edible crab)	112	26
Palaemon serratus (prawn)	120	48
Pleuronectes platessa (plaice)	232	104

*Km: constant of apparent affinity for acetylcholine determined by the equation of Michaelis-Menten.
**Ki: constant of apparent AChE inhibition by carbaryl.

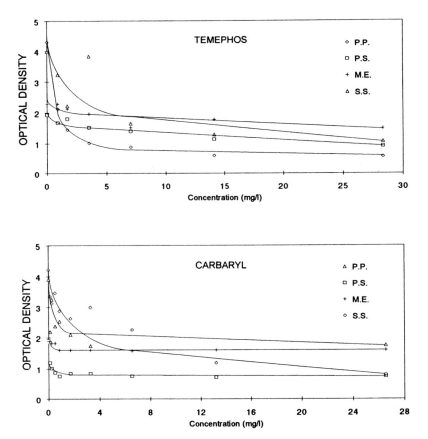

Fig. 3. *In vitro* effects of temephos (OP) and carbaryl (carbamate) on AChE activity of muscle in plaice (*Pleuronectes platessa, P.P.*), prawn (*Palaemon serratus, P.S.*), mussel (*Mytilus edulis, M.E.*), and mackerel (*Scomber scomber, S.S.*) (according to Galgani and Bocquene, 1991)

3.2 *In Vivo* Evaluation of Toxicity of OPs and Carbamates

Generally, the OP compounds show a very high toxicity for marine organisms, particularly in the Crustaceae. Table 6 indicates relations between the effects of pesticides and cholinesterase inhibition in several marine species. Kobayashi et al. (1986) evaluated an LC_{50} of fenitrothion in tiger shrimp (*Penaeus japonicus*) of 1 µg/l after 48 h. After 96 h exposure, LC_{50} of malathion was 1.3 µg/l. Later studies (Bocquené and Galgani, 1991) showed that the inhibitory effects of phosalone were detectable after 29 d in prawn (*Palaemon serratus*) exposed to 0.1 µg/l of this pesticide. The sensitivity of fishes to OPs was lower. Nevertheless, in salmon, lethal effects are observed for concentrations ranging from 120 and 150 µg/l of malathion (LC$_{50}$ at 96 h; Zinkl et al., 1991). Weiss and Gakstatter (1964) proved AChE inhibition in the brain of a fish, *Lepomis macrochirus*, after 30 d of contact with 0.1 µg/l of parathion. Such low lethal or sublethal concentrations show the harmful character of these compounds, the toxicity of which may have been underestimated earlier. Levels much higher than concentrations leading to an effect have been detected in streams. Data are not available on concentrations of these products in marine waters, but concentrations are probably very low, except in aquaculture stocks that have been given antiparasite treatments.

The carbamates, like the OPs, are strong inhibitors of AChE, and the Crustaceae are the most sensitive to them among aquatic species. The LC$_{50}$ (3 h) of carbaryl for daphnia (*Daphnia magna*) was estimated at 50 µg/l (Bocquené, 1991). In this case, the toxicity was also due to one of the products of hydrolysis of carbaryl, 1-naphthol. In a study (Bocquené and Galgani, 1996) on thresholds of induction of AChE inhibition by carbamates, effects were measured in prawn (*Palaemon serratus*) after 29 d of exposure to 0.1 µg/l of carbaryl. Generally, the LC$_{50}$ of carbamates on fish lies between 0.1 and 10 mg/l. Such concentrations are rarely found in marine waters, except in case of accidental release.

3.3 *In Situ* Toxicity of OP Compounds and Carbamates

Examples of action of OP pesticides and carbamates on marine organisms in the natural environment are limited to specific cases, which we have already looked at in the beginning of this chapter.

Table 6. Relations between biological effects and the inhibition of AChE in various species of fish and invertebrates sampled from the sea

Species	Compound	Effect	% inhibition of cholinesterase
Oncorhynchus kisutch[1] (salmon)	Malathion (100 µg/l)	Alteration of movements	28
	Malathion (300 µg/l)	Alteration of movements	48
Penaeus duorarum[2] (prawn)	Malathion	Mortality	75
Palaemon serratus[3] (prawn)	Phosalone (100 µg/l)	Mortality	75
	Carbaryl (100 µg/l)	Mortality	50–70
Loligo pealei[2] (squid)	Fenitrothion	No effect	13–77
	Paraoxon	Mortality	75

[1] Zinkl et al., 1991.
[2] Edwards and Fisher, 1991.
[3] Bocquené and Galgani, 1996.

The most spectacular and economically damaging incidents are linked to massive mortalities of fish or shellfish in farms near dispersal areas. This is the case, for example, of oyster and mussel farms in Charente-Maritime. There were many accidents following treatments with OP insecticides in the marine environment or on the coastal fringe. In 1991, aerial dispersal of fenitrothion for anti-mosquito treatment caused the death of several tons of tiger shrimp (*Penaeus japonicus*) in an aquaculture farm in Languedoc.

Even in cases of voluntary introduction of pesticides in the marine environment, accidents linked to the use of OPs in aquaculture farms were frequent. Treatment of the water for elimination of parasitic copepods is a well-documented example of effects linked to direct introduction of pesticides in the aquatic environment (Cusack and Johnson, 1990). Horsberg et al. (1989) cited three episodes of significant mortality in salmon farms after the use of trichlorfon and dichlorvos as antiparasitic treatments; the dead animals showed inhibitions of 80% in AChE activity and hepatic bioaccumulation of 200 μg/kg.

In such situations, the measurement of cholinesterase activity is a good example of use of a biomarker of effects arising from the presence of a pesticide. The measurement allows us to characterize the spreading of the effect over time and gives indications on dose-effect relationships, as well as return to the normal state (Fig. 4). The results also provide important information for interpretation of effects of pesticides on organisms in the natural environment.

4. USE OF CHOLINESTERASES IN EVALUATING EFFECTS OF OP PESTICIDES AND CARBAMATES IN THE MARINE ENVIRONMENT

The technical and scientific data on cholinesterases described so far allow us to coherently use measurements of cholinesterase activity, the inhibition of which indicates the effect of pesticides, notably OPs and carbamates, on animals. However, certain conditions must be met before this type of biomarker can be applied to the marine environment.

The choice of sites presenting potential risks of effects of pesticides is critical. It is most often made on the basis of the presence of agricultural activities, which are sources of toxic molecules. In the case of the French coast, for example, Britanny and Languedoc-Roussillon are regions located on the coastal border and could be sites of important release of pesticides. In the specific case of Britanny, significant quantities of pesticides, notably carbofuran, were measured in the waters of coastal rivers (Gillet, 1994). The large industrial areas constitute another

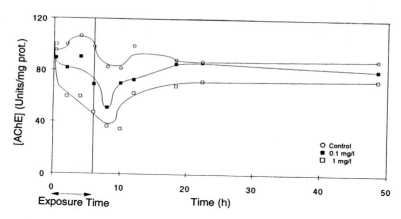

Fig. 4. AChE activity of clam (*Ruditapes documatum*) contaminated *in vivo* by dichlorvos (OP) (according to Le Bris et al., 1995)

example of zones at risk. In this case, the presence of OP compounds or carbamates can be linked to the presence of agrochemical industries. This is the case, for example, of large estuaries and deltas, as well as some gulfs.

The choice of species is also an important criterion in *in situ* evaluation of toxicity. Widely distributed species that have no significant migrations allow us to compare sites that are far apart, which is a fundamental requirement in studies on the oceanographic scale. Moreover, since OP molecules and carbamates generally accumulate in the sediments, a species having benthic behaviour must be used in the study. Finally, organisms that are abundant and easy to catch are most useful for an effective sampling.

In the French coastal waters, no fish species restricted to the continental plateau is present. Thus, for the Atlantic coast and for North European waters, most studies are done on the dragonet (*Callionymus lyra*) and dab (*Limanda limanda*), while the species used in the Mediterranean area are mainly the red mullet (*Mullus barbatus*) for sandy and silty bottoms and the comber (*Serranus cabrilla*) for rocky bottoms. For zones with tidal movements, mussel (*Mytilus edulis* and *M. galloprovincialis*) and oyster (*Crassostrea gigas*) enable us to obtain information relevant to areas with shellfish farms.

Studies on various fish species (Bocquené et al., 1993) have shown that measurements are technically reproducible and that, under standardized sampling conditions, variations in activity are due to the actions of neurotoxic compounds. Measurements done on dab along a transect of 360 km in the North Sea, for example, showed significant inhibitions of different types of cholinesterases. This effect, mainly observed in animals from the coast, is due to compounds transported by the Elbe and Weser estuaries (Fig. 5). The identification of compounds that inhibit AChE is still tricky. The limitations of analytic methods and the diversity of molecules present in the estuarine waters pose problems in the use of measurements of cholinesterase activity as biomarkers. Although the same problems occur for other types of biomarkers (particularly enzymatic biomarkers), in this case, the link between the nature of the molecules involved and the anticholinesterasic activity remains to be demonstrated.

Studies in estuarine waters have nevertheless already shown that, in the natural environment, the presence of insecticides is responsible for cholinesterase inhibition (Zinkl et al., 1991). These results clearly show the advantage of this type of measurement in monitoring biological effects of pesticides. In reality, the use of AChE and other cholinesterases makes good sense in terms of following the neurotoxic effects of pollutants over time and space. Moreover, it allows to evaluate long-term trends. In an experiment, AChE activity was measured on molluscs of

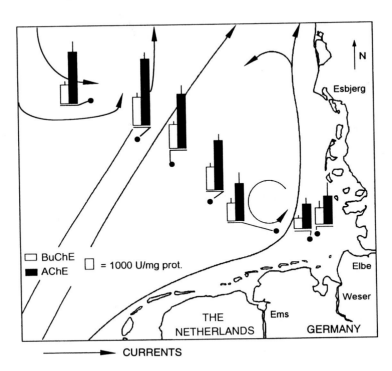

Fig. 5. AChE and BuChE activity in dab (*Limanda limanda*) in the North Sea (according to Galgani et al., 1992)

the French coasts (Fig. 6). The results of this preliminary series of analyses could not take into account variations over time in enzyme activity but clearly indicated the extent of such a monitoring activity. It is possible only if there are appropriate infrastructure for sampling, a standardized sampling procedure, and adequate methods of analysis. These conditions have been met in the case of French coasts, where a coastal surveillance network was created under the aegis of the Environment Ministry, within the framework of the RNO. Similarly, the Conseil International des Mers and the Commission Internationale pour l'Exploration de la Mer Méditerranée envisage the integration of such measures within the framework of the surveillance network of the North Sea (see chapter 3) and the Mediterranean Sea. A reference method and sampling strategy have been established for this purpose. However, the experience gained during the establishment of a network for monitoring

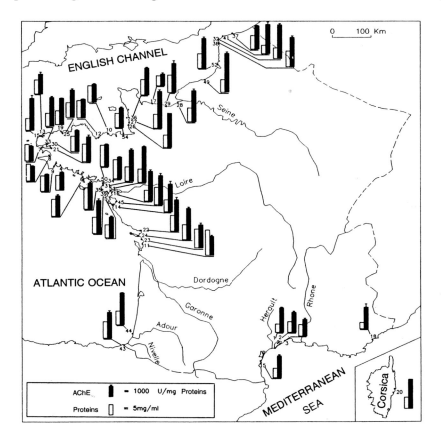

Fig. 6. AChE activity in muscle of mussel (*Mytilus edulis*) from the French coasts (Bocquené et al., 1993)

the effects of pollutants (see chapter 3) has shown that the delay between discovery and refinement of a method for measuring the effect of a pollutant and its application in a structured network is particularly long and involves many research studies.

5. PERSPECTIVES

In such a context, the knowledge of mechanisms of action of pesticides, including those considered secondary, seems necessary to the perfection of new biomarkers that can be used in marine species. Various studies have been able to update the hitherto unknown effects of some pesticides. The molecular analysis of mutagenesis in particular enables us to envisage in the very short term the detection of specific mutations from the effect of pesticides. In a way comparable to the detection of mutations on the gene PSBA of plants in the presence of atrazine (Hirshberg and McIntosh, 1983), the detection of mutations on genes of cholinesterases in the presence of insecticides (Bocquené et al., 1997) seems to be a potential biomarker of effects of pollutants specific to this type of target. Although these measures are presently limited to terrestrial species, the application of biotechnologies, especially molecular biology, seems essential to their extension to marine species. The recent demonstration of mechanisms of resistance to different pesticides such as atrazine, DDT, or carbamates in marine animals (Minier, 1994; Galgani et al., 1995) clearly confirms this trend.

REFERENCES

ACTA (1993). Index phytosanitaire. Association de Coordination Technique Agricole, Paris.

Ashour M.B.A., Gee S.J. and Hammock B.D. (1987). Use of a 96-well microplate reader for measuring routine enzyme activities. *Anal. Biochem.*, 166:353–360.

Ballantyne B. and Marrs T.C. (eds.). (1992). *Clinical and Experimental Toxicology of Organophosphates and Carbamates*. Butterworth/Heinemann.

Barcelo D., Sole M., Durand G. and Albaiges J. (1991). Analysis and behaviour of organophosphorous pesticides in a rice crop field. *Fresenius J. Anal. Chem.*, 339:676–683.

Bocquené G. (1991). L'acétylcholinestérase chez les organismes marins: Outil de surveillance des effets pesticides organophosphorés et carbamates. *Mémoire École Pratique des Hautes Études.*

Bocquené G. and Galgani F. (1996). Cholinesterase inhibition by organophosphorus and carbamate compounds. *Techn. Mar. Env. Scien.* ICES. Times Series.

Bocquené G., Galgani F. and Truquet P. (1990). Characterization and assay conditions for use of AChE activity from several marine species in pollution monitoring. *Mar. Environ. Res.*, 30:75–89.

Bocquené G., Galgani F. and Walker C.H. (1997). Les cholinestérases, biomarqueurs de neurotoxicité. In: Biomarqueurs en écotoxicologie: Aspects fondamentaux. Lagadic L., Caquet Th., Amiard J.C. and Ramade F. (eds.). Masson, Paris, pp. 209–239.

Bocquené G., Galgani F., Burgeot T. et al. (1993). Acetylcholinesterase levels in marine organisms along French coasts. *Mar. Pollut. Bull.*, 26:101–106.

Capel P.D., Giger W., Reichert P. and Wanner O. (1988). Accidental input of pesticides into the Rhine River. *Environ. Sci. Technol.*, 22:992–997.

Chapman P.M., Farrell M. and Brinkhurst R.O. (1982). Relative tolerances of selected oligochaetes to individual pollutants and environmental factors. *Aquat. Toxicol.*, 2:47–67.

Cusack R. and Johnson G. (1990). A study of dichlorvos, a therapeutic agent, for the treatment of salmonids infested with sea lice. *Aquaculture*, 90:101–112.

Durand G. and Barcelo D. (1992). Environmental degradation of atrazine, linuron and fenitrothion in soil samples. *Environ. Toxicol. Chem.*, 36:225–234.

Edwards C.A. and Fisher S.W. (1991). The use of cholinesterase measurements in assessing the impacts of pesticides on terrestrial and aquatic invertebrates. *Chem. Agric.*, 2:255–275.

Ellman G.L., Courtney K.D., Andreas V. Jr. and Featherstone R.M. (1961). A new and rapid colorimetric determination of acetylcholinesterase activity. *Biochem. Pharmacol.*, 7:88–95.

Fairbrother A., Bennett R.S. and Bennett J.K. (1989). Sequential sampling of plasma cholinesterase in mallards as an indicator of exposure to cholinesterase inhibitors. *Environ. Toxicol. Chem.*, 8:117–122.

Fest C. and Schmidt K.J. (1982). *The Chemistry of Organophosphorus Pesticides*, 2nd ed. Springer Verlag.

Fikes J.D. (1990). Organophosphorous and carbamate insecticides. *Vet. Clin. N. Am. Small Anim. Pract.*, 20:353–367.

Franck R., Braun H.E., Chapman N. and Burchat C. (1991). Degradation of parent compounds of nine organophosphorous insecticides in Ontario surface and ground waters under controlled conditions. *Bull. Environ. Contam. Toxicol.*, 47:374–380.

Galgani F. and Bocquené G. (1991). Semi-automated colorimetric and enzymatic assays for aquatic organisms using plate readers. *Water Res.* 25:147–150.

Galgani F., Bocquené G. and Cadiou Y. (1992). Evidence of variation in cholinesterase activity in fish along a pollution gradient in the North Sea. *Mar. Ecol Progr. Ser.*, 91:77–82.

Galgani F., Cornwall R., Toomey B.H. and Epel D.D. (1995). Interaction of environmental xenobiotics with a multixenobiotic defense mechanism in the bay mussel *Mytilus galloprovincialis* from the coast of California. *Environ. Toxicol. Chem.*, 15:325–331.

Gerstl Z. (1990). Estimation of organic chemical sorption by soils. *J. Contam. Hyd.*, 6:357–375.

Getzin L.W. (1985). Factors influencing the persistence and effectiveness of chlorpyrifos in soil. *J. Econ. Entomol.*, 78:412–418.

Gillet H. (1994). Étude de la contamination des eaux superficielles de Bretagne par les pesticides. Results of 1993 follow-up study. Report of the Direction Régionale de l'Agriculture et de la Forêt. Préfecture de la Région Bretagne. Service Régional de la Protection des Végétaux. Rennes.

Gould R.F. (1972). Fate of organic pesticides in the aquatic environment. *Am. Chem. Soc.*, *Adv. in Chem. III*.

Gupta R.C. (1994). Carbofuran toxicity. *J. Toxicol. Environ. Health*, 43:383–418.

Habig C. and Di Giulio R.T. (1988). The anticholinesterase effect of the cotton defoliant, S, S, S,-Tri-*n*-Butyl Phosphorotrithioate (DEF) on channel catfish. *Mar. Environ. Res.*, 24:193–197.

Harrington J.M. and Lew T.S. (1989). Rice pesticide concentrations in the Sacramento river and associated agricultural drains. California Department of Fish and

Game, Environmental Service Division Administrative Report, 89-1, Sacramento, California.

Hirshberg J. and McIntosh L. (1983). Molecular basis of the herbicide resistance in *Amaranthus hybridus*. *Science*, 221:1346–1349.

Hogan J.W. (1970). Water temperature as a source of variation in specific activity of brain acetylcholinesterase of bluegills. *Bull. Environ. Contam. Toxicol.*, 5:347–352.

Horsberg T.E., Hoey T. and Nafstad I. (1989). Organophosphate poisoning of atlantic salmon in connection with treatment against salmon lice. *Acta Vet. Scand.*, 30:385–390.

IFEN (French Institute for Environment) (1999). *Key-numbers for the Environment*. Edition 1999. Paris. Mate.

Kobayashi K., Nakamura Y., Rompas R.M. and Imada N. (1986). Difference in lethal concentration *in vivo* between fenitrothion and its oxo-form in tiger shrimp *Penaeus japonicus*. *Bull. Japan. Soc. Sci. Fisheries*, 52:287–292.

Kuhr R.J. and Dorough H.W. (1976). *Carbamate Insecticides. Chemistry, Biochemistry and Toxicology*. CRC Press, Cleveland, Ohio.

Koelle G.B. (1994). Pharmacology of organophosphates. *J. Appl. Toxicol.*, 14:105–109.

Lartiges S. and Garrigues P. (1993). Determination of organophosphorus and organonitrogen pesticides in water and sediments by GC-NPD and GC-MS. *Analysis*, 21:157–165.

Le Bris H., Maffart P., Boequené G. et al. (1995). Laboratory study on the effect of dichlorvos on two commercial bivalves. *Aquaculture*, 138:139–144.

Logan T.J., Davidson J.M., Baker J.L. and Overcash M.R. (1987). *Effects of Conservation Tillage on Groundwater Quality—Nitrates and Pesticides*. Lewis Publishers.

Long J.P. (1963). Structure-activity relationships of the reversible anticholinesterase agents. In: Cholinesterases and Anticholinesterase Agents. Koelle G.B. (ed.). Springer-Verlag, Berlin, pp. 374–427.

Lores E.M., Moore J.C., Moody P. et al. (1985). Temephos residues in stagnant ponds after mosquito larvicide applications by helicopter. *Bull. Environ. Contam. Toxicol.*, 35:308–313.

Macek K.J., Hutchinson C. and Cope O.B. (1969). The effects of temperature on the susceptibility of bluegills and rainbow trout to selected pesticides. *Bull. Environ. Contam. Toxicol.*, 4:174–183.

Minier C. (1994). Recherche de biomarqueurs de toxicité liés à l'activité estérase non spécifique et à la résistance multixéno biotique chez divers organismes marins. Doctoral thesis, University of Nantes.

Munschy C. (1995). Comportement géochimique des herbicides et de leurs produits de dégradation en millieu estuarien et marin côtier. Doctoral Thesis, University of Paris VI.

Olson D.L. and Christensen G.M. (1980). Effects of water pollutants and other chemicals on fish acetylcholinesterase *in vitro*. *Environ. Res.*, 21:327–335.

Ramade F. (1992). *Précis d'écotoxicologie*. Collection d' Écologie, 22. Masson, Paris.

Rattner B.A. and Fairbrother A. (1991). Biological variability and the influence of stress on cholinesterase activity. *Chem. Agric.*, 2:89–107.

Readman J.W., Liong Wee Kwong L., Mee L.D. et al. (1992). Persistent organophosphorous pesticides in tropical marine environments. *Mar. Pollut. Bull.*, 24:398–402.

Sancho E., Ferrando M.D., Andreu E. and Gamon M. (1993). Bioconcentration and excretion of diazinon by eel. *Bull. Environ. Contam. Toxicol.*, 50:578–585.

Salles C. (1991). Caractérisation biochimique de l'acétylcholinestérase de cinq espèces aquacoles et études des effets *in vitro* des pesticides organophosphorés et carbamates sur ces espèces. *Mémoire École Nationale Vétérinaire de Nantes*.

Tully O. and Morrissey D. (1989). Concentrations of dichlorvos in Beirtreach Bui Bay, Ireland. *Mar. Pollut. Bull.*, 20:190–191.

Tronczynski J. (1990). Programme de recherche sur les produits phytosanitaires en zones littorales et estuariennes. IFREMER.DRO-90-05-MR, Nantes.

Wang T.C., Lenahan R.A. and Tucker J.W. (1987). Deposition and persistence of aerially applied fenthion in a Florida estuary. *Bull. Environ. Contam. Toxicol.*, 38:226–231.

Wanner O., Etgli T., Fleischmann T. et al. (1989). Behavior of the insecticides disulfoton and thiometon in the Rhine river: a chemodynamic study. *Environ. Sci. Technol.*, 23:1232–1242.

Weiss C.M. and Gakstatter J.H. (1964). Detection of pesticides in water by biochemical assay. *J. Wat. Poll. Cont. Fed.*, 36:240–252.

Zinkl J.G., Lockhart W.L., Kenny S.A. and Waard F.J. (1991). The effects of cholinesterase inhibiting insecticides on fish. *Chem. Agric.*, 2:233–254.

7

Biomarkers of Exposure of Birds and Small Mammals to Pollutants

J.-L. Rivière, M.-O. Fouchécourt and C.H. Walker

INTRODUCTION

Terrestrial vertebrates are a heterogeneous group of domesticated and wild species frequenting diverse habitats: rural, suburban, and urban. Like humans, animals can be contaminated by environmental pollutants and suffer consequences such as diminished vitality, reproductive problems, and, at the highest doses, early mortality. If the pollution is extensive, the number of individuals affected will be greater, and the toxic effects will have an impact on numbers in the local populations.

Exposure to toxins is difficult to evaluate in terrestrial species. Unlike truly aquatic species, which are restricted to their ambient medium, amphibians, and most aquatic birds frequent both aquatic and terrestrial environments. Even strictly terrestrial species may drink polluted water. Thus the need to take into account multiple avenues of exposure complicates the evaluation of risks for terrestrial species.

1. ENVIRONMENTAL CONTAMINANTS AND THEIR EFFECTS

The main groups of wild vertebrates, numerically speaking, are the small mammals and birds, but the toxic effects of environmental pollutants have been observed mostly in the latter. Bird mortalities that have accompanied oil spills, such as that resulting from the shipwreck of the Exxon Valdez on the Alaskan coasts, which caused the death of an estimated 100,000 to 300,000 birds (Platt et al., 1990), have raised public awareness about the effects of oil pollution. We must be aware, however, that apart from risks of acute mortality, these pollutions are the source of various embryotoxic effects in birds (Miller et al., 1978; Leighton et al., 1983). The acute toxicity of certain classes of pesticides (e.g., anticholinesterase insecticides, anticoagulant rodenticides) is still the source of fatal accidents, due to poor conditions of use or unsuspected interspecific differences in sensitivity. Risks of secondary toxicity exist in some cases. Fatal accidents have generally a limited impact on the size of bird populations at the global level, but they may contribute to the local extinction of some populations.

The long-term risks posed by sublethal exposure to a wide range of pollutants (metals, organohalogen compounds, and some pesticides) are more difficult to prove. This risk is nevertheless known. Historically, the effects of environmental pollution on reproduction and its role in the decline of animal populations were first brought to public attention with reference to birds (Carson, 1963). The thinning of shell in bird eggs and the decline of certain raptor populations in Great Britain are examples, now classic, of the disastrous consequences of large-scale dissemination of persistent chemical products such as organochlorine insecticides (aldrin, dieldrin, heptachlor, DDT; Cooke, 1973; Newton et al., 1993). The serious contamination of birds of the American Great Lakes by polyhalogenated aromatic hydrocarbons (PHAH) has been known for several years. The PHAH group consists of three types of molecules: the polychlorobiphenyls (PCBs), polychlorodibenzo(p)dioxins (PCDDs), and polychlorodibenzofurans (PCDFs). The family of PCBs comprises 209 different congeners, the family of PCDDs comprises 75 congeners, and the family of PCDFs comprises 135 congeners. (In the remainder of this chapter, the term 'dioxins' signifies PCDD and PCDF together.) These products are highly toxic to birds. Hays and Risebrough (1972) were the first to suggest that dioxins could be the cause of embryo malformations observed in seagull populations around the American Great Lakes. Later studies unequivocally demonstrated this hypothesis (Kubiak et al., 1989; Gilbertson et al., 1991). The toxic effects (embryotoxic and teratogenic) of these products were clearly proved and their responsibility for the decline of certain populations of raptors and water birds has been irrefutably demonstrated.

Most heavy metals are toxic, and lead is one of the most widespread and dangerous. Lead poisoning is very frequent in aquatic birds in wetland areas where hunting is common. In some regions, lead sinkers used by fishermen are responsible for such pollution, as was shown by Sears (1988) in swans of the Thames. The lead is deposited on silty or sandy beds and is easily consumed by birds looking for grit (solid particles, often small stones, that help break down foods in the gizzard). In France, according to Pain (1991), the Camargue is one of the most polluted areas of the world. Lead is extremely toxic to Anatidae, and ingestion of a single lead sinker can kill an adult duck (Pain and Rattner, 1988). Within a single zone, the risk varies as a function of the food habits and lead-sensitivity of different species (Pain, 1990). Apart from its acute toxicity, lead has several sublethal effects that can, in time, contribute to the extinction of certain populations. The bioavailability of lead and its toxic effects on birds and small mammals have been reviewed by various authors (Scheuhammer, 1991; Canters and DeSnoo, 1993).

Mineral pollution, apart from that caused by heavy metals, can affect birds. In the ponds of the Kesterson National Wildlife Refuge (USA), high levels of selenium are found. This element was shown to be highly toxic to local duck populations (Ohlendorf et al., 1988).

Many sites, especially old waste dumps and inactive industrial areas, are polluted by poorly defined mixtures of products, the toxicity of which, for humans and animals that frequent the site, is difficult to judge. Atmospheric pollution in urban areas or near industrial sites is toxic to wild fauna. Llacuna et al. (1993) studied the effects of a coal power station on natural populations of birds and small mammals and on caged animals in Catalonia. The histological examination of the trachea of these animals showed cellular lesions, mostly in birds and especially in caged birds, which could not avoid the atmosphere polluted by emissions from the power station.

We know less about risks of pollutants for terrestrial mammals, such as rodents, lagomorphs (hares and rabbits), carnivorous mammals, large mammals (deer, boar), and insectivores (bats, shrews). There are two reasons for this: the problem has not been considered sufficiently important and the impact of pollutants is very difficult to prove. Certain populations of piscivores, such as otter and mink, seem to be declining, which could be attributed, at least partly, to environmental pollution (Canters and DeSnoo, 1993). Under experimental conditions, PCBs caused reproductive problems in mink, but they have not been clearly demonstrated to have an effect on natural populations (Kihlstrom et al., 1992). Concentrations of dioxin up to 1 µg/kg of soil have not shown noxious effects on reproduction, and more generally on health, of populations of wild mice (Young, 1984, in Nosek et al., 1993), but soil pollution by PCB and heavy metals caused the disappearance of a shrew population (Batty et al., 1990). More detailed information can be found in the bibliographic reviews of Talmage and Walton (1991) and Shore and Douben (1994). The risk that all these pollutants present to amphibians and reptiles is very poorly understood (Hall and Henry, 1992).

2. USE OF BIOMARKERS IN TERRESTRIAL ANIMALS

Biomarkers of exposure to different types of pollutants have been less often studied in birds and other terrestrial species than in aquatic species (Peakall, 1992). The most frequently used in the field are inhibition of cerebral and plasma esterases by organophosphorus insecticides and carbamates, induction of hepatic cytochromes P450 by various pollutants, essentially PHAH, biomarkers of genotoxic effects, and inhibition of δ-aminolevulinate dehydratase activity by lead. Some other indicators, less often used, are mentioned in this chapter. The development of these different biomarkers varies according to the animal group: the inhibition of esterases, for example, has been the subject of many studies in birds and relatively few in small mammals, while the reverse is true for research on genetic lesions. Tables 1 to 3 group different studies on the measurement of biomarkers in wild animals captured on the polluted sites or placed in similar conditions (caged animals, subjected to pollution in field conditions). Most significant studies are reported, but the reader must not consider these tables to be exhaustive. In particular, data before 1980, which can be found in the literature cited, are not listed, with some exceptions. Experimental studies that validate these biomarkers in wild or domesticated animals are not taken into account; nor are measurements of biochemical or physiological parameters not linked to a field pollution.

2.1 Biomarkers of Exposure to Organophosphorus Insecticides and Carbamates: the Esterases

Organophosphorus (OP) insecticides and carbamates, widely used in agriculture, have acute toxicity often higher than that of organochlorines (LD_{50} of parathion in rats: 2 mg/kg; LD_{50} of aldicarb in rat: 0.93 mg/kg; Tomlin, 1994). They pose some risk for small mammals, as they do for birds, but it is essentially the effects on birds that have been studied. Exposure of birds results from direct contact with insecticides during spraying (forest treatments, for example) or secondary intoxication by consumption of contaminated water, plants, or prey. Treatment of lawns (on golf courses, for example) is a frequently reported cause of mortality of geese or swans (Frank et al., 1991; Kendall et al., 1992) and gulls have been fatally poisoned by ingestion of dead crickets after carbofuran treatment (Leighton, 1988, in Mineau, 1991). In raptors, the toxicity may be direct, for example after exposure of birds to insecticide treatments in forests, or secondary, by consumption of contaminated prey (Hunt et al., 1991). Apart from risks of mortality, birds exposed to sublethal doses have diminished reflexes and become easier prey for cats (Galindo et al., 1984) and probably other predators.

Table 1. Response of biomarkers in birds and small mammals: brain and plasma esterases (AChE, acetylcholinesterase activity; BuChE, butyrylcholinesterase activity; CbE, carboxylic esterase activity)

Birds

Pollutant, polluted sites, control sites (location)	Species	Biomarker measured	Results	References
acephate, fenitrothion, mexacarbate (treatment of forests) (New Brunswick, Canada)	Various sparrows	brain AChE	(±)	Ludke et al.,1975 Busby and White, 1991 Fleming et al., 1992, Forsyth and Martin, 1993
acephate (treatment of forests) (Idaho, USA)	Various sparrows	brain AChE plasma BuChE	(+) (+)	Zinkl et al., 1980
fenitrothion (treatment of forests) (Great Britain)	Various sparrows	brain AChE	(±)	Hamilton et al., 1981
malathion (treatment of urban areas) (Winnipeg, Canada)	*P. domesticus* and *C. livia*	brain AChE	(0)	Kucera, 1987
parathion, diazinon, methidathion (winter treatments of orchards) (California, USA)	*B. jamaicensis*	brain AChE plasma AChE	(+) (+)	Hooper et al., 1989
disulphoton/phosalone (pecan farms) (USA)	*C. cristata* and *M. carolinus*	brain AChE	(+, disulphoton) (±, phosalone)	White and Seginak, 1990
fenthion (sparrows previously exposed to fenthion) (USA)	*F. sparverius*	brain AChE plasma AChE	(+) (+)	Hunt et al., 1991
carbofuran (experimental treatment on prairie) (Canada)	*A. platyrhynchos* (caged birds)	brain AChE plasma AChE	(±) (±)	Martin et al., 1991
carbaryl, carbofuran, dimethoate (experimental treatments in paddocks) (Alberta, Canada)	*P. colchicus* and *A. graeca* (caged birds)	brain AChE	(0)	Somers et al., 1991
chlorpyriphos, diazinon, ethyl parathion, methidathion? (winter treatments of orchards) (California, USA)	*B. jamaicensis, B. lineatus, F. sparverius* and *S. mexicana*	plasma AChE	(±)	Wilson et al., 1991

Contd.

Table 1 contd.

Birds

Pollutant, polluted sites, control sites (location)	Species	Biomarker measured	Results	References
demeton S-methyl (wheat fields) (Boxworth project, Great Britain)	*P. domesticus*	brain AChE plasma BuChE	(±) (+)	Tarrant et al., 1992
diazinon (golf course) (USA)	*A. americana*	brain AChE	(+)	Kendall et al., 1992
acephate, diazinon (treatment of ornamental trees) (Laval city, Canada)	*T. migratorius*	brain AChE plasma AChE	(0) (−)	Decarie et al., 1993
methiocarb (experimental treatment on cherry trees) (Great Britain)	*P. domesticus, C. chloris, C. palumbus, S. vulgaris* and *T. merula*	brain AChE, plasma AChE, liver CbE	(±) (±) (±)	Hardy et al., 1993
? (cotton fields) (Israel) azinphos-methyl (cherry orchards) (Modene region, Italy)	*H. spinosus* and *B. ibis* *P. montanus*	brain AChE, plasma BuChE brain BuChE plasma CbE	(±) (±) (0) (+)[a]	Yawetz et al., 1993 Lari et al., 1994

Small mammals

Pollutant, polluted sites, control sites (location)	Species	Biomarker measured	Results	References
acephate (treatment of forests) (Idaho, USA)	*S. columbianus* and *T. hudsonicus*	brain AChE	(+)	Zinkl et al., 1980
methiocarb (experimental treatment of fruit trees) (Kent, Great Britain)	*A. sylvaticus, C. glareolus* and *M. agrestis*	brain AChE plasma AChE liver CbE	(±) (±) (±)	Hardy et al., 1993

(+), inhibition;
(±), variable effect;
(0), no effect;
?, not specified.
[a] increase in activity.

Table 2. Response of biomarkers in birds and small mammals: hepatic cytochromes P450[a] (AHH, aryl hydrocarbon hydroxylase; BROD, benzoxyresorufin O-dealkylase; EROD, ethoxyresorufin O-deethylase; NADPH-R, NADPH-cytochrome P450 reductase; P450, total cytochrome P450 content; PCB, polychlorobiphenyls; PCDD, polychlorodibenzo(p)dioxins; PCDF, polychlorodibenzofurans; PROD, pentoxyresorufin O-depentylase; TEQs, dioxine equivalents)

Birds

Pollutant, polluted sites, control sites (location)	Species	Biomarker measured	Results	References
organochlorine insecticides, PCB[c] (coasts of England, Scotland, Ireland)	*F. arctica, A. torda, U. aalge, P. aristotelis, P. carbo, P. puffinus* (adults)	aldrine epoxydase	(± corr)	Walker and Knight, 1981
pentachlorobenzene, TCDD[b] (Great Lakes, Canada)	*L. argentatus* (adults)	AHH	(+)	Ellenton et al., 1985
pentachlorobenzene, DDE, mirex, HCB TCDD[c] (9 polluted sites and 1 control site) (Great Lakes, Canada)	*L. argentatus* (embryos from eggs incubated in laboratory)	EROD aniline hydroxylase, aminopyrine demethylase AHH	(± corr)	Boersma et al., 1986
DDE, dieldrin, heptachlore epoxyde, oxychlordane, HCB, PCB[c] (Newfoundland, Canada)	*L. argentatus* (embryos and young birds from eggs incubated *in situ*)	AHH, EROD aminopyrine demethyiase P450	(± corr)	Peakall et al., 1986
organochlorines, PCB[b] (San Francisco Bay, USA)	*N. nycticorax* (fledglings)	AHH	(0)	Hoffman et al., 1986
DDE, HCB, PCB[c] (Marano lagoon, Italy)	*P. nigricollis* (adults)	EROD aldrine epoxydase	(± corr)	Fossi et al., 1986
PCB and PCDD[b] (Green Bay, polluted site; Lake Poygan, control site) (Great Lakes, USA)	*S. forsteri* (embryos from eggs incubated *in situ*)	AHH	(+)	Hoffman et al., 1987
PCDD (pine plantations) (Wisconsin, USA)	*T. migratorius* (adults)	AHH EROD	(+)	Martin et al., 1987, in Rattner et al., 1989
PCB[c] (four colonies of birds) (north and centre of Italy)	*L. cachinnans* (adults)	aldrine epoxydase, EROD	(± corr)	Fossi et al., 1988
PCB[c] (Great Britain)	Various aquatic species	aldrine epoxydase antibodies[d]	(+ corr)	Ronis et al., 1989

Contd.

Table 2 contd.

Birds

Pollutant, polluted sites, control sites (location)	Species	Biomarker measured	Results	References
HCB, DDE, organochlorine insecticides and PCB[c] (a control site, Nicomekl, a polluted site, Vancouver, and a highly polluted site, Crofton) (British Columbia, Canada)	A. herodias (fledglings from eggs incubated in laboratory)	P450 EROD PROD	(0 corr) (+ corr) (0 corr)	Bellward et al., 1990
lindane, HCB, DDT, PCB[c] (Venice lagoon and 2 waste sites in Tuscany) (Italy)	L. ridibundus (adults)	aldrine epoxydase EROD NADPH-R	(+) (+) (0)	Fossi et al., 1991
Industrial region (Rhine banks, France)	A. fuligula, A. platyrhynchos and P. cristatus (adults)	EROD PROD BROD P450 alkoxycoumarin O-dealkylases metabolism of parathion	(0 corr)	Rivière, 1992
PCB[c] (coasts of Japan)	L. ridibundus and L. crassirostris (adults)	P450 EROD AHH aldrine epoxydase NADPH-R	(0 corr) with levels of PCB; (+ corr) with levels of TEQs in L. ridibundus	Yamashita et al., 1992
Two polluted sites (Green Bay and Saginaw Bay) and a control site (Patuxent Wildlife Research Center) (Great Lakes, USA)	N. nycticorax (fledglings from eggs incubated in laboratory)	AHH	(+)	Hoffman et al., 1993
DDE, PCB and Hg[c] (2 polluted sites, Green Bay and Saginaw Bay; 3 control sites, Cut River, Lime Island, and Pointe aux Chenes) (Great Lakes, USA)	S. hirundo (fledglings from eggs incubated in laboratory or taken from nest)	AHH	(+)	Hoffman et al., 1993
PCB, PCDD and PCDF[c] (8 colonies of birds) (Netherlands)	S. hirundo (embryos from eggs incubated in laboratory)	EROD PROD	(+ corr) (+ corr)	Bosveld et al., 1993

Birds

Pollutant, polluted sites, control sites (location)	Species	Biomarker measured	Results	References
PCB, PCDD and PCDF[c] (2 polluted sites, Vancouver and Crofton) (British Columbia, Canada)	A. herodias (embryos from eggs incubated in laboratory)	EROD	(−)	Sanderson et al., 1994

Small mammals

Pollutant, polluted sites, control sites (location)	Species	Biomarker measured	Results	References
Dioxins (253-27 800 ng TEQs/kg soil; polluted soil) (Germany)	M. arvalis	EROD	(+)	Schrenk et al., 1991
'Pryor' site (Aroclor 1254, 0.3-863 mg/kg) (Superfund classified site, Oklahoma, USA)	R. fulvescens	EROD CYP1A1[d]	(+) males, (0) females (+)	Lubet et al., 1992
'MOTCO' site (chlorinated alcane/alcenes, benzene, naphthalene, styrene, PCB, dieldrine), 'French Limited' site (benzene and naphthalene) and 'Crystal Co.' site (As 1000 mg/kg) (Superfund classified sites, USA)	S. hispidus	AHH aniline hydroxylase P450	(±) according to the site	Rattner et al., 1993
Industrial pollution (3 industrial sites; power stations, refineries, incinerators, vehicular traffic) (Vienna, Austria)	A. flavicollis	EROD PROD BROD ethoxycoumarin O-dealkylase P450	(±) according to the site and sex	Bhatia et al., 1994

(+), induction; (±), variable result; (0), no induction; (−) decrease over time; (+ corr), correlation between pollutant concentrations and enzyme induction; (± corr) variable degree of correlation; (0 corr), no correlation.

[a] This table integrates the table presented by Rattner et al. (1989).
[b] Products identified/measured on the polluted site.
[c] Products identified/measured in animal tissues.
[d] Use of antibodies directed against purified forms of cytochromes P450.

Table 3. Response of biomarkers in small mammals: biomarkers of genotoxicity (HCB, hexachlorobenzene, PCB, polychlorobiphenyls; PAH, polycyclic aromatic hydrocarbons)

Birds

Pollutant, polluted sites, control sites (location)	Species	Biomarker measured	Results	References
Industrial zone (Ontario, Canada)	*M. musculus*	sister chromatids exchange	(+)	Nayak and Petras, 1985
mixes of oils, fats, partly burned hydrocarbons	*P. leucopus* and *S. hispidus*	chromosomal aberrations (karyotype)	(+)	McBee et al., 1987
PCB, HCB and metals (Brazos County, Texas, USA)	*P. leucopus*	cellular content of DNA, number of aberrant cells and lesions per cell (flow cytometry)	(+)	McBee and Bickham, 1988
	P. leucopus	chromosomal aberrations (G bands)	(+)	McBee, 1991
'MOTCO' site (chlorinated alcane/ alcenes, benzene, naphthalene, styrene, PCB, dieldrin), 'French Limited' site (benzene and naphthalene) and 'Crystal Co.' site (arsenic 1000 mg/kg) (Superfund classified sites, USA)	*S. hispidus*	chromosomal aberrations (G bands)	(+)	Thompson et al., 1988
PAH (pollution of sediments, Elizabeth River; less polluted site, Nansemond River) (Virginia, USA)	*O. zibethicus*	DNA adducts	(±)	Halbrook et al., 1992
PCB (0.3-863 mg/kg soil) (Oklahoma, USA)	*P. leucopus* *S. hispidus* *R. fulvescens*	chromosomal aberrations (karyotypes)	(+)	Shaw-Allen and McBee, 1993

The biochemical mode of action of OPs and carbamates is well known. They are powerful inhibitors of a group of enzymes, the B esterases, which hydrolyse esters of aliphatic and aromatic acids. These enzymes, widespread in animal tissues, are easy to measure by a colorimetric method (Ellman et al., 1961; Fairbrother et al., 1991), but their exact functions are still poorly understood (Walker, 1993). Esterases of type B are classified in two main groups, the cholinesterase (ChE) group, which comprises acetylcholinesterases (AChE) and butyrylcholinesterases (BuChE), and the carboxyl esterases (CbE). In most studies, the enzymatic activity of these esterases is measured with different substrates (e.g., acetylthiocholine and butyrylthiocholine for cholinesterases and naphthyl acetate for carboxyl esterase), but the biochemical characteristics of enzymes that catalyse these activities in wild species are not known.

AChE activity is present in the erythrocytes of most vertebrates, with the exception of birds; it has not been found in any bird species studied so far. AChE and BuChE are present in the plasma in a ratio that is highly variable according to the species. The nervous system has a significant AChE activity, the essential physiological role of which is to hydrolyse acetylcholine, the principal mediator in transmission of nerve impulses at cholinergic synapses. The CbEs are present in all tissues, but their exact physiological role is not known. There has been a hypothesis that their presence in large quantity in plasma and tissue serves an important function of detoxification in binding (and therefore inactivating) the anticholinesterasic compound present (Chambers and Carr, 1993).

The measurement of esterase inhibition is an excellent biomarker of exposure to OP compounds and carbamates. The ChEs are sensitive biomarkers. Inhibition is apparent well before the appearance of symptoms of neurotoxicity. In birds, cerebral AChE activities were first measured and then esterases in tissues were examined. The two methods have their advantages and disadvantages, which have been summarized by Walker (1992). Measurement of cerebral activities requires the sacrifice of the animals, while measurement of activities in blood is non-destructive and can be done repetitively, for example, in laboratory study of the development of these parameters in a single animal following a sublethal intoxication. Conversely, it is easier to sample the brain of a dead bird in the field than to sample its blood. The measurement of activities in brain has the advantage of dealing directly with inhibition in the target organ, and therefore acting as a biomarker of toxic effect, particularly useful in identifying the causes of poisoning. The CbE of blood can be more sensitive to inhibition than brain AChE (Thompson, 1991), but this is not a general rule. The main disadvantage of plasma CbE is that it is more variable than brain AChE, which makes it difficult to establish standard conditions and controls (Thompson, 1993).

The activity of brain esterases, which are not the most thoroughly studied, varies with numerous factors, which were reviewed in detail by Rattner and Fairbrother (1991). Variations among individuals have been proved in many species of birds and mammals. In birds of a single species, if sex does not seem to be an important factor of variation, the activity changes between hatching and the adult stage. For example, plasma BuChE increases with age in tree sparrows (*Passer montana*). To measure the inhibition of the enzyme in a group of birds of different ages, one must compare the slopes of regression lines linking the activity to a morphological parameter representative of the age, (e.g., wing length) (Greig-Smith et al., 1992; Thompson and Walker, 1993). Sometimes, seasonal variations are also observed, some of which are linked to the stage of sexual maturity, and circadian variations. The influence of physiological and environmental factors has also been studied: food regime, external temperature, handling of animals in laboratory assays, and pathologies are sources of potential variation, the importance of which is difficult to establish.

Inhibition of brain AChE, after exposure to lethal doses of OPs or carbamates, varies considerably according to the chemicals used (Holmes and Boag, 1990). In two exceptional cases, birds died of OP poisoning when inhibition of brain AChE was low but inhibition of plasma AChE was high. This has been shown by Hooper et al. (1991) in northern bobwhite given a lethal dose of terbufos, an OP, and Booth et al. (1992) in Japanese quail dosed with fenamiphos (OP). In these cases, the plasma activities are more useful biomarkers than cerebral AChE. In these two unrepresentative cases, the biochemical and physiological cause of bird mortality remained to established. On some occasions, overcompensation also occurs, the enzymatic activity being higher in birds after the initial phase of exposure. This phenomenon has been observed for cerebral AChE after experimental treatments (Westlake et al., 1981a, b) and for plasma CbE in birds sampled immediately after a treatment with low doses of demeton S-methyl (Thompson et al., 1991) or parathion (Yawetz et al., 1993). Important differences have also been observed, for a single inhibitor, in sensitivity to OPs according to the species (Kemp and Wallace, 1990).

The response of esterases is very useful as a biomarker of exposure of birds to insecticide treatments by OPs and carbamates. The Service Canadien de la Faune has used it systematically following studies by Ludke et al. (1975). In Canada, very large areas of coniferous forests are treated, generally by fenitrothion, to control tortricid lepidoptera (*Choristoneura* spp.) and other pest species. In the method used there, birds were shot down on the day following the treatment or some days later. The small mammals were trapped. The birds were located by their calls, but efforts were made to capture birds that did not call, in order to avoid a sampling bias. In some cases, caged birds were subjected to treatments (Mineau and Peakall, 1987).

Whatever the mode of exposure, the birds are taken on dry ice to the laboratory, where they are dissected and their AChE activities measured. The rate of inhibition is calculated by comparison with activities measured in individuals of the same species, captured at the same time, in a nearby untreated area. For raptors, birds coming from

rescue centres have been used. The phosphorylase or carbamylase enzyme can be reactivated *in vitro*, thus allowing direct measurement of the rate of inhibition (Stansley, 1993). Lastly, reference values for brain AChE activity were provided by Hill (1988) for 48 bird species, representing 11 orders and 23 families. To evaluate the degree of exposure and risk for the birds, the proportion of birds showing a level of inhibition higher than 20% or higher than 50% is calculated (Lu'dke et al., 1975). Inhibition higher than 20% is considered a certain index of exposure and inhibition higher than 50% an index of a near-certain fatal outcome. The use of this method allows comparisons, for example between different techniques of spraying (Busby and White, 1991) or between different active ingredients. It is also sometimes acknowledged that there is significant exposure when the reduction in brain activities is higher than or equal to twice the standard deviation of control values.

The degree of inhibition of esterases can also be diagnostic of acute intoxication for animals found dead (Greig-Smith, 1991), and can confirm suspicions of intoxication by these products in living animals. In some cases, interspecific differences that were hitherto unknown can be proved, thus establishing a high risk for certain species. For example, several fatal cases showed that carbophenothion (OP) can be highly toxic for some species of geese (*Anser anser* and *A. brachyrhynchus*). The initial evaluation of toxicity and the determination of doses of use were done on a similar species (*Branta canadensis*) that was much less sensitive to this product (Tucker and Haegele, 1971, in Mineau, 1991).

There are other methods for measuring degree of exposure of birds to anticholinesterase OPs, for example, measurement of metabolites (alkyl-phosphates) in droppings. These methods were used with success to evaluate the exposure of raptors (Wilson et al., 1991).

2.2 Biomarker of Exposure to Polyhalogenated and Non-halogenated Polycyclic Aromatic Hydrocarbons (PHAH and PAH): Cytochrome P450, Plasma Porphyrins and Retinoids

2.2.1 Cytochrome P450

Cytochrome P450-dependent monooxygenase activities have been useful for some years as biomarkers of exposure of aquatic organisms to water pollution. Cytochrome P450-dependent enzymes are a family of enzymes that exist in all living organisms and catalyse oxidations of organic substrates (monooxygenation). Present in most tissues and organs of vertebrates, they are particularly abundant in the liver, where

they metabolize most toxic organic molecules, thus playing a fundamental role in their elimination (Nelson, 1998). Hepatic monooxygenase activities correspond to several forms, of various specificity of substrate and level of expression (Lewis, 1996). In addition to constitutive forms, certain forms of the family of CYP1A are strongly induced by a large variety of industrial pollutants (PAH and PHAH of planar conformation, such as benzo(a)pyrene, certain congeners of PCB or dioxins). Other forms (CYP2B) are induced by treatments with medicines (phenobarbital) or non-planar congeners of PCB. The induction of CYP1A is easily proved by an increase of hepatic enzymatic activities, such as that of aryl hydrocarbon hydroxylase (AHH) and ethoxyresorufin O-deethylase (EROD), while the induction of CYP2B brings a very high increase in pentoxyresorufin O-depentylase (PROD). These activities are measured in homogenates of liver, post-mitochondrial supernatants, and most often microsomal fractions obtained by ultracentrifugation. There are immunochemical techniques directly quantifying the form of cytochrome P450 catalysing this activity, but they are available only for a very small number of species, essentially laboratory rodents and some fishes.

Cytochrome P450 induction is a good marker of exposure to PHAHs, dangerous products for which it is important to be able to evaluate the exposure and risk to wild animals (Nosek et al., 1993). All these molecules are present in the environment in varying proportions and quantities and their measurement poses difficult analytic problems because of the large number of congeners present. Even when these problems are resolved, it is difficult to link the measurements to a potential environmental toxicant, because of the specific toxicity of each congener and the possibilities of interactions of these molecules with very similar structures. Measurement of induction of EROD (or AHH) activity in species at risk or in sentinel animals is thus an interesting alternative. This biomarker is particularly appreciated in cases of exposure to complex mixtures that are difficult to characterize analytically and for which an overall assay is often faster, more economical, and more meaningful than a series of analytic measurements.

The use of hepatic cytochrome P450-associated enzymes of birds and small mammals as biomarkers of pollution has been reviewed by Rattner et al. (1989), who concluded that this approach is useful. However, because it is used much less frequently than in aquatic species, they emphasized the need for supplementary studies for validation. Most data on birds come from Canadian and US research teams, who have studied colonies of aquatic birds that inhabit the Great Lakes. The measurements are made on embryos at the late embryonic stage, hatchlings of one or two days, or adults from contaminated and control sites. In some cases, eggs are incubated in the laboratory and the activities measured at hatching or a few days afterwards. Monooxygenase

activities are more difficult to measure than esterase activities because the enzymes are more rapidly inactivated after the animals die, and because of the need to prepare subcellular fractions, use a fluorimeter (less often available in laboratories than a colorimeter), and use more costly reactive agents. Direct immunochemical quantification of the induced proteins, generally with antibodies directed against the corresponding forms in the rat, have been tried in some cases (Ronis et al., 1989; Yamashita et al., 1992), but the necessary materials are difficult to get and costly to use.

The study of some purified cytochrome P450 in the hen or chicken (Sinclair and Sinclair, 1993) showed that their structure differs clearly from those of isolated forms in rodents. Laboratory assays are therefore needed to precisely identify the inducer pollutants and sources of variation in enzymatic activity other than pollutants. The expression of monooxygenase activities in birds has been reviewed by Walker and Ronis (1989). Interspecific differences are very important, even between similar species. Physiological variations linked to age or sex are insufficiently known, even in the few species that have been most often studied from this point of view. Birds are subject to seasonal cycles that modify their behaviour (migrations) and physiology (sexual activities, state of energy reserves) and that can, as in fishes, modulate monooxygenase activities. This phenomenon has been found in some studies (Fossi et al., 1988) and not proved in others (Rivière, 1992).

The relationship between induction of EROD activity and presence of PCB and/or dioxins is not absolute. Other pollutants also are inducers: organochlorine insecticides, for example, and perhaps other natural substances present in the food. Moreover, it has been shown that in duck, EROD activity is effectively induced by β-naphthoflavone, a model inducer of EROD activity in rat, but also by phenobarbital, an inducer of CYP2B and PROD activity in rat (Leffin and Rivière, 1992), which indicates profound differences either in the nature of the forms induced or in the catalytic activities of the forms induced. The correlation between EROD and PROD activities was very high in natural populations of mallard duck and tufted duck (respectively, $r = 0.829$ and $r = 0.813$), which suggests a single mechanism of regulation for the two activities or even, what is less likely, the simultaneous presence of two types of inducers (Rivière, 1992). This high correlation was not found in natural populations of grebes ($r = 0.403$), which confirms the existence of significant differences between species.

Experimental studies on rats showed a very high correlation between the toxicity of dioxins and induction of cytochromes of the family 1A, the different biological effects being mediated by a common molecular mechanism, interaction with the receptor *Ah* (Safe, 1986). In birds, the correlation between induction and toxicity of dioxins and the significance of interspecific differences have been highlighted in experimental studies by Brunstrom (1991). The analysis of data obtained on natural populations showed also that there is some correlation between induction of monooxygenase activities, notably AHH activity or EROD activity, and the existence of embryotoxic and teratogenic effects. Induction of EROD activity in natural bird populations can be a good biomarker of effects on reproduction, but for all that, it must be demonstrated unequivocally that all the inducers can cause these same effects.

The induction potential of most environmental pollutants has been tested in laboratory rodents, but the transposition of these data to species of small wild mammals requires a prior validation study verifying analytically the presence of inducer pollutants in the organism, as was done by Hincks and Brindley (1986) for *Microtus montanus* and Novak and Qualls (1989) for *Sigmodon hispidus*. Also, the induction of cytochrome P450-associated enzymes is an indicator of the bioavailability of a product, as has been shown by Roos et al. (1994), who measured the rate of induction of EROD activity in laboratory rats when 5% plain soil contaminated by PAH was added to their food. The induction of CYP1A and/or CYP2B was proportional to the corporal levels of PCB for laboratory colonies and for several wild species (Lubet et al., 1992; Simmons and McKee, 1992; DeJongh et al., 1993; Dragnev et al., 1994; Henneman et al., 1994). However, the level of induction obtained was often variable (Hincks and Brindley, 1986; Miura et al., 1991; Watkins, 1991; Henneman et al., 1994), depending on the species and the sex (Elangbam et al., 1991; Bhatia et al., 1994). Some species are very sensitive. Schrenk et al. (1991) showed that hepatic EROD activity in terrestrial voles was significantly increased after exposure to contaminated soil containing 253 ng of dioxin equivalents per kg of soil. As for birds, most of these studies have been done by US research teams to contribute to risk evaluation of polluted sites identified by the Environmental Protection Agency (EPA). These are generally waste sites contaminated by PCB and petroleum and/or oil products, which are sometimes associated with metals, insecticides, or solvents. The species most often mentioned are *S. hispidus*, *P. leucopus*, and various voles (*Microtus* spp.). All these studies demonstrated the induction of one or several monooxygenase activities, hepatic EROD activity being the most frequently measured and the most strongly induced (Lubet et al., 1992). EROD activity of other organs can be a very useful biomarker, for example, of induction of pulmonary activity (Fouchécourt and Rivière, 1995, 1996).

Apart from measurements of hepatic activity in birds or small mammals captured on a polluted site, induction of EROD activity can also serve to quantify directly the concentrations of dioxins or dioxin equivalents present in a contaminated medium (a soil, for example) or in the

tissues of an exposed animal. The technique used for this bioassay is based on the response of a cellular line of hepatoma of rat, expressing CYP1A, in the presence of an organic extract of the matrix to be analysed. With this bioassay, it was possible to measure the concentration of dioxin equivalents in birds (Tillitt et al., 1991) and to demonstrate the bioamplification of these pollutants (Jones et al., 1993).

2.2.2 Plasma Porphyrins and Retinoids

Porphyrins are a group of pathologies that are congenital or caused by chemical products. They are characterized biochemically by modifications in activity of enzymes of heme biosynthesis and accumulation of various porphyrins in the tissues and products of excretion. Several classes of pollutants and toxicants (hexachlorobenzene, PHAH) are porphyrogenic agents. A team headed by G.A. Fox studied in detail the different porphyrins present in the liver of gulls of the Great Lakes and suggested that the presence of highly carboxylated porphyrins could be used as a biomarker of exposure to these products (Fox et al., 1988). This biomarker does not seem to have been studied in small mammals.

The works of Spear et al. (1992) showed that concentration of hepatic retinoids—vitamin A and its various precursors—is greatly reduced in birds of the Great Lakes, in direct correlation with contents of dioxins, and that can be the cause of teratogenic effects of these products. According to these authors, the induction or inhibition of enzymes of biotransformation (cytochromes P450 and enzymes of glucuronoconjugation) is thought to be the cause of these metabolic irregularities (see chapter 10).

2.3 Biomarkers of Exposure to Genotoxic Products

Environmental exposure to genotoxic products can cause genetic disturbances in the germinal or somatic cells of exposed individuals. Apart from the consequences for reproduction (embryotoxic and teratogenic effects resulting from lesions of gametes or zygotes), a simple genome modification can cause development of malignant tumours or physiological disorders, which may or may not be genetically transmitted. Genotoxic compounds can act on DNA at the molecular level (adducts, bonding between bases), at the gene level (mutation), or at the structural level (chromosomal aberrations, aneuploidy).

Mutagenic products are generally electrophilic molecules that directly attack DNA. These electrophilic entities can be generated by monooxygenases that contain cytochrome P450 and transform chemically inert molecules, such as PAH, into reactive metabolites capable of attacking nucleophilic centres such as DNA (Thompson et al., 1989; Degawa et al., 1994; Lewis, 1996).

Genetic lesions are detectable in the laboratory (*in vivo* or *in vitro*) but the techniques are not applicable to wild animals *in situ*. However, DNA adducts can be detected after tagging with ^{32}P; this technique, used on fishes and rats (Degawa et al., 1994), has also been used in the field to study DNA adducts in the muskrat (Halbrook et al., 1992).

In the field, chromosomal alterations are much more often studied. Generally simple, inexpensive and sensitive techniques can be applied to wild animals that provide complete information on the clastogenic effects of pollutants. Flow cytometry is a promising technique enabling the study of micronuclei (index of clastogeny) in blood cells (polychromatic erythroblasts; Tometsko et al., 1993). With this technique, variations in the cellular content of DNA can be detected. The quantity of cellular DNA can vary following a chromosomal aberration leading to unequal distribution of DNA in the daughter cells during cellular divisions. This method enables us to measure a variation rate of the order of 2–3% in cellular DNA content, a rate that has been observed in mice captured on a site polluted by PCB (McBee and Bickham, 1988). Aneuploidies of erythrocytes were proved in domesticated ducks exposed to radioactive pollution during an original experiment. The ducks were taught to respond to a call, then released on the experimental site, the reservoir of an old cooling system of a nuclear reactor (George et al., 1991).

Flow cytometry is a simple and rapid technique. More difficult techniques, such as research of sister chromatids, study of karyotypes, or detection of abnormal G bands, enable more powerful investigations into the mode of action of toxicants.

The rate of exchange of sister chromatids in bone narrow cells or circulatory lymphocytes (non-invasive method) is a sensitive biomarker of exposure, validated on animals (mice, bovines) living on sites with high industrial pollution (Nayak and Petras, 1985; Parada and Jaszczak, 1993). The sensitivity and excellent dose-effect correlation of this biomarker makes it a good warning signal of an increase in genotoxic risk for terrestrial species. Several studies have shown the advantage of comparing the karyotypes of animals captured on polluted and reference sites (cf. Table 3). McBee et al. (1987) and Thompson et al. (1988) showed an increase in number of aberrant cells and number of lesions per aberrant cell in bone marrow cells of *Sigmodon hispidus* and *Peromyscus leucopus*, in individuals that showed no apparent morphological alteration. The most frequent lesions were breaks in chromatids and translocations. More sensitive than *S. hispidus*, *P. leucopus* has been the subject of complementary studies. The study of abnormal G bands enabled identification of internal aberrations in chromosomes (inversions, deletions, fusions) and the phenomenon of aneuploidy (Thompson et al., 1988; McBee, 1991). The study of karyotype of *P. leucopus* enabled selection of chromosome 21 as the most sensitive to clastogenic effects. Study of this single chromosome sufficed to establish exposure to clastogenic products (McBee, 1991).

Similarity of response of these biomarkers when measured *in situ* with laboratory genotoxicity tests was evaluated by Shaw-Allen and McBee (1993) with animals (*P. leucopus, S. hispidus*, and *R. fulvescens*) captured on a site contaminated by PCB (Aroclor 1254®). The number of aberrant cells and the rate of lesions per cell were not significantly higher than those observed in individuals of the same species captured on three reference sites. The results correspond to those obtained in the laboratory following administrations of Aroclor 1254®. These biomarkers of genotoxicity are particularly sensitive. McBee et al. (1987) verified that positive responses that they obtained were not the result of a genetic variability between populations.

2.4 Biomarkers of Exposure to Heavy Metals: δ-aminolevulinate Dehydratase (ALAD), Metallothioneins and Porphyrins

For evaluating the degree of lead poisoning in birds and small mammals, the methods used are the search of lead sinkers in bird gizzards and measurement of lead levels in various organs, mainly liver and kidney, but sometimes also in hard tissues (bone) or phanera (bird feathers). Other methods have been suggested. According to Way and Schroder (1982), measurement of lead in faeces of *R. norvegicus* was an effective means of estimating corporal contamination. Inhibition of an erythrocyte enzyme, δ-aminolevulinate dehydratase (ALAD), is an early and sensitive indicator of exposure to lead. This enzyme is part of the cascade of enzymatic activities that lead to biosynthesis of heme. The inhibition is often quite well correlated with lead poisoning. In a sample of 141 wood ducks (*Aix sponsa*) captured in a mining region in the United States, the coefficient of correlation (r) found by Blus et al. (1993) was 0.50. ALAD inhibition was measured in pigeons exposed to urban pollution (Gonzalez and Tejedor, 1992) and in black ducks (*Anas rubripes*; Pain, 1989). According to Scheuhammer (1989), the measurement of erythrocyte ALAD activity is relatively simple, precise, and economical, and could be used much more widely. However, it does not have decisive advantages over measurement of lead levels and lead concentrations in tissues. The temporal scales of response of these different biomarkers are not the same. Lead levels and ALAD inhibition reflect recent exposures, whereas concentrations in the phanera and bone integrate exposure over a longer period. The chief advantage of measuring lead level in tissues and ALAD activity is that the animal need not be killed. The rate of urinary elimination of δ-aminolevulinic acid has sometimes been used to evaluate the contamination of small mammals (*Peromyscus* spp.) near a lead treatment plant (Lower and Tsutakawa, 1978).

Metallothioneins are cytosolic proteins of low molecular weight that play a role in homeostasis and detoxification of metals. Cosson (1989) determined metallothionein levels in the greater flamingoes and little egrets of the Camargue. Higher concentrations of metallothioneins were found in birds from lakes with low pH, polluted by acid rain (St Louis et al., 1993), than in those from unpolluted lakes. Correlations between renal concentrations of metallothioneins and cadmium were proved in aquatic birds (Elliott et al., 1992) but, as emphasized by Peakall and Walker (1994), the use of metallothioneins as biomarker does not have particular advantages over chemical measurements. Metallothioneins have only rarely been studied in wild rodents. Seasonal variations, parallel to zinc, copper, and cadmium levels, were observed in the bank vole (*Clethrionomys glareolus*), but their toxicological significance is yet to be established (Wlostowski, 1992).

The presence of plasma porphyrins is studied for all the products that interfere with biosynthesis of heme. Lead is an inhibitor of ferrochelatase, the final enzyme in the cycle, which ensures the insertion of iron in the protoporphyrin. Exposure to lead causes an increase in free protoporphyrin. It is a potentially useful indicator, because this compound is quite easy to measure in plasma by fluorimetry (O'Halloran and Myers, 1988; Passer et al., 1989; Blus et al., 1993). According to different authors, the correlation with lead poisoning could be higher (Scheuhammer, 1989) or lower (Blus et al., 1993) than for ALAD. This biomarker does not seem, for the present, to be clearly superior to measurement of lead levels.

2.5 Biomarkers of Exposure to Anticoagulant Rodenticides

Second generation anticoagulant rodenticides are products that bind strongly to certain cellular proteins and have a high acute toxicity. Poisoned rodents develop fatal hemorrhages, but, whatever the dose absorbed, mortality occurs only after several days. During this period, the weakened rodents become easy prey for raptors and other predators. The risk of secondary toxicity is therefore high, but difficult to assess. At the biochemical level, these products act on the vitamin K cycle, which is essential for the carboxylation of precursors of coagulation proteins. Attempts have been made to develop antibodies against certain precursors of these proteins in order to study increase in their plasma levels (Walker, 1990, 1992).

2.6 Other Biomarkers

Many toxic effects have been measured at the molecular, cellular, or functional level in birds and small mammals. Most cannot be used as biomarkers of pollution for terrestrial wildlife for various reasons: lack of specificity, low sensitivity, or lack of correlation with toxic effects. Some offer interesting potential, but the absence of data, especially field data, prevents their validation.

The activity of certain plasma enzymes that are markers for liver damage has sometimes been measured. However, the present trend is to consider that variations observed are produced generally at very high doses and with a low specificity and are therefore not considered useful biomarkers of exposure (Peakall and Walker, 1994). The state of energy reserves has a physiological and toxicological importance but is often hard to link to a chemical pollution. Normal variation can be considerable. The clearest example is the decrease of lipid reserves during migratory flights of birds, which make it very difficult to establish biomarkers on this basis. Hormone levels could provide useful indications, primarily as biomarkers of toxic effects, particularly hormones that control reproduction. Although hormone regulation in rodents (rats) and birds (domesticated species, such as hen or duck) is well known, that of wild species is not as well understood. Pioneering experiments (Peakall, 1967) have shown that certain pollutants disturb hormone metabolism *in vitro*, but the exact consequences in terms of *in vivo* regulation of hormonal levels have yet to be established. Some data have shown how difficult these studies are, even to only obtain a correct value for plasma hormonal rates (Le Maho et al., 1992). Brouwer (1991) showed that hydroxylated metabolites of PCB interfere with plasma transport of vitamin A and thyroxin. This was demonstrated experimentally in different animals (seals and cormorants) and the effects were observed in seals fed with PCB-contaminated fish from the North Sea. An analysis and detailed bibliography on these different biomarkers can be found in Mayer et al. (1992).

3. CONCLUSION: USE OF BIOMARKERS FOR EVALUATION OF ENVIRONMENTAL RISK OF POLLUTANTS

The study of biomarkers of exposure is motivated by the concern for evaluating long-term risks resulting from prolonged exposure to pollutants present in low concentrations in the environment. In the United States, with regard to environmental pollution and human risk, the EPA prioritized the development of studies aiming to obtain reliable biomarkers enabling evaluation of risk of cancer, reproductive and growth problems, neurotoxicity, pulmonary toxicity, and changes in the immune system (Fowle and Sexton, 1992). It will be useful to study in parallel the same biomarkers in sentinel animals and wild species at risk. However, we must recognize that most of these biomarkers cannot be applied to wild species and can be developed only at the cost of intensive research on the mode of action of toxicants at the molecular and cellular level. Some biomarkers of exposure have potential and can be used as biomarkers of toxic effect. This is the case especially with biomarkers of genotoxic effects. The immunotoxic potential of environmental pollutants has been proved in laboratory rodents (Vecchi et al., 1986; Kerkvliet et al., 1990; Hardin et al., 1992) but, to our knowledge, the integrity of the immune system has been little studied in wild species (with the exception of fish) that have been naturally exposed to pollutants.

REFERENCES

Batty J., Leavitt R.A., Biondo N. and Polin D. (1990). An ecotoxicological study of a population of the white footed mouse (*Peromyscus leucopus*) inhabiting a polychlorinated biphenyls-contaminated area. *Arch. Environ. Contam. Toxicol.*, 19:283–290.

Bellward G.D., Norstrom R.J., Whitehead P.E. et al. (1990).Comparison of polychlorinated dibenzodioxin levels with hepatic mixed-function oxidase induction in great blue herons. *J. Toxicol. Environ. Health*, 30:33–52.

Bhatia A., Tobil F., Lepschy G. et al. (1994). Biomonitoring of pollution: the hepatic cytochrome P-450 enzyme system in the feral mouse *Apodemus flavicollis* as indicator. *Chemosphere*, 28:1525–1537.

Blus L.J., Henny C.J., Hoffman D.J. and Grove R.A. (1993). Accumulation and effects of lead and cadmium on wood ducks near a mining and smelting complex in Idaho. *Ecotoxicology*, 2:139–154.

Boersma D.C., Ellenton J.A. and Yagminas A. (1986). Investigation of the hepatic mixed-function oxidase system in herring gull embryos in relation to environmental contaminants. *Environ. Toxicol. Chem.*, 5:309–318.

Booth G.M., Chambers E.W. and Carter M.W. (1992). *In vivo* evidence that Nemacur 10 g and 15 g do not inhibit brain cholinesterase of birds. Abst. 13th Annual Meeting, Society of Environmental Toxicology and Chemistry, Cincinnati, 8-12 November 1992, p. 386.

Bosveld A.T.C., Gradener J. and Van Kampen M. et al. (1993). Occurrence and effects of PCBs, PCDDs and PCDFs in hatchlings of the common tern (*Sterna hirundo*). *Chemosphere*, 27:419–427.

Brouwer A. (1991). Role of biotransformation in PCB-induced alterations in vitamin A and thyroid hormone metabolism in laboratory and wildlife species. *Biochem. Soc. Trans.*, 19:731–737.

Brunstrom B. (1991). Toxicity and EROD-inducing potency of polychlorinated biphenyls (PCBs) and polycyclic aromatic hydrocarbons (PAHs) in avian embryos. *Comp. Biochem. Physiol.*, 100C:241–243.

Busby D.G. and White L. (1991). Factors influencing variability in brain acetyl-cholinesterase activity in songbirds exposed to aerial fenitrothion spraying. In:

Cholinesterase-inhibiting Insecticides. Their Impact on Wildlife and the Environment. Mineau P. (ed.). Elsevier, Amsterdam, pp. 89–109.

Canters K.J. and DeSnoo G.R. (1993). Chemical threat to birds and mammals in the Netherlands. *Rev. Environ. Contam. Toxicol.*, 130:1–29.

Carson R. (1963). *Printemps silencieux*. Plon, Paris.

Chambers J.E. and Carr R.L. (1993). Inhibition patterns of brain acetylcholinesterase and hepatic and plasma aliesterases following exposures to three phosphorothionate insecticides and their oxons in rats. *Fund. Appl. Toxicol.*, 21:111–119.

Cooke A.S. (1973). Shell thinning in avian eggs by environmental pollutants. *Environ. Pollut.*, 4:85–152.

Cosson R.P. (1989). Relationships between heavy metal and metallothionein-like protein levels in the liver and kidney of two birds: the greater flamingo and the little egret. *Comp. Biochem. Physiol.*, 94C:243–248.

Decarie R., Desgranges J.L., Lepine C. and Morneau F. (1993). Impact of insecticides on the American robin (*Turdus migratorius*) in a suburban environment. *Environ. Pollut.*, 80:231–238.

Degawa M., Kojima M., Yoshinari K. et al. (1994). DNA adduct formation of hepatocarcinogenic aromatic amines in rat liver: effect of cytochrome P450 inducers. *Cancer Lett.*, 79:77–81.

De Jongh J., Wondergem F., Seinen W. and Van Den Berg M. (1993). Toxicokinetic interactions between chlorinated aromatic hydrocarbons in the liver of the C57BL/6J mouse: I. Polychlorinated biphenyls (PCBs). *Arch. Toxicol.*, 67:453–460.

Dragnev K.H., Beebe L., Jones C.R. et al. (1994). Subchronic dietary exposure to Aroclor 1254 in rats: accumulation of PCBs in liver, blood and adipose tissue and its relationship to induction of various hepatic drug-metabolizing enzymes. *Toxicol. Appl. Pharmacol.*, 125:111–122.

Elangbam C.S., Qualls C.W. and Lochmiller R.L. (1991). O-Dealkylation of resorufin ethers as an indicator of hepatic cytochrome P-450 isoenzyme induction in the cotton rat (*Sigmodon hispidus*): a method for monitoring environmental contamination. *Bull. Environ. Contam. Toxicol.*, 47:23–28.

Ellenton J.A., Brownlee L.J. and Hollebone B.R. (1985). Aryl hydrocarbon hydroxylase levels in herring gull embryos from different locations on the Great Lakes. *Environ. Toxicol. Chem.*, 4:615–622.

Elliott J.E., Scheuhammer A.M., Leighton F.A. and Pearce P.A. (1992). Heavy metal and metallothionein concentration in Atlantic Canadian seabirds. *Arch. Environ. Contam. Toxicol.*, 22:63–73.

Ellman G.L., Courtney K.D., Andres V. and Featherstone R.M. (1961). A new and rapid colorimetric determination of acetylcholinesterase activity. *Biochem. Pharmacol.*, 7:88–95.

Fairbrother A., Marden B.T., Bennett J.K. and Hooper M.J. (1991). Methods used in determination of cholinesterase activity. In: Cholinesterase-inhibiting Insecticides. Their Impact on Wildlife and the Environment. Mineau P. (ed.). Elsevier, Amsterdam, pp. 35–71.

Fleming R.A., Holmes S.B. and Busby D.G. (1992). An interlaboratory comparison of data on brain cholinesterase activity in forest songbirds exposed to aerial application of zectran. *Arch. Environ. Contam. Toxicol.*, 22:228–237.

Forsyth D.J. and Martin P.A. (1993). Effects of fenitrothion on survival behavior, and brain cholinesterase activity of white-throated sparrows (*Zonotrichia albicollis*). *Environ. Toxicol. Chem.*, 12:91–103.

Fossi C., Leonzio C. and Focardi S. (1986). Increase of organochlorines and MFO activity in water birds wintering in an Italian lagoon. *Bull. Environ. Contam. Toxicol.*, 37:538–543.

Fossi C., Leonzio C., Focardi S. and Renzoni A. (1988). Seasonal variations in aldrin epoxidase (MFO) activity of yellow-legged herring gulls: the relationship to breeding and PCB residues. *Bull. Environ. Contam. Toxicol.*, 41:365–370.

Fossi C., Leonzio C., Focardi S. et al. (1991). Modulation of mixed-function oxidase activity in black-headed gulls living in anthropic environments—biochemical acclimatization or adaptation? *Environ. Toxicol. Chem.*, 10:1179–1188.

Fouchécourt M.O. and Rivière J.L. (1995) Activities of cytochrome P450-dependent monooxygenases and antioxidant enzymes in different organs of Norway rats (*Rattus norvegicus*) inhabiting reference and contaminated sites. *Chemosphere*, 31:4375–4386.

Fouchécourt M.O. and Rivière J.L. (1996). Activities of liver and lung cytochrome P450-dependent monooxygenases and antioxidant enzymes in laboratory and wild Norway rats exposed to reference and contaminated soils. *Arch. Environ. Contam. Toxicol.*, 30:513–522.

Fowle J.R. and Sexton K. (1992). EPA priorities for biologic markers research in environmental health. *Environ. Health Persp.*, 98:235–241.

Fox G.A., Kennedy S.W., Norstrom R.J. and Wigfield D.C. (1988). Porphyria in herring gulls: a biochemical response to chemical contamination of Great Lakes food chains. *Environ. Toxicol. Chem.*, 7:831–839.

Frank R, Mineau P., Braun H.E. et al. (1991). Deaths of Canada geese following spraying of turf with diazinon. *Bull. Environ. Contam. Toxicol.*, 46:852–858.

Furness R.W., Greenwood J.J.D. (1993). *Birds as Monitors of Environmental Change*. Chapman and Hall, London.

Galindo J., Kendall R.J, Driver C.J. and Lacher T.E. (1984). The effects of methyl parathion on susceptibility of bobwhite quail to domestic house cat predation. *Behav. Neur. Biol.*, 43:21–36.

Gebauer M.B. and Weseloh D.V. (1993). Accumulation of organic contaminants in sentinel mallards utilizing confined disposal facilities at Hamilton Harbour, Lake Ontario, Canada. *Arch. Environ. Contam. Toxicol.*, 25:234–243.

George L.S., Dallas C.E., Brisbin I.L. Jr. and Evans D.L. (1991). Flow cytometry DNA analysis of ducks accumulating cesium[137] on a reactor reservoir. *Ecotoxicol. Environ. Saf.*, 2:337–347.

Gilbertson M., Kubiak T., Ludwig J. and Fox G. (1991). Great Lakes embryo mortality, edema and deformities syndrome (glemeds) in colonial fish-eating birds similarity to chick edema-disease. *J. Toxicol. Environ. Health*, 33:455–520.

Gonzalez M. and Tejedor M.C. (1992). Delta-ALAD activity variations in red blood cells in response to lead accumulation in rock doves (*Columba livia*). *Bull. Environ. Contam. Toxicol.*, 49:527–534.

Greig-Smith P.W. (1991). Use of cholinesterase measurements in surveillance of wildlife poisoning in farmland. In: Cholinesterase-inhibiting Insecticides. Their Impact on Wildlife and the Environment. Mineau P. (ed.). Elsevier, Amsterdam, pp. 89–109.

Greig-Smith P.W., Frampton G. and Hardy A.R. (1992). Pesticides, Cereal Farming, and the Environment: The Boxworth Project. HMSO Publication.

Halbrook R.S., Kirkpatrick R.L., Bevan D.R. and Dunn B.P. (1992). DNA adducts detected in muskrats by ^{32}P-postlabeling analysis. *Environ. Toxicol. Chem.*, 11:1605–1613.

Hamilton G.A., Hunter K. and Ruthven A.D. (1981). Inhibition of brain acetylcholinesterase activity in songbirds exposed to fenitrothion during aerial spraying of forests. *Bull. Environ. Contam. Toxicol.*, 27:856–863.

Hall R.J. and Henry P.F.P. (1992). Assessing effects of pesticides on amphibians and reptiles: status and needs. *Herpetol. J.*, 2:65–71.

Hardin J.A., Hinoshita F. and Sherr D.H. (1992). Mechanisms by which benzo[a] pyrene, an environmental carcinogen, suppresses B cell lymphopoiesis. *Toxicol. Appl. Pharmacol.*, 117:155–164.

Hardy A.R, Westlake G.E., Lloyd G.A. et al. (1993). An intensive field trial to assess hazards to birds and mammals from the use of methiocarb as a bird repellent on ripening cherries. *Ecotoxicology*, 2:1–31.

Hays H. and Risebrough R.W. (1972). Pollutant concentrations in abnormal terns from Long Island Sound. *Auk*, 89:19–35.

Henneman J.R., Fox S.D., Lubet R.A. et al. (1994). Induction of cytochrome P-450 in *Sigmodon hispidus* (cotton rats) exposed to dietary Aroclor 1254. *J. Toxicol. Environ. Health*, 41:369–386.

Hill E.F. (1988). Brain cholinesterase activity of apparently normal wild birds. *J. Wildl. Dis.*, 24:51–61.

Hincks J.R. and Brindley W. (1986). Effects of varying inducers type and dose on hepatic monooxygenase activities in the mountain vole *Microtus montanus*. *Comp. Biochem. Physiol.*, 85C:385–389.

Hoffman D.J., Rattner B.A., Bunck C.M. et al. (1986). Association between PCBs and lower embryonic weight in black-crowned night-herons in San Francisco Bay. *J. Toxicol. Environ. Health*, 19:383–391.

Hoffman D.J., Rattner B.A., Sileo L. et al. (1987). Embryotoxicity,teratogenicity, and aryl hydrocarbon hydroxylase activity in forster's terns on Green Bay, Lake Michigan. *Environ. Res.*, 42:176–184.

Hoffman D.J., Smith G.J. and Rattner B.A. (1993). Biomarkers of contaminant exposure in common terns and black-crowned night herons in the Great Lakes. *Environ. Toxicol. Chem.*, 12:1095–1103.

Holmes S.B. and Boag P.T. (1990). Inhibition of brain and plasma cholinesterase activity in zebra finches orally dosed with fenitrothion. *Environ. Toxicol. Chem.*, 9:323–334.

Hooper M.J., Detrich P.J., Weisskopf C.P. and Wilden B.W. (1989). Organophosphorus insecticide exposure in hawks inhabiting orchards during winter dormant-spraying. *Bull. Environ. Contam. Toxicol.*, 42:651–659.

Hooper M.J., Brewer L.W., Cobb G.P. and Kendall R.J. (1991). An integrated laboratory and field approach for assessing hazards of pesticide exposure to wildlife. In: Pesticide Effects on Terrestrial Wildlife. Somerville S. and Walker C.H. (eds.). Taylor and Francis, London, pp. 271-283.

Hunt K.A., Bird D.M., Mineau P. and Shutt L. (1991). Secondary poisoning hazard of fenthin to American kestrels. *Arch. Environ. Contam. Toxicol.*, 21:84–90.

Jokinen M.P., Clarkson T.B. and Prichard R.W. (1985). Animal models in atherosclerosis research. *Exp. Mol. Pathol.*, 42:1–28.

Jones P.D., Ankley G.T., Best D.A. et al. (1993). Biomagnification of bioassay derived 2,3,7,8-tetrachlorodibenzo-p-dioxin equivalents. *Chemosphere*, 26:1203–1212.

Karnofsky D.A. (1965). The chick embryo in drug screening: Survey of teratological effects observed in the 4-day old chick embryo. In: Teratology: Principles and Techniques. Wilson J.G. and Warkany J.K. (eds.). University of Chicago Press, Chicago, pp. 185–215.

Kemp J.R. and Wallace K.B. (1990). Molecular determinants of the species-selective inhibition of brain acetylcholinesterase. *Toxicol. Appl. Pharmacol.*, 104:246–258.

Kendall R.J., Brewer L.W., Hitchock R.R. and Mayer F.L. (1992). American wigeon mortality associated with turf application of diazinon AG500. *J. Wildl. Dis.*, 28:263–267.

Kerkvliet N.I., Baecher-Steppan L., Smith B.B. et al. (1990). Role of the *Ah* locus in suppression of cytotoxic T lymphocyte activity by halogenated aromatic hydrocarbons (PCBs and TCDD): structure-activity relationships and effects in C57Bl/6 mice congenic at the *Ah* locus. *Fundam. Appl. Toxicol.*, 14:532–541.

Kihlstrom J.E., Olsson M., Jensen S. et al. (1992). Effects of PCB and different fractions of PCB on the reproduction of the mink (*Mustela vison*). *Ambio*, 21:563–569.

Kubiak T.J., Harris H.J., Smith L.M. et al. (1989). Microcontaminants and reproductive impairment of the Forster's tern on Green Bay, Lake Michigan—(1983). *Arch. Environ. Contam. Toxicol.*, 18:706–727.

Kucera E. (1987). Brain cholinesterase activity in birds after a city-wide aerial application of malathion. *Bull. Environ. Contam. Toxicol.*, 38:456–460.

Lari L., Massi A., Fossi M.C. et al. (1994). Evaluation of toxic effects of the organophosphorus insecticide azinphos-methyl in experimentally and naturally exposed birds. *Arch. Environ. Contam. Toxicol.*, 26:234–239.

Leffin M. and Riviere J.L. (1992). Effect of some inducers on hepatic and extrahepatic drug metabolism in the mallard duck (*Anas platyrhynchos*). *Comp. Biochem. Physiol.*, 102C:471–476.

Leighton F.A., Peakall D.B. and Butler R.G. (1983). Heinz-body hemolytic anemia from the ingestion of crude oil: a primary toxic effect in marine birds. *Science*, 220:871–873.

Le Maho Y., Karmann H., Briot D. et al. (1992). Stress in birds due to routine handling and a technique to avoid it. *Am. J. Physiol.*, 263:R775–R781.

Lewis D.F.V. (1996). *Cytochromes P450: Structure, Function and Mechanism*. Taylor and Francis.

Llacuna S., Gorriz A., Durfort M. and Nadal J. (1993) Effects of air pollution on passerine birds and small mammals. *Arch. Environ. Contam. Toxicol.*, 24:59–66.

Lower W.R. and Tsutakawa R.K. (1978). Statistical analysis of urinary delta-aminolevulinic acid (ALAU) excretion in the whitefooted mouse associated with lead smelting. *J. Environ. Pathol. Toxicol.*, I:551–560.

Lubet R.A., Nims R.W., Fox S.D. et al. (1992). Induction of hepatic CYPIA activity as a biomarker for environmental exposure to Aroclor 1254 in feral rodents. *Arch. Environ. Contam. Toxicol.*, 22:339–344.

Ludke J.L., Hill E.F. and Dieter M.P. (1975). Cholinesterase (ChE) response and related mortality among birds fed ChE inhibitors. *Arch. Environ. Contam. Toxicol.*, 3:1–21.

Ma T.E. and Harris M.M. (1985). *In situ* monitoring of environmental mutagens. *Curr. Dev.*, 4:77–106.

Martin P.A., Solomon K.R., Forsyth D.F. et al. (1991). Effects of exposure to carbofuran-sprayed vegetation on the behavior, cholinesterase activity and growth of mallard ducklings (*Anas platyrhynchos*). *Environ. Toxicol. Chem.*, 10:901–909.

Mayer F.L., Versteeg D.J., McKee M.J. et al. (1992). Physiological and non specific biomarkers. In: Biomarkers: Biochemical, Physiological and Histological Markers of Anthropogenic Stress. Huggett R.J., Kimerle R.A., Mehrle Jr. P.M. and Bergam H.L. (eds.). Lewis Publishers, Boca Raton, pp. 5–85.

McBee K. (1991). Chromosomal aberrations in native small mammals (*Peromyscus leucopus* and *Sigmodon hispidus*) at a petrochemical waste site: II. Cryptic and inherited aberrations detected by G-band analysis. *Environ. Toxicol. Chem.*, 10:1321-1329.

McBee K. and Bickham J.W. (1988). Petrochemical-related DNA damage in wild rodents detected by flow cytometry. *Bull. Environ. Contam. Toxicol.*, 40:343–349.

McBee K. and Bickham J.W. (1990). Mammals as bioindicators of environmental toxicity *Curr. Mammal.*, 2:37–88.

McBee K., Bickam J.W., Brown K.W. and Donnelly K.C. (1987). Chromosomal aberrations in native small mammals (*Peromyscus leucopus* and *Sigmodon hispidus*) at a petrochemical waste site: I. Standard karyology. *Arch. Environ. Contam. Toxicol.*, 16:681–688.

McMillan D.E. (1990). The pigeon as a model for comparative behavioral pharmacology and toxicology. *Neurotoxicol. Teratol.*, 12:523–529.

Miller D.S., Peakall D.B. and Kinter W.B. (1978). Ingestion of crude oil: sublethal effects in herring gull chicks. *Science*, 199:315–317.

Mineau P. (1991). Difficulties in the regulatory assessment of cholinesterase-inhibiting insecticides. In: Cholinesterase-inhibiting Insecticides. Their Impact on Wildlife and the Environment. Mineau P. (ed.). Elsevier, Amsterdam, pp. 277–299.

Mineau P. and Peakall D.B. (1987). An evaluation of avian impact assessment techniques following broad-scale forest insecticide sprays. *Environ. Toxicol. Chem.*, 6:781–791.

Miura Y., Hisaki H., Fukushima B. et al. (1991). Comparative aspects of microsomal cytochrome-P-450-dependent monooxygenase activities in musk shrew (*Suncus murinus*), mongolian gerbil (*Meriones unguiculatus*), harvest mouse (*Micromys minutus*) and rat. *Comp. Biochem. Physiol.*, 100B:249–252.

National Research Council (1991). *Animals as Sentinels of Environmental Health Hazards.* National Academy Press, Washington.

Nayak B.N. and Petras M.L. (1985). Environmental monitoring for genotoxicity: *in vivo* sister chromatid exchange in the house mouse (*Mus musculus*). *Can. J. Genet. Cytol.*, 27:351–356.

Nelson D.R. (1998). Metazoan cytochrome P450 evolution. *Comp. Biochem. Physiol.*, Part C, 121C:15–22.

Newman J.R. and Schreiber R.K. (1984). Animals as indicators of ecosystem responses to air emissions. *Environ. Manag.*, 8:309–324.

Newton I., Wyllie I. and Asher A. (1993). Longterm trends in organochlorine and mercury residues in some predatory birds in Britain. *Environ. Pollut.*, 79:143–151.

Nosek J.A., Craven S.R., Karasov W.H. and Peterson R.E. (1993). 2,3,7,8 Tetrachlorodibenzo-*p*-dioxin in terrestrial environments: implications for resource management. *Wildl. Soc. Bull.*, 21:179–187.

Novak J. and Qualls C.W. (1989). Effects of phenobarbital and 3-methylcholanthrene on the hepatic cytochrome P-450 metabolism of various alkoxyresorufin ethers in the cotton rat (*Sigmodon hispidus*). *Comp. Biochem. Physiol.*, 94C:543–545.

Olhendorf H.M., Kilness A.W., Simmons J.L. et al. (1988). Selenium toxicosis in wild aquatic birds. *J. Toxicol. Environ. Health*, 24:67–92.

O'Halloran J. and Myers A.A. (1988). Blood lead levels and free red blood cell protoporphyrin as measure of lead exposure in mute swans. *Environ. Pollut.*, 52:19–38.

Pain D.J. (1989). Hematological parameters as predictors of blood lead and indicators of lead poisoning in the black duck, *Anas rubripes. Environ. Pollut.*, 60:67–82.

Pain D.J. (1990). Lead shot ingestion by waterbirds in the Camargue, France: an investigation of levels and interspecific differences. *Environ. Pollut.*, 66:273–285.

Pain D.J. (1991). Lead shot densities and settlement rates in Camargue marshes, France. *Biol. Conserv.*, 57:273–286.

Pain D.J. and Rattner B.A. (1988). Mortality and hematology associated with the ingestion of one number four lead shot in black ducks. *Bull. Environ. Contam. Toxicol.*, 40:159–164.

Parada R. and Jaszczak K. (1993). A cytogenetic study of cows from a highly industrial or an agricultural region. *Mutat. Res.*, 300:259–263.

Passer E.L., Leinaeng R.H., Birmingham L.W. et al. (1989). Effect of lead on blood protoporphyrin levels of a group of ring teal ducks, *Callonetta leucophrys. Zoo Biol.*, 8:357–366.

Peakall D.B. (1967). Pesticide-induced breakdown of steroids in birds. *Nature*, 216:505–506.

Peakall D.B. (1992). *Animal Biomarkers as Pollution Indicators.* Chapman and Hall. London.

Peakall D. and Walker C.H. (1994). The role of biomarkers in environmental assessment (3) Vertebrates. *Ecotoxicology*, 3:173–179.

Peakall D.B., Norstrom R.J., Rahimtula A.D. and Buteer R.D. (1986). Characterization of mixed-function oxidase systems of the nestling herring gull and its implications for bioeffects monitoring. *Environ. Toxicol. Chem.*, 5:379–386.

Platt J.F., Lensink C.J.K, Butler W. et al. (1990). Immediate impact of the Exxon Valdez oil spill on marine birds. *Auk*, 107:387–397.

Rattner B.A. and Fairbrother A. (1991). Biological variability and the influence of stress on cholinesterase activity. In: Cholinesterase-inhibiting Insecticides. Their Impact on Wildlife and the Environment. Mineau P (ed.). Elsevier, Amsterdam, pp. 89–109.

Rattner B.A., Hoffman D.J. and Marn C.M. (1989). Use of mixed-function oxygenases to monitor contaminant exposure in wildlife. *Environ. Toxicol. Chem.*, 8:1093–1102.

Rattner B.A., Flickinger E.L. and Hoffman D.J. (1993). Morphological, biochemical, and histopathological indices and contaminant burdens of cotton rats (*Sigmodon hispidus*) at three hazardous waste sites near Houston, Texas, U.S.A. *Environ. Pollut.*, 79:85–93.

Rivière J.L.(1992). Hepatic microsomal monooxygenase activities in natural populations of the mallard duck *Anas platyrhynchos*, the tufted duck *Aythya fuligula* and the great crested grebe *Podiceps cristatus*. *Ecotoxicology*, 1:117–135.

Rivière J.L. (1993). Les animaux sentinelles. *Courrier de l'Environnement I.N.R.A*, 20:59–67.

Ronis M.J.J. and Walker C.H. (1989). The microsomal monooxygenases of birds. *Rev. Biochem. Toxicol.*, 10:381–384.

Ronis M.J.J., Borlakoglu J., Walker C.H. et al. (1989). Expression of orthologues to rat cytochromes P-450IA1 and I.I.B.I. in sea birds: evidence for environmental induction. *Mar. Environ. Res.*, 28:123–130.

Ronis M.J.J., Hansson T., Borlakoglu J. and Walker C.H. (1989). Cytochromes P-450 of sea birds: cross-reactivity studies with purified rat cytochromes. *Xenobiotica*, 19:1167-1173.

Roos P.H., Hanstein W.G., Strotkamp D.et al. (1994). Risk assessment and bioavailability for mammals of soil bound polycyclic aromatic hydrocarbons. In: Cytochrome P450—8th International Conference. Lechner P (ed.). John Libbey Eurotext, Paris, pp. 785-788.

Safe S. (1986). Comparative toxicology and mechanism of action of polychlorinated dibenzo-p-dioxins and dibenzofurans. *Annu. Rev. Pharmacol. Toxicol.*, 26:371–399.

Sanderson J.T., Elliott J.E., Norstrom R.J. et al. (1994). Monitoring biological effects of polychlorinated dibenzo-p-dioxins, dibenzofurans, and biphenyls in great blue heron chicks (*Ardea herodias*) in British Columbia. *J. Toxicol. Environ. Health*, 41:435–450.

Scheuhammer A.M. (1989). Monitoring wild bird populations for lead exposure. *J. Wildl. Manag.*, 53:759–765.

Scheuhammer A.M. (1991). Effects of acidification on the availability of toxic metals and calcium to wild birds and mammals. *Environ. Pollut.*, 71:329–375.

Schrenk D., Lipp H.P., Brunner H. et al. (1991). Induction of hepatic P450-dependent monooxygenase in feral mice from a PCDD/PCDF-contaminated area. *Chemosphere*, 22:1011-1018.

Sears J. (1988). Regional and seasonal variations in lead poisoning in the mute swan, *Cygnus olor*, in relation to the distribution of lead and lead weights in the Thames area. *Biol. Conserv.*, 46:115–134.

Shaw-Allen P.L. and McBee K. (1993). Chromosome damage in wild rodents inhabiting a site contaminated with Aroclor 1254. *Environ. Toxicol. Chem.*, 12:677–684.

Shore R.F. and Douben P.E.T. (1994). Predicting ecotoxicological impacts of environmental contaminants on terrestrial small mammals. *Rev. Environ. Contam. Toxicol.*, 134:49–89.

Simmons G.J. and McKee M.J. (1992). Alkoxyresorufin metabolism in white-footed mice at relevant environmental concentrations of Aroclor 1254. *Fund. Appl. Toxicol.*, 19:446–452.

Sinclair J.F. and Sinclair P.R. (1993). Avian cytochrome P450. In: Cytochrome P450. Schenkman J.B. and Greim H (eds.). Springer-Verlag, Berlin, pp. 259–277.

Somers J.D., Khan A.A., Kumar Y. and Barrett M.W. (1991). Effects of simulated field spraying of carbofuran, carbaryl and dimethoate on pheasant and partridge chicks. *Bull. Environ. Contam. Toxicol.*, 46:113–119.

Spear P.A., Bilodeau A.Y. and Branchaud A. (1992). Retinoids: from metabolism to environmental monitoring. *Chemosphere*, 25:1733–1738.

Stansley W. (1993). Field results using cholinesterase reactivation techniques to diagnose acute anticholinesterase poisoning in birds and fish. *Arch. Environ. Contam. Toxicol.*, 25:315–321.

StLouis V.L., Breebaart L., Barlow J.C. and Klaverkamp J.F. (1993). Metal accumulation and metallothionein concentrations in tree swallow nestlings near acidified lakes. *Environ. Toxicol. Chem.*, 12:1203–1207.

Talmage S.S. and Walton B.T. (1991). Small mammals as monitors of environmental contaminants. *Rev. Environ. Contam. Toxicol.*, 119:47–145.

Tarrant K.A., Thompson H.M. and Hardy A.R. (1992). Biochemical and histological effects of the aphicide demeton S-methyl on house sparrows (*Passer domesticus*) under field conditions. *Bull. Environ. Contam. Toxicol.*, 48:360–366.

Thompson C., Andries M., Lundgren K. et al. (1989). Humans exposed to polychlorinated biphenyls (PCBs) and polychlorinated dibenzofurans (PCDFs) exhibit increased SCE frequencies in lymphocytes when incubated with alpha-naphthoflavone: involvement of metabolic activation by P-450 isozymes. *Chemosphere*, 18:687–694.

Thompson H.M. (1991). Serum B esterases as indicators of exposure to pesticides. In: Cholinesterase-inhibiting Insecticides. Their Impact on Wildlife and the Environment. Mineau P. (ed.). Elsevier, Amsterdam, pp. 109–125.

Thompson H.M. (1993). Avian serum esterases: species and temporal variations and their possible consequences. *Chem. Biol. Interact.*, 87:329–338.

Thompson H.M. and Walker C.H. (1994). Serum 'B' esterases as indicators of exposure to pesticides. In: Non-destructive Biomarkers in Vertebrates. Fossi M.C. and Leonzio C. (eds.). Lewis Publishers, Boca Raton, pp. 35–60.

Thompson H.M., Walker C.H. and Hardy A.R. (1991). Changes in activity of avian serum esterases following exposure to organophosphorus insecticides. *Arch. Environ. Contam. Toxicol.*, 20:514–518.

Thompson R.A., Schroder G.D. and Connor T.H. (1988). Chromosomal aberrations in the cotton rat, *Sigmodon hispidus*, exposed to hazardous waste. *Environ. Mol. Mutagen.*, II:359–367.

Tillitt D.E., Ankley G.T., Verbrugge D.A. et al. (1991). H4IIE rat hepatoma cell bioassay-derived 2,3,7,8-tetrachlorodibenzo-*p*-dioxin equivalents in colonial fish eating waterbird eggs from the Great Lakes. *Arch. Environ. Contam. Toxicol.*, 21:91–101.

Tometsko A.M., Dertinger S.D. and Torous D.K. (1993). Analysis of micronucleated cells by flow cytometry. 2. Evaluating the accuracy of high-speed scoring. *Mutat. Res.*, 292:137–143.

Tomlin C. (1993). *The Pesticide Manual*, 10th ed.. C.P.C. Publications, The Bath Press, Bath.

Vecchi A., Sironi M., Sfreddo-Gallotta E. et al. (1986). Effect of inducers of P-450 cytochrome isoenzymes on T.C.D.D. immunosuppressive activity. *Chemosphere*, 15:1707–1714.

Walker C.H. (1990). Biochemical effects of pesticides exploitable in field testing. In: Pesticide Effects on Terrestrial Wildlife. Somerville S. and Walker C.H. (eds.). Taylor and Francis, London, pp. 153–164.

Walker C.H. (1992). Biochemical responses as indicators of toxic effects of chemicals in ecosystems. *Toxicol. Lett.*, 64-65:527–533.

Walker C.H. (1993). The classification of esterases which hydrolyse organophosphates: recent developments. *Chem. Biol. Interact.*, 87:17–24.

Walker C.H. and Knight G.C. (1981). The hepatic microsomal enzymes of sea birds and their interaction with liposoluble pollutants. *Aquat. Toxicol.*, 1:343–354.

Walker C.H. and Ronis M.J.J. (1989). The monooxygenases of birds, reptiles and amphibians. *Xenobiotica*, 19:1111–1121.

Watkins J.B. (1991). Effect of microsomal enzyme inducing agents on hepatic biotransformation in cotton rats (*Sigmodon hispidus*): comparison to that in Sprague-Dawley rats. *Comp. Biochem. Physiol.*, 98C:433–439.

Way C.A. and Schroder G.D. (1982). Accumulation of lead and cadmium in wild populations of the commensal rat, *Rattus norvegicus*. *Arch. Environ. Contam. Toxicol.*, 11:407–417.

Westlake G.E., Bunyan P.J., Martin A.D. et al. (1981 a). Carbamate poisoning. Effects of selected carbamate pesticides on plasma enzymes and brain esterases of Japanese quail (*Coturnix coturnix japonica*). *J. Agric. Food Chem.*, 29:779–785.

Westlake G.E., Bunyan P.J., Martin A.D. et al. (1981b). Organophosphate poisoning. Effects of selected organophosphate pesticides on plasma enzymes and brain esterases of Japanese quail (*Coturnix coturnix japonica*). *J. Agric. Food Chem.*, 29:772–778.

White D.H. and Seginak J.T. (1990). Brain cholinesterase inhibition in songbirds from pecan groves sprayed with phosalone and disulfoton. *J. Wildl. Dis.*, 26:1-3-106.

Wilson B.W., Hooper M.J., Littrell E.E. et al. (1991). Orchard dormant sprays and exposure of red-tailed hawks to organophosphates. *Bull. Environ. Contam. Toxicol.*, 47:717–724.

Wlostowski T. (1992). Seasonal changes in subcellular distribution of zinc, copper, cadmium and metallothionein in the liver of bank vole (*Clethrionomys glareolus*)—a possible essential role of cadmium and metallothionein in the hepatic metabolism of copper. *Comp. Biochem. Physiol.*, 101C:155–162.

Yamashita N., Shimada T., Tanabe S. et al. (1992). Cytochrome P-450 forms and its inducibility by P.C.B. isomers in black headed gulls and black tailed gulls. *Mar. Pollut. Bull.*, 24:316–321.

Yawetz A., Zook-Rimon Z. and Dotan A. (1993). Cholinesterase profiles in two species of wild birds exposed to insecticide sprays in their natural habitat. *Arch. Insect Biochem. Physiol.*, 22:501–509.

Zinkl J.G., Roberts R.B., Henny C.J. and Lenhart D.J. (1980). Inhibition of brain cholinesterase activity in forest birds and squirrels exposed to aerially applied acephate. *Bull. Environ. Contam. Toxicol.*, 24:676–683.

8

Biomarkers of Exposure of Terrestrial Plants to Pollutants: Application to Metal Pollution

J. Vangronsveld, M. Mench, B. Mocquot and H. Clijsters

INTRODUCTION

As primary producers, chlorophyll-containing plants constitute the basis of almost all terrestrial ecosystems. Photosynthesis represents the unique entry of bioavailable (solar) energy into ecosystems. All animals, bacteria, fungi, and the plants themselves rely to some extent on this energy. Moreover, plants contribute to many other important ecological functions:

- The physical structure of individual and groups of plants defines the habitat for animals.
- Plants are involved in nutrient dynamics, especially through the production of organic matter that plays a key role in soil fertility.
- They supply protection against soil erosion.

In spite of their crucial function in ecosystems, the use of plants in the evaluation of actual and potential adverse consequences of human activities has been underestimated. Several processes such as germination, growth, biochemical responses, adaptations, and mortality of plants can be used to evaluate the quality of the terrestrial environment.

Generally, the potential harmful effects of metal-polluted soils on plants is predicted on the basis of chemical soil analysis. However, these results can lead to misinterpretations since metal availability is a function of the chemical form of the element in the soil (metal speciation) and of several soil parameters, e.g., pH, organic matter content, and cation exchange capacity. Moreover, soils are frequently contaminated by a mixture of metals. Each of these metals may be phytotoxic, or they may interact in a synergistic, antagonistic, or cumulative way.

Toxicity tests are among the most frequently used tools to evaluate the effects of soil contamination. The most widely used acute phytotoxicity tests on terrestrial plants are the seed germination test (a direct exposure method) and the root elongation test (typically performed with soil elutriates). However, seed germination, although often presented as a sensitive and critical stage in the plant life cycle, is rather insensitive to many toxic substances, including metals. Metal uptake in the embryonic plant is limited since it mainly derives nutrients from the seed reserves. The first stages of seed development are therefore almost independent of the environment. Life-cycle bioassays are used to assess sublethal responses of plants to toxic chemicals. Exposure may be either acute or chronic. Biomass, shoot length, leaf number and area, days to flower, allocation to reproduction, and other parameters are investigated during growth under controlled environmental conditions in greenhouse or growth chamber. Whatever test is used, the exposure of test plants must be long enough for its effects to become measurable, which limits the advantage of such tests as diagnostic tools.

Biomarkers may be used as an early warning system of specific or general stress. In this context, the term 'biomarker' can be defined as any plant biochemical, physiological, or morphological parameter that can be used to quantify the exposure of plants to stress (Ernst and Peterson, 1994). In this chapter, we review the potential of metabolites, enzymatic processes, fluorescence, and changes in plant growth as biomarkers for the evaluation of soil contamination by metals.

It is theoretically desirable to use biomarkers that are specific for the stressor under investigation, metals in this case. In reality, however, only a few potential biomarkers seem to be really metal-specific. In the first part of this chapter, metal-specific biomarkers are discussed. Less specific biomarkers are treated in the second part. In the third part, important precautions required in using these tools are indicated.

1. METAL-SPECIFIC BIOMARKERS

1.1 Phytochelatins

In many plant species exposure to increasing levels of metals appear to induce the production of metal-binding polypeptides, usually designated as phytochelatins (for a review, see Rauser, 1990 and Steffens, 1990).

These polypeptides probably play a central role in the homeostatic control of metal ions in plants (Steffens, 1990). They may also be involved in the mechanism of metal tolerance of selected cell lines and whole plants (Rauser, 1984; Tomsett et al., 1989). The similarity of phytochelatins to glutathione indicates that they cannot be primary gene products and that their biosynthesis shares a common pathway with glutathione (Steffens, 1990). They can be considered as linear polymers of the γ-glutamylcysteine portion of glutathione. Evidence for the linkage of glutathione metabolism (Fig. 1) to phytochelatin synthesis (Fig. 2) has been shown by many authors in both *in vivo* and *in vitro* studies (Rauser, 1987; Scheller et al., 1987; Grill et al., 1989; Rügsegger et al., 1990). In *in vitro* studies, Grill et al. (1989) showed that phytochelatins were formed from glutathione in the presence of cadmium ions by an enzyme (purified from *Silene cucubalus*) they characterized as γ-glutamylcysteine dipeptidyl transpeptidase (phytochelatin synthase).

Under laboratory conditions, a dose- and time-dependent relationship between metal exposure and phytochelatin concentration was shown during short-term exposure to copper (Schat and Kalff, 1992), cadmium (De Knecht et al., 1993), and zinc (Harmens et al., 1993). As could be expected, phytochelatin concentration depends not only on the exposure level, but also on the tolerance level of the investigated plants. In a comparison between Cd-sensitive and Cd-tolerant individuals of *Silene vulgaris*, the former produced significantly more phytochelatins than the latter at the same external Cd concentration (De Knecht et al., 1992; Fig. 3). When Cd-tolerant and Cd-sensitive plants were exposed at external Cd concentrations that caused a 50% inhibition in root growth in both populations (EC_{50}), the phytochelatin concentration was nearly the same (De Knecht et al., 1994). This result indicates that the cellular, and not the environmental, Cd concentration is responsible for the

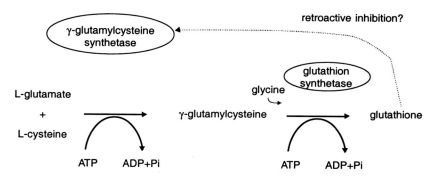

Fig. 1. Pathways of glutathione synthesis deduced from *in vitro* enzyme analysis. The dotted line indicates the possibility of regulation by retroinhibition (Foyer et al., 1995).

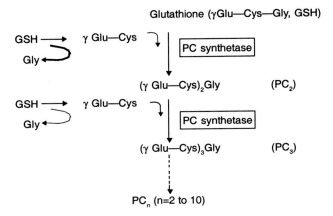

Fig. 2. Pathways of phytochelatin synthesis (GSH = glutathione; PC = phytochelatin)

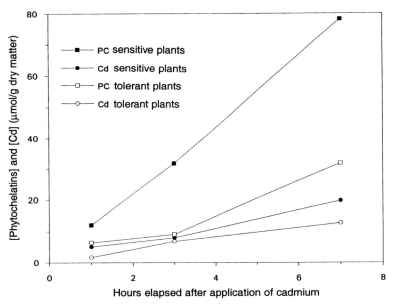

Fig. 3. Levels of phytochelatins and cadmium (μmoles/g dry matter) in roots of *Silene vulgaris* sensitive and tolerant to cadmium

synthesis of this biomarker. Phytochelatin synthesis has also been shown to be more or less metal-specific. Low concentrations (10 μM) of Ag, Cd, Hg, and Te induced the same phytochelatin concentration as a concentration 100 times as high (1000 μM) of Pb (Grill et al., 1987).

Another possible biomarker of metal exposure may be the sulphide-containing phytochelatin complex (Verkleij et al., 1990; Reese et al., 1992;

Speiser et al., 1992); the biosynthesis of this component requires adenylosuccinate synthetase and succino amino imidarole carboxamide ribonucleotide synthetase (Juang et al., 1993).

Phytochelatin synthase (Fig. 2), which is activated by free metal ions (see above), is a third potential specific biomarker for free metal ions in the cell. So far, however, its potential use as a biomarker has not been evaluated.

1.2 Selenoproteins

The production of a particular metabolite has been reported following selenium exposure that is different in tolerant and sensitive plants. Selenium-sensitive plants cannot differentiate between S and Se and they incorporate selenium in sulphur-containing amino acids such as selenomethionine and selenocysteine (Fig. 4). Incorporation of these seleno amino acids into proteins may cause a decrease in the activity of some enzymes (Brown and Shrift, 1981). Selenium-tolerant plants synthesize and accumulate non-protein seleno amino acids such as selenocystathioneine and methylselenocysteine (Peterson and Butler, 1967, 1971; Burnell, 1981). The occurrence of selenoproteins in selenium-sensitive plants may therefore be an excellent biomarker for selenium-induced stress in some plants.

2. NON–METAL-SPECIFIC BIOMARKERS

Generalist biomarkers, which respond to a variety of environmental stressors, are undoubtedly useful to indicate that something in the environment is hazardous to plant life. In spite of their non-specificity, such biomarkers can be used to evaluate metal phytotoxicity of soils if other stressors are absent or remain constant. Phytotoxicity responses of test plants grown on polluted substrates under controlled environmental conditions (growth chamber) should only reflect the interference of metals assimilated through the roots with metabolic processes. Activity of certain stress-induced enzymes, membrane permeability and composition, polyamine levels, sugar ratio in roots, and fluorescence are some of the parameters that may be used as biomarkers.

2.1 Enzymatic Changes

The correlation between the induction of various enzymes (peroxidase, malic enzyme, isocitrate dehydrogenase, glutamate dehydrogenase, glucose-6-phosphate dehydrogenase; Vangronsveld et al., 1997) and the metal concentration was so high that precise toxic threshold value for

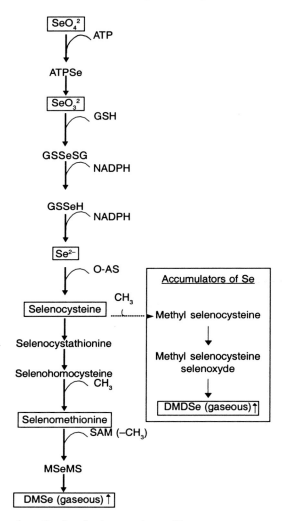

Fig. 4. Pathway of synthesis of seleno amino acids

zinc and cadmium in the primary leaf of *Phaseolus vulgaris* could be calculated from the enzyme induction curves in this organ (Van Assche et al., 1988). Because of the metal-specific responses in different organs (Van Assche et al., 1986), isoperoxidase activity appears to be particularly useful for the evaluation of the phytotoxicity of metal-polluted soils.

The quantitative relationship observed in *P. vulgaris* between enzyme induction and metal assimilation (Van Assche et al., 1988; Fig. 5) in combination with the metal-specific changes in the isoperoxidase

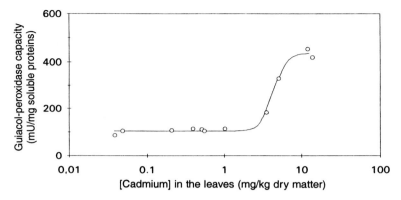

Fig. 5. Dose-effect relationship between cadmium concentration (mg/kg dry matter) and the capacity of guiacol-peroxidase enzymes (milli-units/mg soluble protein) in primary leaves of bean (*Phaseolus vulgaris* cv. Limburgse vroege)

pattern (Van Assche et al., 1986; Vangronsveld et al., 1997) and data on morphological parameters (length, weight, leaf area) provide a reliable biological test system for the evaluation of soil phytotoxicity (Van Assche and Clijsters, 1990a; Vangronsveld and Clijsters, 1992). Cultivation of the test plants under controlled environmental conditions is very important for these studies to avoid interference of other stressors. Attention must also be paid to the test plant species itself as the sensitivity to metals varies between species and even between cultivars (ecotypes, populations) of the same species (Marschner, 1983). For routine screening of soil phytotoxicity, a *P. vulgaris* cultivar (Limburgse vroege) with moderate sensitivity to zinc, cadmium, and copper has been used (Van Assche and Clijsters, 1990a; Vangronsveld and Clijsters, 1992; Mench et al., 1994; Vangronsveld et al., 1995). More sensitive species such as *Spinacea oleraceae* and *Lycopersicon esculentum* and less sensitive species such as *Zea mays* are now included in this test system (Mocquot et al., 1996; Lagriffoul et al., 1998).

The values obtained for the different parameters examined can be transformed into phytotoxicty classes, each parameter being used to classify the substrate in a given phytotoxicity class. Such a classification system was developed for the parameters studied in *P. vulgaris* cv. Limburgse vroege (Van Assche and Clijsters 1990a; Vangronsveld and Clijsters, 1992) and it is presented in Table 1. The 'phytotoxicity index' for a given substrate represents the mean value of the toxicity class numbers obtained for each parameter separately.

Enzyme induction and/or isoenzyme variation proved also to be a useful criterion for air pollution monitoring (Keller, 1974; Wellburn et al., 1976; Flückiger et al., 1978; Rabe and Kreeb, 1979).

Table 1. Phytotoxicity classification scheme using bean (*Phaseolus vulgaris* L., cv Limburgse vroege) as test plant for diagnostic of phytotoxicity of metal-contaminated substrates. Phytotoxicity level for each parameter depends on the effect of stress (expressed as percentage of morphological or enzymatic parameters measured in control plants) (Van Assche and Clijsters, 1990; Vangronsveld and Clijsters, 1992).

	Level of phytotoxicity (class)			
Parameters	Non-toxic (class 1)	Mildly toxic (class 2)	Moderately toxic (class 3)	Highly toxic (class 4)
Height of aerial parts	>85	85–70	70–50	<50
Leaf area of primary leaves	>85	85–70	70–50	<50
Root weight	>85	85–70	70–50	<50
Enzymatic activity				
leaves				
POD	<150	150–325	325–500	>500
ME, ICDH	<125	125–175	175–250	>250
roots				
POD, ME, GDH	<125	125–175	175–250	>250
D-isoperoxydases/total isoperoxydases (× 100)				
leaves	0	0–25	35–50	>50
roots	0	0–15	15–30	>30

POD, peroxydases; ME, maleic enzyme; ICDH, isocitrate dehydrogenase; GDH, glutamate dehydrogenase.

2.2 Metabolites

In addition to or instead of enzymes, metabolites can be used as general stress biomarkers. An increase in putrescine was observed after exposure of barley and rapeseed to a toxic concentration of chromium (Jacobsen et al., 1992). In mung bean plants treated with Cd, Cu, Ni, Zn, Pb, Hg, and As, strong elevations in putrescine content were observed, especially in the roots (Geuns et al., 1992; Geuns and Colpaert, 1993). Increased levels of putrescine were also observed under the influence of SO_2 (Priebe et al., 1978) and UV-B treatment (Kramer et al., 1992). In roots of mung beans, it was reported that the ratio of glucose plus fructose/saccharose significantly increased after exposure to toxic amounts of Cd, Cu, Ni, Zn, Pb, Hg, and As (Geuns and Colpaert, 1993).

2.3 Membrane Parameters

Membrane permeability (quantified on the basis of potassium leakage) and membrane degradation products (e.g., thiobarbaturic acid reactive compounds) significantly increase under conditions of metal phytotoxicity (Fig. 6) and several other forms of stress (Passow and Rothstein, 1960; Wainwright and Woolhouse, 1978; De Filippis, 1979; De Vos et al., 1989, 1992; Weckx et al., 1993; Vangronsveld et al., 1993).

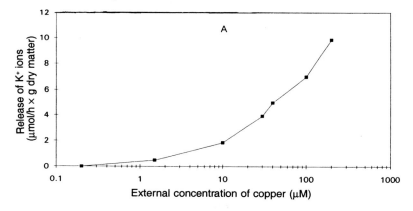

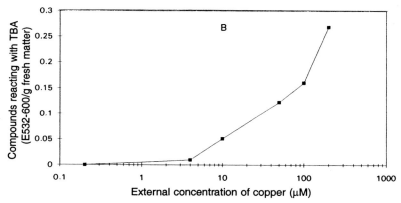

Fig. 6 A. Release of K$^+$ ions (μmoles per h and per g fresh matter).
B. Formation of compounds reactive with thiobarbituric acid (E$_{532-600}$ per g fresh matter) in roots of bladder campion (*Silene vulgaris*) (De Vos et al., 1989).

In roots of mung beans the ratio of stigmasterol to sitosterol significantly increased after exposure to toxic amounts of Cd, Cu, Ni, Zn, Pb, Hg, and As (Geuns and Colpaert, 1993). Ros et al. (1990) also found an increase in the Mg^{2+}-ATP-ase activity of plasmatic membranes of rice shoots after an *in vivo* exposure to cadmium and nickel. They associated this increase with changes in lipid composition of the plasmatic membrane, namely reduction in the campesterol/sitosterol ratio.

2.4 Genotoxicity Tests

Effects at the level of the genome are other possible biomarkers for metal toxicity. The toxic effects of pesticides and fungicides on the mitosis of root meristems has been well documented (e.g., Sharma and

Panneerselvam, 1990). However, studies at that level require some experience in karyotyping.

Most reports about interactions of metals with nucleic acids and their metabolism are based on studies on bacteria and animals. Information from higher plants is rather poor. Ernst (1980) cited an increase of the number of structural chromosome aberrations after exposure of *Crepis capillaris* seeds to cadmium.

Metals can interact directly or indirectly with nucleic acids. The numerous nucleophylic centres in nucleic acids constitute potential binding sites for metal ions. The nature and localization of bonds apparently depend on the metal considered. Induction of crosslinks between both DNA strands, single-strand DNA breaks, and chelation or formation of complexes between DNA and metals have been described (see Gebhart and Rossman, 1991, for a review). Similar reactions may also occur between metals and RNA, thereby affecting the specific functions of the latter.

Besides these direct effects, metals may indirectly induce breaks in the DNA strand: depurination of DNA, which may have mutagenic effects, was reported to be induced by copper (II), nickel (II), and chromium (IV) (Schaaper et al., 1987). Metal-supported generation of oxygen free radicals can also cause DNA damage. Exposure of DNA to oxygen free radical generating systems was found to induce extensive strand breakage and degradation of DNA (Halliwell and Gutteridge, 1984).

Effects of metals on the function of various enzymes involved in the metabolism of nucleic acids can also indirectly contribute to alterations in the genetic information by affecting replication fidelity.

The activity of hydrolytic enzymes such as ribonuclease and deoxyribonuclease was found to increase after *in vivo* cadmium exposure of *Glycine max* seedlings. These effects were considered to be symptoms of cadmium-accelerated senescence (Lee et al., 1976a).

2.5 Fluorescence

Fluorometric analysis of photosynthesis has gained wide acceptance as a method to evaluate the biochemical and physiological condition of plants (Jones and Winchell, 1984; Kramer et al., 1987; Walker, 1988; Ogren, 1991). Toxicological data specific to photosynthetic systems have been collected for hundreds of chemicals and on several plant species. This large collection of information makes chlorophyll fluorescence one of the most promising biomarkers for detection of exposure and effects.

Fluorescence of chlorophyll is a good biomarker to evaluate ecologically significant environmental stress. The test is sensitive, reliable, and feasible. The underlying mechanism of plant fluorescence is well-documented. An important advantage of the use of the fluorescence method is that it is non-destructive, easily applicable in the field, and less time-consuming than other techniques, provided it is standardized.

Sampson and Popovic (1988) described the use of fluorescence to determine toxicity of heavy metals and pesticides in the alga *Dunaliella tertiolecta*. They presented dose-response data illustrating the sensitivity of the technique and identified stressor-specific differences in fluorescence profiles. Inhibition of photosynthesis by copper was studied by fluorescence on *Oryza sativa* (Lidon et al., 1993) and on a copper-tolerant ecotype of *Silene compacta* (Ouzounidou et al., 1995). The inhibition of photosynthesis by cadmium has also been studied in bean (*Phaseolus vulgaris*; Krupa et al., 1993; Ciscato et al., 1999).

3. LIMITS OF THE USE OF BIOMARKERS

It is very important to note that the degree of adaptation to environmental stress (in this case metals) and the homogeneity of the plant population are critical factors for the sensitivity of a plant species and therefore for the efficiency and reliability of a biomarker (i.e. its accuracy, sensitivity, and reproducibility). Like any other taxonomic group, plants have the ability to adapt to environmental conditions that may cause metabolic and functional injury to non-adapted organisms. In the case of heavy metals, adaptation may be the result of an exposure of plants to metal concentrations that are unusually high compared to the normal concentrations in soils, for example in the vicinity of metal processing industrial plants. However, adaptation to elevated metal concentration may also be the result of long-term exposure, as on soils with naturally elevated metal contents or on old non-ferrous mining sites. Various mechanisms have been developed by plants to alleviate the stress due to exposure to toxic metal concentrations (Ernst et al., 1990, 1992; Peterson, 1993). These adaptations may mitigate the impact of metals on plants. Therefore, the use of growth responses, biochemical responses, or fluorescence as biomarkers for metal pollution, necessitates a precise knowledge of the adaptation level already present in the population of the test plant species.

The physiological and metabolic responses of test plants as well as their growth constitute biological criteria for the overall phytotoxic effect, since they are the result of interactions of the metals with other soil factors and with the plant. In general, growth responses are rather poor biomarkers since they may be strongly influenced by differences in nutrient status of the soils. Moreover, biomarkers should exhibit a more sensitive response than visible symptoms of metal exposure. They should allow early recognition of stress effects in a dose- or time-dependent manner before effects on growth and development are visible. Biomarkers should also be used to predict the consequences of metal pollution for the environment. Examples of metal-induced adverse effects at the (sub)cellular level are: an increase of biomembrane permeability (Passow and Rothstein, 1960; Wainwright and Woolhouse, 1978; De Filippis, 1979; De Vos et al., 1989, 1992; Weckx et al., 1993; Vangronsveld et al., 1993), depletion of the glutathione pool (De Vos et al., 1992), a diminished uptake of nutrients (Weber et al., 1991), and inhibition of photosynthesis (Clijsters and Van Assche, 1985). It is only after a sufficiently long exposure to elevated metal concentrations that disturbances of these processes may cause visible growth responses.

As previously mentioned, it is desirable to use biomarkers that are specific for metal stress. However, most of the biomarkers presently known are not specific. Effects on membrane permeability, on photosynthesis (and fluorescence), on glutathione pool, on antioxidative enzymes, and other parameters have been described for many environmental stressors (e.g., metals, UV, ozone, SO_2, drought, heat, cold). Such non-specific parameters can be used as biomarkers for metal stress only if other stressors are absent or possibly constant.

It is important to emphasize that most of the potential biomarkers presented here reflect only the level of free metal ions in the cell, i.e., those metal ions that interfere with the cellular metabolism. In consequence, no reliable information can be obtained about the total metal uptake and load of a plant (and possible transfer to the consumer compartment of an ecosystem) without simultaneous metal quantification in the plant tissues.

4. CONCLUSIONS

Experiments on using plant characteristics as biomarkers for metal pollution are still in an initial stage. Additional basic research needs to be done on the choice of plant species, organs and parameters that have to be analysed, on standardization of sampling, sample preparation and analysis, and on the quantitative relationship between metal concentration and biological effect.

Research on biomarkers in plants has mainly consisted of short-term exposures under laboratory conditions. When metabolites, enzymatic activities, or fluorescence were measured, the different laboratories mostly followed their own procedures. The methods have still not been standardized. Therefore, comparison of data is very difficult. Whatever parameter is used as biomarker, the question of specificity of the response arises. Do the changes in the analysed parameter(s) arise from the stressor under investigation or from (an)other stressor(s), or do the values measured reflect normal variation due to climatic, edaphic, or life-cycle differences? When soil metal phytotoxicity is measured, these questions can be answered by standardizing the test for example by growing selected plant species during a given period under strictly controlled environmental conditions (growth chamber). Such a test gives reliable and comparative information about the potential phytotoxicity of soils. However, the ultimate goal of ecotoxicologists is to monitor environmental stress (metals in this case) in the field or on plant material collected in the field. Little has been done on plant growth in the field. The aim of future research should be to fill this gap in order to understand the impact of various environmental conditions and the role of genotypic adaptation in the response of biomarkers.

Another important criticism is that single species tests were usually considered up to now and that measurements are at the level of organs, tissues or cells. Are such tests sufficiently predictive for effects on higher levels of biological organization, such as populations or communities?

REFERENCES

Brown T.A. and Shrift A. (1981). Exclusion of selenium from proteins of selenium-tolerant *Astragalus* species. *Plant Physiol.*, 67:1051–1053.

Burnell J.N. (1981). Selenium metabolism in *Neptunia amplexicaulis*. *Plant Physiol.*, 67:316–324.

Ciscato M., Valcke R. and Vangronsveld J. (1999). Effects of heavy metals on the fast chlorophyll fluorescence induction kinetics of photosystem II: a comparative study. 54c: *Z. Naturforsch.*, 735–739.

Clijsters H. and Van Assche F. (1985). Inhibition of photosynthesis by heavy metals. *Photosynth. Res.*, 7:31–40.

De Filippis L.F. (1979). The effect of heavy metal compounds on the permeability of *Chlorella* cells. *Z. Pflanzenphysiol.*, 92:39–49.

De Knecht J.A., Koevoets P.L.N., Verkleij J.A.C. and Ernst W.H.O. (1992). Evidence against a role for phytochelatins in naturally selected increased cadmium tolerance in *Silene vulgaris* (Moench) Garcke. *New Phytol.*, 122:681–688.

De Knecht J.A., Van Dillen M., Koevoets P.L.N. et al. (1993). Phytochelatins in cadmium-sensitive and cadmium-tolerant *Silene vulgaris*: chain length distribution and sulfide incorporation. *Plant Physiol.*, 104:255–261.

De Vos C.H.R., Vonk M.J., Vooijs H. and Schat H. (1992). Glutathione depletion due to copper-induced phytochelatin synthesis causes oxidative stress in *Silene cucubalus*. *Plant Physiol.*, 98:853–858.

De Vos C.H.R., Vooijs H., Schat H. and Ernst W.H.O. (1989). Copper-induced damage to the permeability barrier in roots of *Silene cucubalus*. *J. Plant Physiol.*, 135:165–169.

Ernst W.H.O. (1980). Biochemical aspects of cadmium in plants. In: Cadmium in the Environment, Part I, Ecological Cycling. Nriagu J.O. (ed.). John Wiley and Sons, New York, pp. 639–653.

Ernst W.H.O. and Peterson P.J. (1993). The role of biomarkers in environmental assessment (4). Terrestrial plants. *Ecotoxicology*, 3:180–192.

Ernst W.H.O., Schat H. and Verkleij J.A.C. (1990). Evolutionary biology of metal resistance in *Silene vulgaris*. *Evolut. Trends Plants*, 4:45–51.

Ernst W.H.O., Verkleij J.A.C. and Schat H. (1992). Metal tolerance in plants. *Acta Bot. Neerl.*, 41:229–248.

Flückiger W., Flückiger-Keller H. and Oertli J.J. (1978). Der Einfluss verkehrsbedingter Luftverunreinigingen auf die Peroxydaseaktivität, das ATP-Bildungsvermogen isolierter Chloroplasten und das Längenwachstum van Mais. *J. Plant Dis. Protect.*, 85:41–47.

Foyer H., Jouannin L., Souriau N. et al. (1993). The molecular biochemical and physiological function of glutathion and its action in poplar. In: Contribution to Forest Tree Physiology. Eurosilva. Les Colloques de l'INRA, 76:141–170.

Gebhart E. and Rossman T.G. (1991). Mutagenicity, carcinogenicity, teratogenicity. In: Metals and their Compounds in the Environment. Occurrence, Analysis and Biological Relevance. Merian E. (ed.). VCH Verlagsgesellschaft, Weinheim, pp. 617–640.

Geuns J. and Colpaert J. (1993). Cadmium toxicity in mung bean seedlings. Proc. Environmental Platform KU Leuven, pp. 193–195.

Geuns J.M.C., Colpaert J., Asard H. and Caubergs R. (1992). Cadmium effects in mung bean seedlings. *Physiol. Plant.*, 85:371.

Grill E., Winnackers E.L. and Zenk M.H. (1987). Phytochelatins, a class of heavy-metal-binding peptides from plants, are functionally analogous to metallothioneins. *Proc. Nat. Acad. Sci. USA*, 84:439–443.

Grill E., Loffler S., Winnacker E.L. and Zenk M.H. (1989). Phytochelatins, the heavy-metal-binding peptides of plants, are synthesized from glutathione by γ-glutamylcysteine dipeptidyl transpeptidase (phytochelatin synthase). *Proc. Nat. Acad. Sci. USA*, 86:6838–6842.

Halliwell B. and Gutteridge J.M.C. (1984). Oxygen toxicity, oxygen radicals, transition metals and disease. *Biochem. J.*, 219:1–14.

Harmens H., Cornelisse E., Den Hartog P.R. et al. (1993). Phytochelatins do not play a key role in naturally selected zinc tolerance in *Silene vulgaris*. *Plant Physiol.*, 103:1305–1309.

Jacobsen S., Hausschild M.Z. and Rasmussen U. (1992). Induction by chromium of chitinases and polyamines in barley (*Hordeum vulgare* L.) and rapeseed (*Brassica napus* ssp. *oleifera*). *Plant Sci.*, 84:119–128.

Jones T.W. and Winchell L. (1984). Uptake and photosynthetic inhibition by atrazine and its degradation products on four species of submerged vascular plants. *J. Environ. Qual.*, 13:250–300.

Juang R.H., McCue K.F. and Ow D.W. (1993). Two purine biosynthetic enzymes that are required for cadmium tolerance in *Schizosaccharomyces pombe* utilize cysteine sulfinate *in vitro*. *Arch. Biochem. Biophys.*, 403:392–401.

Keller T. (1974). The use of peroxidase activity for monitoring and mapping air pollution areas. *Europ. J. Forest Pathol.*, 4:11–19.

Kramer D., Adawi O., Morse P.D. and Crofts A.R. (1987). A portable double-flash spectrophotometer for measuring the kinetics of electron transport components in intact leaves. Progress in photosynthesis research: Proceedings of the 7th International Congress on Photosynthesis, Vol. 2. Elsevier, Amsterdam, pp. 665–668.

Kramer G.F., Krizek D.T. and Mirecki R.M. (1992). Influence of photosynthetically active radiation and spectral quality on UVB-induced polyamine accumulation in soybean. *Phytochemistry*, 31:1119–1125.

Krupa Z., Oquist G. and Hunger N.P.A. (1993). The effect of cadmium on photosynthesis of *Phaseolus vulgaris*—a fluorescence analysis. *Physiol. Plant.*, 88:626–630.

Lagriffoul A., Mocquot B., Mench M. and Vangronsveld J. (1998). Cadmium toxicity effects on growth, mineral and chlorophyll contents, and activities of stress-related enzymes in young maize plants (*Zea mays* L.). *Plant Soil*, 200:241–250.

Lee K.C., Cunningham B.A., Paulsen G.M. et al. (1976). Effects of cadmium on respiration rate and activities of several enzymes in soybean seedlings. *Physiol. Plant.*, 36:4–6.

Lidon F.C., Ramalho J.C. and Henriques F.S. (1993). Copper inhibition of rice photosynthesis. *J. Plant Physiol.*, 142:12–17.

Maksymiec W., Russa R., Urbanik-Symniewska T. and Baszynski T. (1993). Effect of excess Cu on the photosynthetic apparatus of runner bean leaves treated at two different growth stages. *Physiol. Plant.*, 91:715–721.

Marschner H. (1993). General introduction to the mineral nutrition of plants. In: Encyclopaedia of Plant Physiology, Vol. 15A: Inorganic plant nutrition. Lauchli A. and Bieleski R.L. (eds.). Springer-Verlag, Berlin, pp. 5–60.

Mench M., Vangronsveld J., Didier V. and Clijsters H. (1993). Evaluation of metal mobility, plant availability and immobilization by chemical agents in a limed-silty soil. *Environ. Pollut.*, 86:279–286.

Mocquot B., Vangronsveld J., Clijsters H. and Mench M. (1996). Copper toxicity in young maize (*Zea mays* L.) plants: effects on growth, mineral and chlorophyll contents, and enzyme activities. *Plant Soil*, 182:287–300.

Ogren E. (1991). Prediction of photoinhibition of photosynthesis from measurements of fluorescence quenching components. *Planta*, 184:538–544.

Ouzounidou G., Moustakas M. and Lannoye R. (1993). Chlorophyll fluorescence and photoacoustic characteristics in relation to changes in chlorophyll and Ca^{2+} content of a Cu-tolerant *Silene compacta* ecotype under Cu treatment. *Physiol. Plant.*, 93:551–557.

Passow H. and Rothstein A. (1960). The binding of mercury by the yeast cell in relation to changes in permeability. *J. Gen. Physiol.*, 43:621–633.

Peterson P.J. (1993). Metal pollution tolerance. In: Plant Adaptation to Environmental Stress. Mansfield T.A., Fowden L. and Stoddart J. (eds.). Chapman and Hall, London, pp. 171–188.

Peterson P.J. and Butler G.W. (1967). Significance of selenocystathionine in an Australian selenium-accumulating plant *Neptunia amplexicaulis*. *Nature*, 219:599–600.

Peterson P.J. and Butler G.W. (1971). The occurrence of selenocystathionine in *Morinda reticulata* Bench. a toxic seleniferous plant. *Aust. J. Biol. Sci.*, 24:175–177.

Priebe A., Klein H. and Jager H.J. (1978). Role of polyamines in SO_2-polluted pea plants. *J. Exp. Bot.*, 39:1045–1050.

Rabe R. and Kreeb K.H. (1979). Enzymatic activities and chlorophyll and protein content in plants as indicators of air pollution. *Environ. Pollut.*, 19:119–137.

Rauser W.E. (1984). Copper-binding protein and copper tolerance in *Agrostis gigantea*. *Plant Sci.*, 33:239–247.

Rause W.E. (1987). Changes in glutathione content of maize seedlings exposed to cadmium. *Plant Sci.*, 51:171–175.

Rauser W.E. (1990). Phytochelatins. *Annu. Rev. Biochem.*, 59:61–86.

Reese R.N., White C.A. and Winge D.R. (1992). Cadmium-sulfide crystallites in Cd-(γEC) nG peptide complexes from tomato. *Plant Physiol.*, 98:225–229.

Ros R., Cooke D.T., Burden R.S. and James C.S. (1990). Effects of the herbicide MCPA and the heavy metals cadmium and nickel on the lipid composition, Mg^{2+}-ATPase activity and fluidity of plasma-membranes from rice *Oryza sativa* (cv Bahia) shoots. *J. Exp. Bot.*, 41:457–462.

Rüegsegger A., Schmutz D. and Brunold C. (1990). Regulation of glutathione synthesis by cadmium in *Pisum sativum* L. *Plant Physiol.*, 93:1579–1584.

Samson G. and Popovic R. (1988). Use of algal fluorescence for determination of phytotoxicity of heavy metals and pesticides as environmental pollutants. *Ecotoxicol. Environ. Saf.*, 16:272–278.

Schaaper R.M., Koplitz R.M., Tkeshelashvili L.K. and Loeb L.A. (1987). Metal-induced lethality and mutagenesis: possible role of apurinic intermediates. *Mutat. Res.*, 177:179–188.

Schat H. and Kalff M.A.A. (1992). Are phytochelatins involved in differential metal tolerance or do they merely reflect metal imposed strain? *Plant Physiol.*, 99:1475–1480.

Scheller H.V., Huang B., Hatch E. and Goldsbrough P.B. (1987). Phytochelatin synthesis and glutathione levels in response to heavy metals in tomato cells. *Plant Physiol.*, 85:1031–1035.

Sharma C.B.S.R. and Panneerselvam M. (1990). Genetic toxicology of pesticides in higher plant systems. *Plant Sci.*, 9:409–442.

Speiser D.M., Abrahamson S.L., Banuelos G. and Ow D.W. (1992). *Brassica juncea* produces a phytochelatin-cadmium-sulfide complex. *Plant Physiol.*, 99:81–821.

Steffens J.C. (1990). The heavy metal-binding peptides of plants. *Annu. Rev. Plant Physiol. Plant Mol. Biol.*, 41:553–575.

Tomsett A.B., Salt D.E., De Miranda J. and Thurman D.A. (1989). Metallothioneins and metal tolerance. *Aspects Appl. Biol.*, 22:365–372.

Van Assche F. and Clijsters H. (1990). A biological test system for the evaluation of the phytotoxicity of metal-contaminated soils. *Environ. Pollut.*, 66:157–172.

Van Assche F., Cardinaels C. and Clijsters H. (1988). Induction of enzyme capacity in plants as a result of heavy metal toxicity: dose-response relations in *Phaseolus vulgaris* L. treated with zinc and cadmium. *Environ. Pollut.*, 52:103–115.

Van Assche F., Put C. and Clijsters H. (1986). Heavy metals induce specific isozyme patterns of peroxidase in *Phaseolus vulgaris* L. *Arch. Intern. Physiol. Biochim.*, 94:60.

Vangronsveld J. and Clijsters H. (1992). A biological test system for the evaluation of metal phytotoxicity and immobilization by additives in metal contaminated soils. In: Metal Compounds in Environment and Life. Interrelation between Chemistry and Biology. Merian E. and Haerdi W. (eds.). Science and Technology Letters, Northwood, pp. 117–126.

Vangronsveld J., Van Assche F. and Clijsters H. (1993). Reclamation of a bare industrial area, contaminated by non-ferrous metals: *in situ* metal immobilization and revegetation. *Environ. Pollut.*, 87:51–59.

Vangronsveld J., Mocquot B., Bench M. and Clijsters H. (1997). Biomarqueurs du stress oxydant chez les végétaux. In: Lagadic L., Caquet Th., Amiard J.C. and Ramade F. (eds.). Biomarqueurs en écotoxicologie: Aspects fondamentaux. Masson, Paris, pp. 165–184.

Vangronsveld J., Weckx Y., Kubacka-Zebalska M. and Clijsters H. (1993). Heavy metal induction of ethylene production and stress enzymes: II. Is ethylene involved in the signal transduction from stress perception to stress responses? In: Cellular and Molecular Aspects of the Plant Hormone Ethylene. Pech J.C., Latche A. and Balagué C. (eds.). Kluwer Academic Publishers, Dordrecht, pp. 240–246.

Verkleij J.A.C., Koevoets P., Vain 'T., Riet J. et al. (1990). Poly (γ-glutamylcysteinyl) glycine or phytochelatin and their role in cadmium tolerance of *Silene vulgaris*. *Plant Cell Environ.*, 13:912–921.

Wainwright S.J. and Woolhouse H.W. (1978). Inhibition by zinc of cell wall acid phosphatases from roots in zinc-tolerant and non-tolerant clones of *Agrostis tenuis* Sibth. *J. Exp. Bot.*, 29:525–531.

Walker D. (1988). Measurement of O_2 and chlorophyll fluorescence. In: Techniques in Bioproductivity and Photosynthesis. Coombs J., Hall D., Long S. and Scurlock J. (eds.). Pergamon Press, Oxford, pp. 95–106.

Weber M.B., Schat H. and Ten Bookum-Van Der Maarel D.C. (1991). The effect of copper toxicity on the contents of nitrogen compounds in *Silene vulgaris* (Moench) Garcke. *Plant Soil*, 133:101–109.

Weckx Y., Vangronsveld J. and Clijsters H. (1993). Heavy metal induction of ethylene production and stress enzymes: I. Kinetics of the responses. In: Cellular and Molecular Aspects of the Plant Hormone Ethylene. Pech J.C., Latché A. and Balagué C. (eds.). Kluwer Academic Publishers, Dordrecht, pp. 238–239.

Wellburn A.L., Capron T.M., and Chan H.S. and Horsman D.C. (1976). Biochemical effects of atmospheric pollutants on plants. In: Effects of Air Pollutants on Plants. Mansfield T.A. (ed.). Cambridge University Press, Cambridge, pp. 105–114.

9

Endocrine Biomarkers: Hormonal Indicators of Sublethal Toxicity in Fishes

A. Hontela

1. PHYSIOLOGICAL INTEGRITY AND RESISTANCE TO ENVIRONMENTAL XENOBIOTICS IN FISH

The environment is a multifactorial matrix in which an organism must, in order to ensure its survival, constantly prove its capacity for physiological homeostasis. Nychthemeral, seasonal, and annual changes in factors such as temperature, photoperiod, and availability of food imply a precise synchronization and adjustment of vital physiological functions. The endocrine system plays an important role in maintenance of physiological integrity by allowing, through its hormonal effectors, the regulation of homeostatic response of the organism in reaction to fluctuations of environmental factors. Osmoregulation, metabolism, growth, and reproduction are hormone-dependent physiological functions that are constantly modulated in response to environmental stimuli (Hontela and Stacey, 1990; Rankin and Jensen, 1993). Studies in environmental toxicology have proved the harmful effects of certain pollutants on the physiological performance of fishes and other aquatic organisms (Heath, 1987). The consensus that emerges, supported by an increasing number of experimental studies, is that certain pollutants disturb the endocrine function and cause physiological anomalies through a hormonal imbalance (Colborn and Clement, 1992). The sensitivity of the hormonal response, which generally precedes the expression of pathologies in tissues and organs or mortality (Hontela et al., 1993a), is a key characteristic of early indicators of damage induced by xenobiotics. Moreover, the hormonal action is an integrated process. It is characterized by the action of specific hormones on multiple cellular targets and by the regulation of hormone levels in blood at several levels (hypothalamus, hypophysis, and target glands). Therefore, any significant alteration of hormonal balance induced by xenobiotics may alter the physiological performance of the organism as well as its capacity to ensure its survival and transmission of its genetic patrimony. A pollutant can alter endocrine function by acting on processes of hormone synthesis and release in the hypothalamus, the anatomical and functional link with the nervous system, and similarly in the hypophysis and target glands, those that depend on hypophysial trophic hormones as well as others (Fig. 1). By its toxic action on the kidney or liver, a pollutant can also alter the hormonal balance by modifying the metabolism of hormones, their half-life in blood, their activities, and their rate of elimination.

In this chapter, two major themes are addressed. First, we consider how the hormonal status of an organism challenged by environmental pollutants might provide biomarkers of its early physiological response. Second, we examine how the hormonal biomarkers can be used to evaluate the physiological competence and health of the organism. To

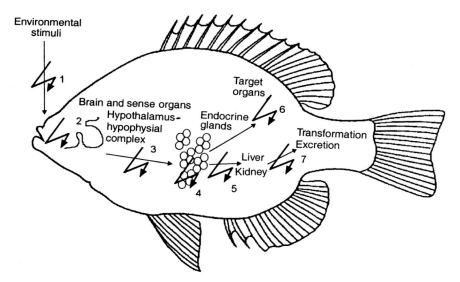

Fig. 1. Potential sites of toxic action of pollutants on the endocrine system in fish. 1, perception of environmental stimuli by the nervous system; 2, synthesis and release of hypothalamic hormones (RH/IH); 3, synthesis and release of hypophysial hormones (e.g., GTH, ACTH, TSH); 4, response of endocrine glands under the control of the hypophysis (e.g., gonads, interrenal tissue, thyroid gland) and other glands (e.g., endocrine pancreas); 5, synthesis and release of hormones by the endocrine glands; 6, response of target organs to hormones; 7, metabolism and elimination of hormones by excretory organs.

illustrate these themes, three endocrine systems will be considered, using teleost fishes as a model: osmoregulatory hormones and their response to acid pH, reproductive hormones and their response to industrial pollutants, and finally cortisol and thyroidal hormones and their response to heavy metals and organic pollutants. Standardized measurement protocols and statistical analysis of data are discussed in the last section in order to facilitate the use of hormonal biomarkers *in situ*.

2. HORMONAL BIOMARKERS IN FISHES

2.1 Use of Hormonal Indicators in Detection of Physiological Anomalies Caused by Acid pH

2.1.1 Experimental Laboratory Studies

Acidification of surface waters has caused a decline in fish stocks in Canada as well as Scandinavia, and species of the Salmonidae are the most vulnerable from this point of view (Harvey and Lee, 1982). Partly because of the enormous economic impact of this phenomenon, the physiological response to acid pH in teleosts has been rapidly elucidated owing to the intensive efforts of several researchers in experimental physiology. Laboratory studies, including ours, have demonstrated that the vital function altered by exposure to acid water is osmoregulation and that, once a toxic pH threshold has been exceeded, the homeostatic mechanisms activated in response to this stress of the osmo-ionic type are overloaded (Wood, 1987). Acid pH causes mortalities mainly by increase of losses of plasma electrolytes, especially Na^+ and Cl^-, reduction of gaseous exchanges across the gills, and modification of ventilation rate. The cause of mortality in fishes exposed to acid pH is therefore an osmo-ionic failure.

Osmoregulatory hormones play an important role in the regulation of hydric and ionic equilibrium in fish living in fresh water, a hypoosmotic medium that tends to dilute the internal medium of the fish by favouring the entry of water and loss of electrolytes (Rankin and Jensen, 1993). The elimination of water and active reabsorption of ions are determined by hormones that regulate exchanges of ions and water across the membranes of the osmoregulatory organs, especially gills and kidney. Acid pH impairs the homeostatic efforts of fishes mainly because it reduces the active reabsorption of ions by the gills (Wood, 1987). The secretion of prolactin increases in fishes exposed to acid pH in the laboratory (Wendelaar Bonga et al., 1988) and plasma cortisol also increases (Brown et al., 1984). Chevalier et al. (1986) reported a significant increase in urotensins, osmoregulatory hormones stored in the urophysis of the caudal neurosecretory system, in brook trout (*Salvelinus fontinalis*) from acidified lakes of the Canadian Shield (Quebec). Moreover, they established a link between aluminium levels in the gills, metal bioavailability by acidification of soils, and the severity of pathologies in the gills (Chevalier et al., 1985). A laboratory study further established, from dose-response relationships, strong correlations between urotensin I content in the urophysis and water pH, on the one hand, and between urotensin II content and total concentration of aluminium in water, on the other (Hontela et al., 1989). An increase has also been detected in hypothalamic and hypophysial contents of another osmoregulatory hormone, the neurohormone arginine vasotocine (AVT), in *Salvelinus fontinalis* exposed to acid pH and aluminium in the laboratory (Hontela et al., 1991). This hormone, linked to specific branchial receptors (Guibboline et al., 1989), induces a strong vasoconstriction of branchial lamellae (Bennett and Rankin, 1986) and reduces the entry of water through the gills (Lahlou and Giordan, 1970). The synthesis, storage, and secretion of AVT increase with time in freshwater fishes (Perrot et al., 1991) and the net effect of AVT is to prevent dilution of the internal medium. Since acid pH induces a dilution of the extracellular fluid by favouring a loss of electrolytes and reducing their active reabsorption by the gills, it has been possible to formulate the hypothesis that increase in AVT level constitutes a hormonal biomarker of stress caused by acid pH.

2.1.2 Validation of AVT as Biomarker of Toxic Stress Caused by Acid pH *in Situ*

AVT levels have been determined in the hypophysis and hypothalamus of *Salvelinus fontinalis* taken from six lakes of the Canadian Shield, representing an acidity gradient ranging from a nearly neutral pH (6.9) to a pH of 5.0, which is considered sublethal for that species (Menendez, 1976). The lakes were chosen on the basis of physico-chemical data such as hardness of water (mg $CaCO_3$/l), colour (expressed in ng Pt/l), pH, and total aluminium. Data were obtained from the Ministère des Loisirs, de la Chasse et de la Pêche (Quebec), which performs environmental monitoring of this region. The levels of total aluminium in the water are covariant with the pH, the acid lakes being rich in aluminium, as has been shown by other authors (Schofield and Trojnar, 1980). Since photoperiod and temperature, two environmental variables that change with the season, modify AVT levels in fishes (Hontela and Lederis, 1985), all the lakes were sampled in spring, summer, and autumn to detect possible seasonal variations in AVT levels, which could mask the effect of pH. The number of fish sampled was 25–45 fish/lake/campaign.

The effects of factors contributing to the variation in AVT, especially the season, size of fish, and pH, were quantified by covariance analysis, and it was demonstrated that the AVT levels in fish from lakes with pH lower than 5.6 were higher (Table 1) than in fish from lakes with pH higher than or equal to 5.6 (Hontela, 1993b). Moreover, the results obtained in fish from acidified lakes confirmed the data obtained in laboratory studies, since the AVT levels in fish exposed to a pH of 5.6 in laboratory and in the field were similar (Hontela et al., 1991, 1993b). Our studies thus demonstrated, by means of laboratory experiments and field studies, that increase in hypothalamic AVT constitutes a hormonal biomarker of toxic stress caused by acid pH and aluminium in *Salvelinus fontinalis*. The response constitutes an early warning since the fish sampled did not show evident external pathologies. The AVT was measured in endocrine tissues and the fish had to be sacrificed to sample these tissues in our studies, because the technique of radioimmunological assay (RIA) used was not sensitive enough to measure AVT in the blood. Future studies must allow measurement of plasma or seric AVT, and the release of fish after taking a blood sample, in order to minimize the effect of sampling on vulnerable populations.

2.2 Use of Hormonal Markers for the Detection of Anomalies Caused by Organic Pollutants on Reproduction

2.2.1 Experimental Laboratory Studies

Among the various biotic and abiotic conditions that must be met to ensure reproduction of a species (e.g., production of viable gametes,

Table 1. Effect of pH and size on levels of arginine vasotocine (AVT) in brook trout, *Salvelinus fontinalis*, in acidified lakes of the Canadian Shield (Hontela et al., 1993b).

Lake	Slope of AVT-size relation[1]	pH	Al monomer (μg/L)	Colour (ng Pt/L)	Parr (CPUE[2])
Welly	1.809	5.3	65	10	0.04
Brochard	1.668	5.0	75	16	0
Pyrolle	1.389	5.5	38	38	0
Franciscain	1.167	5.8	29	50	2.2
de la Loutre	1.184	6.9	7.5	17	2.0
Demarest	0.960	6.3	8.5	29	0.2

1) log AVT = 1.849 + [slope × length (m)], from Hontela et al. (1993b).
2) number (expressed in units of CPUE, capture per unit of effort) of Parr captured in a snare of 15 m × 2 m/h.

presence of a substrate adequate for the eggs or their oxygenation), normal synthesis and secretion of hormones of the hypothalamo-hypophysio-gonadal axis are the most important. The levels of hypothalamic neurohormones (GnRH and GRIF, gonadotropin releasing and inhibitory hormones), hypophysial hormones (GTH, gonadotropins), and sexual steroids of the androgen, estrogen, and progesterone type, are subject to a precise regulation, and changes in the available concentrations at the level of their receptors significantly modify the function of target tissues. Several studies have reported anomalies at the hypothalamo-hypophysio-gonadal axis in fish exposed to pollutants. It has been observed that synthesis of steroid hormones and/or the capacity to produce viable gametes are reduced in different species of teleosts after exposure to cadmium (Sangalang and Freeman, 1974; Das, 1988), malathion and hexachlorocyclohexane (Singh and Singh, 1987), cyanide (Ruby et al., 1987), or polychlorobiphenyls (PCB; Chen et al., 1986). The laboratory studies of Thomas (1989, 1990) helped to identify the cellular targets of pollutants in the hypothalamo-hypophysio-gonadal axis. Pollutants such as PCB, lead, or benzo(a)pyrene interfere with the functioning of the axis. These effects may occur in the hypothalamus as well as in the hypophysis or gonads. The studies cited above were all conducted in the laboratory, under controlled experimental conditions. They enabled the establishment of mechanisms of toxic action of pollutants and the demonstration that several xenobiotics decrease the secretion of hormones required for reproduction.

2.2.2 Validation of Sexual Steroids as Biomarkers of Reproductive Anomalies *in Situ*

A large number of studies have been dedicated to evaluation of the impact of chlorine-bleached kraft mill effluents on fish reproduction. The importance of this industry in Canada as well as Scandinavia, the quantities of effluents discharged into water courses, and the presence of highly toxic organochlorine pollutants in these effluents (Hodson et al., 1992) justify this *in situ* evaluation of the effects of organochlorine substances on fish physiology. Many effects have been listed during these studies. They range from increasing mortality in exposed fish populations (Sandstrom and Thoresson, 1990) to reduction in their fertility (Munkittrick et al., 1991). Studies that demonstrated the existence of anomalies in the synthesis of sexual steroids in fish exposed to organochlorine effluents in the Canadian Great Lakes are particularly useful because they apply to *in situ* studies, as well as laboratory experiments. Thus, it has been demonstrated that the components of kraft mill effluents, some of which have a high estrogenic activity, act on the hypothalamo-hypophysio-gonadal axis at several levels and significantly reduce the synthesis of specific sexual steroids (Van der Kraak et al., 1992; McMaster et al., 1992). Other studies have linked these endocrine anomalies to a reduction in fertility, measured by production of viable gametes and survival of juvenile stages, in the same populations. Apart from harmful effects on fertility, alterations in synthesis of sexual steroids can also have consequences on the sexual development of fishes. The feminization of males in various species of the family Poecilidae, associated with an abnormal synthesis of sexual steroids, was diagnosed in fish caught downstream of a paper factory in Florida (Davis and Bortone, 1992).

To validate sexual steroids as hormonal biomarkers of reproductive anomalies in fishes, it is necessary to quantify more precisely the relationship between decrease in plasma levels of sexual steroids in a population and reduction of its reproductive success. It is also essential to determine the threshold of plasma sexual steroids needed to ensure normal reproduction. It should be noted that there are fewer studies relating to effects of pollutants on gonadotropin secretion than on sexual steroids. The great specificity of gonadotropins in different species of teleosts is a limiting factor, since specific antibodies needed for the conventional measurement method (RIA) are not available for many families, other than Salmonidae and Cyprinidae. On the other hand, the similarity of the chemical structure of sexual steroids in different species is an important characteristic of these biomarkers since the same methods of detection can be used in all the species.

2.3 Use of Plasma Cortisol and Thyroid Hormones in Detection of Physiological Anomalies Caused by Organic Pollutants and Heavy Metals

Hormonal biomarkers can be used not only to detect early response to a very specific pollutant, as in the case of osmoregulatory hormones and acid pH, for example, but also to evaluate the physiological performance associated with a specific function. This is the case of reduction in levels of plasma sex steroids in fishes exposed to organochlorine substances. The last series of investigations that will be mentioned here concerns the studies that have demonstrated an endocrine anomaly in the process of secretion of cortisol and thyroid hormones in fishes exposed *in situ* to organic pollutants and heavy metals.

2.3.1 Experimental Laboratory Studies

Cortisol is a corticosteroid hormone synthesized in teleosts by interrenal tissue in response to a stimulation by adrenocorticotropic hormone (ACTH). Many experimental studies in comparative endocrinology have demonstrated that a physiologically healthy fish has the capacity to increase its plasma cortisol in response to general stress such as handling or capture (Schreck, 1990; Barton and Iwama, 1991; Pickering, 1993). Plasma cortisol also increases following acute exposure to a chemical stress due to exposure to certain pollutants such as heavy metals and polycyclic aromatic hydrocarbons (PAH; Thomas, 1990; Gill et al., 1993; Folmar, 1993). In our laboratory, we exposed rainbow trout, *Oncorhynchus mykiss*, to sublethal doses of inorganic mercury ($HgCl_2$) or organic mercury (CH_3HgCl). Mercury induced a significant increase in plasma cortisol, organic mercury being more toxic than inorganic mercury (Bleau et al., 1996). Similar results were obtained with cadmium. Fish exposed up to 7 d at sublethal doses showed an increase in level of plasma cortisol (Hontela et al., 1996). These studies show that the increase in plasma cortisol is one of the primary stages in the neuroendocrine and metabolic sequence activated in reaction to a stress (Donaldson et al., 1984). Cortisol, a metabolic hormone involved in neoglucogenesis and lipolysis (Sheridan, 1986; Vijayan and Moon, 1992), plays a role in the maintenance of plasma glucose, one of the energy substrates that can be used in reaction to stress, and mediates secondary responses to stress (Mazeaud and Mazeaud, 1981).

A question the ecotoxicologist must address is the role of cortisol in chronic exposures to sublethal chemical stress that, in a sedentary fish, can last for months, even years. There is very little information available on the cortisolic status of fish from contaminated environments. Cortisol is implicated, either by synergy with other hormones, or by direct effects, in regulation of several vital physiological functions such as growth, reproduction, and resistance to disease, functions that are all potentially disturbed by pollutants. Cortisol produces antigonadal effects (Carragher and Sumpter, 1990) and is a known immunosuppressant in mammals as well as fish (Maule et al., 1989). It plays an important role in metabolism, because it can stimulate the catabolism of proteins and decrease muscular mass as well as somatic growth in fishes (Freeman and Idler, 1974; Davis et al., 1986; Barton et al., 1987). Significant synergies have been demonstrated, particularly for cortisol and thyroid hormones. Thyroxine (T4) stimulates secretion of cortisol in interrenal tissue (Young and Lin, 1984) and cortisol increases the clearance of triiodothyronine (T3) in rainbow trout (Brown et al., 1991).

The thyroid gland is involved in the hormonal response to pollutants, sometimes in the same way as interrenal tissue. Anomalies of thyroid function, characterized by a decrease in plasma levels of thyroid hormones, have been detected in fish exposed to mercury (Kirubagaran and Joy, 1989), pesticides (Yadav and Singh, 1987; Ram et al., 1989; Sinha and Singh, 1991), and industrial pollutants (Bhattacharya et al., 1989). Studies conducted in our laboratory demonstrated that acute exposure to heavy metals (cadmium and mercury) increased plasma levels of thyroid hormones, whereas subchronic exposures (up to 30 d) reduced them (Ricard et al., 1998; Hontela, 1998) in rainbow trout. The mechanisms responsible for this reduction of plasma levels of thyroid hormones have not been elucidated in fishes. It has been demonstrated in mammals that certain industrial pollutants modify the transport of thyroid hormones in blood by acting against transport proteins (van den Berg et al., 1991) and favouring thus the elimination of non-bound hormones. Moreover, stimulation of metabolism of thyroid hormones by hepatic enzymes such as UDP-GT (uridyl diphosphate glucuronyl transferase) may be induced by pollutants and accelerate excretion of metabolized thyroid hormones (Collins and Capen, 1980; Saito et al., 1991).

2.3.2 *In Situ* Validation of Plasma Cortisol and Thyroid Hormones as Biomarkers of Physiological Anomalies Caused by Organic Pollutants and Heavy Metals

Many studies on fish have demonstrated that hormonal response to an acute stress caused by capture and sampling is characterized by an increase in cortisol and plasma glucose (Haux et al., 1985; Barton and Iwama, 1991). This hormonal response (Fig. 2) has been used by our team to evaluate the functional integrity of the hypothalamo-hypophysio-interrenal axis in fish living in environments contaminated by industrial pollutants to test the hypothesis that chronic exposure to pollutants induces anomalies at the level of cortisol secretion. Individuals of two species, yellow perch, *Perca flavescens*, and northern pike, *Esox lucius*, were collected, following standardized procedures of capture and handling, in sites contaminated by a mixture of PCB, PAH, and heavy metals (essentially mercury and cadmium) in the Saint Lawrence River and in reference sites. The capacity to increase plasma cortisol in response to the acute stress of capture and sampling proved to be higher in fish from reference sites (Fig. 3; Hontela et al., 1992, 1995). Cortisol failure was detected in sexually mature male and female fish, as well as in immature fish. These fish also showed very low plasma levels of thyroxine (Hontela et al., 1995). Cortisol impairment was also diagnosed in individuals of both species sampled by the same method downstream of a kraft mill located on the Saint Maurice River in Quebec (Fig. 3; Hontela et al., 1997). Lockhart et al. (1972) reported a similar cortisol impairment in pike sampled in a lake highly contaminated with mercury (16–25 mg total Hg/kg muscle). Fish from this site showed lower levels of plasma cortisol and glucose than fish from a reference lake. In the same study, contaminated fish transferred into a reference lake and recaptured one year later had a lower increase in their cortisol level than fish that were never subjected to contamination.

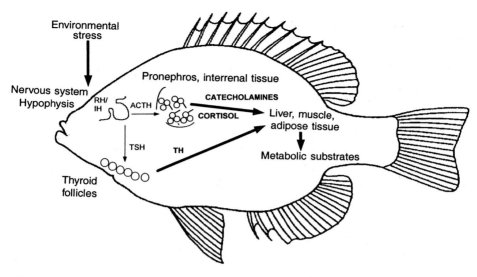

Fig. 2. Neuro-endocrine and metabolic response to an environmental stress in fish. Discharge of cortisol by interrenal tissue and increase in plasma levels of this hormone constitute the normal response of a healthy fish. Cortisol, catecholamines, and thyroid hormones (TH) mobilize energy substrates by stimulating gluconeogenesis, lipolysis, and glycogenolysis.

Some laboratory studies confirm the hypothesis that chronic exposures to pollutants induce cortisol failure in fishes. Kirubagaran and Joy (1991) exposed catfish, *Clarias batrachus*, to mercury for 90 d. They observed a very low secretion of cortisol and morphological alterations in hypophysis and interrenal tissue. The cortisol failure, characterized by a reduced capacity to increase plasma cortisol following an acute stress, was associated with specific pathologies in the hypophysial corticotropic cells that secrete ACTH and in the interrenal steroidogenic cells (Hontela et al., 1992, 1997). Using digitalized image analysis, we demonstrated that in fish from contaminated sites, the interrenal tissue and the hypophysial corticotropic zone have atrophied cells with small nuclei and large intercellular spaces (Fig. 4).

The mechanisms responsible for cortisol impairment in fish exposed to pollutants *in situ* have not yet been identified. The atrophy of endocrine cells involved in the regulation of plasma cortisol can be caused by the functional exhaustion of the hypothalamo-hypophysio-interrenal axis due to exposure to pollutants and prolonged stimulation. A direct toxic action on cells of this axis can also cause cortisol impairment. DDT and its metabolites seem to have such a cyto-specific action. Ilan and Yaron (1983) exposed *Sarotherodon mossambicus* to DDT and to its metabolite, DDE. They observed a reduction of cortisol secretion in the

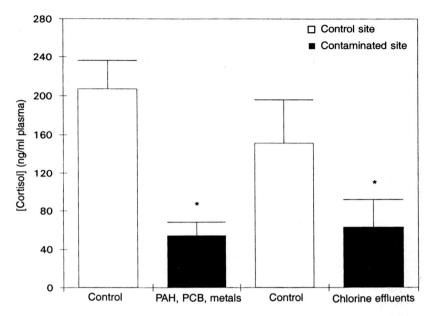

Fig. 3. Concentration of plasma cortisol (mean ± standard error) in the teleost *Perca flavescens*, sampled according to standardized protocols in sites contaminated by a mixture of organic pollutants and heavy metals, in sites contaminated by kraft mill bleached effluents, and in respective control sites. Fish from contaminated sites showed cortisol impairment characterized by a dramatic decrease in the capacity to increase plasma cortisol levels in reaction to a standardized stress of capture and handling.

treated fish. Later studies on birds and mammals reported relevant data to elucidate the mode of action of certain pollutants on the process of cortisol secretion. Metabolites of DDT, especially DDE, bind by covalent bonds to cells of the *zona fasciculata*, the equivalent of interrenal tissue of fish, and cause microhemorrhages (Jonsson et al., 1993). We have examined the possibility that certain pollutants, including DDT, that are frequently detected in the Saint Lawrence River at very low levels can exert direct toxic effects on the interrenal tissue of fish. *In vitro* studies provided evidence that DDT and its metabolites can impair cortisol secretion (Benguira and Hontela, 2000).

In characterizing a biomarker, it is important to quantify the link between its response and exposure, in the case of a marker of exposure, or an effect, in the case of a marker of effect. Since cortisol and thyroid hormones have well-established functions in metabolism, especially gluconeogenesis and lipolysis for cortisol (Sheridan, 1986; Vijayan and Moon, 1992), we have studied certain metabolic parameters in fish with cortisol failure and in control fish. We have thus discovered anomalies in metabolism of carbohydrates, specifically a hepatic glycogen level

Interrenal tissue with cortisol impairment   Normal interrenal tissue

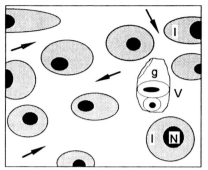
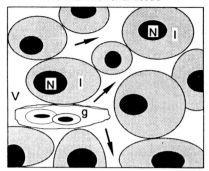

Fig. 4. Steroidogenic interrenal cells in a fish with cortisol failure sampled from a contaminated site and in a healthy fish sampled from a control site. The interrenal cells (I) and their nuclei (N) are smaller, and the intercellular spaces (indicated by arrows) are larger in the fish with cortisol failure than in fish having a normal capacity to increase the plasma cortisol. V, blood vessel; g, red blood cell.

higher in fish with cortisol impairment than in control fish (Hontela et al., 1995). Similar problems were detected in perch exposed to kraft mill effluents in the Baltic Sea (Andersson et al., 1988). Studies comparing growth rate and efficiency of metabolic conversion in fish from contaminated sites and fish from reference sites revealed lower growth efficiency in fish exposed to contaminants (Sherwood et al., 2000).

3. ELEMENTS FOR SETTING UP A STRATEGY FOR USE OF HORMONAL BIOMARKERS *IN SITU*

3.1 Physiological and Ecological Parameters to be Considered

The use of endocrine markers is justified by the significant role that the endocrine system plays in the maintenance of physiological homeostasis (Mayer et al., 1992). It is well established by clinical empirical studies and by experimental studies on different types of vertebrates that a significant hormonal imbalance can lead to health problems. Because of its sensitivity and integrating role, hormonal response represents an early response of the organism, which precedes the expression of visible pathologies and serious physiological problems. The great challenge that ecotoxicologists face is the detection of endocrine anomalies for environmentally realistic pollutant concentrations, in a complex and variable environment. A multidisciplinary approach and the close

˙collaboration of ecologists, chemists, and ecophysiologists will permit the implementation of effective tools and the planning of adequate sampling campaigns (Kendall and Lacher, 1991). It could be important to choose a sedentary species that accumulates pollutants independently of the temporal variation of contamination level.

The yellow perch is a good sentinel species for ecotoxicological studies in the temperate zone of Canada, since its behaviour favours prolonged stays in well-defined sites (Fortin and Magnin, 1972; Aalto and Newsome, 1990). It is critical to choose a species sufficiently sensitive to pollutants. The yellow perch seems to satisfy this requirement since a study of the dynamics of a population of *Perca fluviatilis* exposed to kraft mill effluents indicated increased mortality of juvenile stages (Karas et al., 1991). Similarly, the pike is a convenient species for ecotoxicological studies since it is relatively sedentary and it is a carnivore located at higher levels of the food web (Toner and Lawler, 1969).

3.2 Statistical Treatments

The use of biomarkers relies on the use of sampling protocols that allow the detection of a signal in spite of interference and of the natural variability of the response. This consideration applies to endocrine markers used *in situ*. The great number of species, their diversified responses to environmental conditions such as photoperiod, temperature, or pH, and differences in sexual maturity at the time of sampling are variables that may modify the response to pollutants and may also modify the hormonal response. Several experimental approaches have been presented in this chapter. We have seen that the levels of the osmoregulatory hormone AVT vary not only with water pH but also with the season and the size of the fish. Sampling of fish of a particular size for three seasons and statistical treatment of data based on the use of covariance analysis (ANCOVA) gave us the opportunity to quantify the variation of AVT levels due to pH and to identify AVT as a hormonal biomarker of stress caused by acidity.

The use of cortisol impairment as a biomarker must rely on very specific sampling protocols since several factors influence cortisol secretion. The fish must be captured in the same way in all the sites, preferably with nets to avoid injury. They must be sampled at the same time in all the sites, since the existence of nychthemeral variations in plasma cortisol level has been demonstrated (Peter et al., 1978). The time of sampling, the period between capture and confinement in an enclosure before a blood sample is taken, the time spent in the enclosure during the sampling of blood, and the means by which the blood sample is taken (with or without anaesthesia) are variables that are important but easily controlled. It is preferable to sample a very small number of fish each day and to ensure that the protocols are precisely followed. Using such protocols, it has been possible to predict an endocrine anomaly in fish exposed to pollutants, which leads to a decline in the physiological competence of the fish, particularly in its metabolism.

3.3. Non-invasive Methods

Non-invasive methods, especially the use of blood parameters, are particularly important in evaluating the state of health of vulnerable and threatened species (Fossi and Leonzio, 1993). Endocrine biomarkers can be determined by means of blood parameters and the health of organisms can thus be evaluated without sacrificing them. Moreover, it gives the opportunity to study the same organisms over a long period. This method can be very useful, especially for evaluation of procedures to decontaminate and restore polluted sites.

ACKNOWLEDGEMENTS

I thank my colleagues and collaborators who have contributed to the research presented in this chapter, particularly Professor Gaston Chevalier of the TOXEN Laboratory of Université du Quebec at Montreal, in which most of our studies were conducted, and Professor Joe Rasmussen of McGill University. I thank Mr Claude Daniel and Ms Anne Ricard for their technical assistance and Ms Michelle Dard-Girard for revising the text.

REFERENCES

Aalto S.K. and Newsome G.E. (1990). Additional evidence supporting endemic behaviour of a yellow perch *(Perca flavescens)* population. *Can. J. Fish. Aquat. Sci.,* 47:1959–1962.

Andersson T., Förlin L., Härdig J. and Larsson A. (1988). Physiological disturbances in fish living in coastal water polluted with bleached kraft pulp mill effluents. *Can. J. Fish. Aquat. Sci.,* 45:1525–1536.

Barton B.A. and Iwama G.K. (1991). Physiological changes in fish from stress in aquaculture with emphasis on the response and effects of corticosteroids. *Ann. Rev. Fish Dis.,* 1:3–26.

Barton B.A., Schreck C.B. and Barton L.D. (1987). Effects of chronic cortisol administration and daily acute stress on growth, physiological conditions, and stress responses in juvenile rainbow trout. *Disease Aquat. Org.,* 2:173–185.

Benguira S. and Hontela A. (2000). Adrenocorticotrophin-and cyclicadenosine 3′, 5′-monophosphate, stimulated cortisol secretion in interrenal tissue of rainbow trout exposed *in vitro* to DDT compounds. Environ. *Toxicol. Chem.,* 19: 842-847.

Bennett M.B. and Rankin J.C. (1986). The effects of neurohypophysial hormones on the vascular resistance of the isolated perfused gill of the European eel, *Anguilla anguilla. Gen. Comp. Endocrinol.,* 64:60–66.

Bhattacharya T., Bhattacharya S., Ray A.K. and Dey S. (1989). Influence of industrial pollutants on thyroid function in *Channa punctatus*. *Ind. J. Exp. Biol.*, 27:65–68.

Bleau H., Daniel C., Chevalier G. et al. (1996). Effects of acute exposure to mercury chloride and methyl mercury on plasma cortisol, T3, T4, glucose and liver glycogen in rainbow trout (*Oncorhynchus mykiss*). *Aquat. Toxicol.*, 34:221–235.

Brown S.B., Eales J.G., Evans R.E. and Hara T.J. (1984). Interrenal, thyroidal, and carbohydrate responses of rainbow trout (*Salmo gairdneri*) to environmental acidification. *Can. J. Fish. Aquat. Sci.*, 41:36–45.

Brown S.B., MacLatchy D.L., Hara T.J. and Eales J.G. (1991). Effects of cortisol on aspects of 3,5,3'-triiodo-L-thyronine metabolism in rainbow trout (*Oncorhynchus mykiss*). *Gen. Comp. Endocrinol.*, 81:207–216.

Carragher J.F. and Sumpter J.P. (1990). The effect of cortisol on the secretion of sex steroids from cultured ovarian follicles of rainbow trout. *Gen. Comp. Endocrinol.*, 77:403–407.

Chen T.T., Reid P.C., Van Beneden R. and Sonstegard R.A. (1986). Effect of Aroclor 1254 and Mirex on estradiol-induced vitellogenin production in juvenile rainbow trout (*Salmo gairdneri*). *Can. J. Fish. Aquat. Sci.*, 43:169–173.

Chevalier G., Gauthier L. and Moreau G. (1985). Histopathological and electron microscopic studies of gills of brook trout, *Salvelinus fontinalis*, in acidified lakes. *Can. J. Zool.*, 63:2062–2070.

Chevalier G., Gauthier L., Lin R. et al. (1986). Effect of chronic exposure to an acidified environment on the urophysis of the brook trout, *Salvelinus fontinalis*. *J. Exp. Biol.*, 45:291–299.

Colborn T. and Clement C. (1992). *Chemically-induced Alterations in Sexual and Functional Development: the Wildlife/Human Connection*, vol. XXI. Advances in Modern Environmental Toxicology. Princeton Scientific Publishing Co., Inc.

Collins W.T. Jr. and Capen C.C. (1980). Biliary excretion of [125]-I-thyroxine and fine structural alterations in the thyroid glands of Gunn rats fed polychlorinated biphenyls (PCB). *Lab. Invest.*, 43:158–164.

Das R.C. (1988). Cadmium toxicity to gonads in a freshwater fish, *Labeo bata* (Ham.) *Arch. Hydrobiol.*, 112:467–474.

Davis K.B., Torrance P., Parker N.C. and Shuttle M.A. (1985). Growth, body composition and hepatic tyrosine aminotransferase activity in cortisol-fed channel catfish, *Ictalurus punctatus*. *J. Fish Biol.*, 29:177–184.

Davis W.P. and Bortone S.A. (1992). Effects of kraft mill effluent on the sexuality of fishes: an environmental early warning? In: Chemically-induced Alterations in Sexual and Functional Development: the Wildlife/Human Connection, vol XXI. Advances in modern environmental toxicology. Colborn T. and Clement C. (eds.). Princeton Scientific Publishing Co., Princeton, pp. 113–127.

Donaldson E.M., Fagerlund U.H.M. and McBride J.R. (1984). Aspects of the endocrine stress response to pollutants in salmonids. In: Contaminant Effects on Fisheries. Adv. Environ. Sci. Technol., Vol. 16. Cairns V.V.W. (ed.).

Folmar L.C. (1993). Effects of chemical contaminants on blood chemistry of teleost fish: a bibliography and synopsis of selected papers. *Environ. Toxicol. Chem.*, 12:337–375.

Fortin R., and Magnin E. (1972). Dynamique d'un groupement de perchaude, *Perca flavescens* (Mitchill) dans la Grande Anse de l'Ile Perrot, au Lac Saint Louis. *Natur. Can.*, 99:36–380.

Fossi, C., and Leonzio C. (1993). *Nondestructive Biomarkers in Vertebrates*. Lewis Publishers, Boca Raton.

Freeman H.C. and Idler D.R. (1973). Effects of corticosteroids on liver transaminases in two salmonids, the rainbow trout (*Salmo gairdneri*) and the brook trout (*Salvelinus fontinalis*). *Gen. Comp. Endocrinol.*, 20:69–76.

Gill T.S., Leitner G., Porta S. and Epple A. (1993) Response of plasma cortisol to environmental cadmium in the eel, *Anguilla rostrata* Lesueur. *Comp. Biochem. Physiol.*, 104C:489–495.

Guibbolini M.E., Henderson I.W., Mosley W. and Lahlou B. (1989). Arginine vasotocin binding to isolated branchial cells of the eel: effect of salinity. *J. Mol. Endocrinol.*, 1:125–130.

Harvey H.H. and Lee C. (1982). Population responses of fishes to acidified waters. In: Acid rain/fisheries: Proceedings of an International Symposium on Acid Precipitation and Fishery Impacts in Northeastern North America. Johnson R.E. (ed.). American Fisheries Society, Bethesda, pp. 227–242.

Haux C., Sjöbeck M. and Larsson A. (1985). Physiological stress responses in a wild fish population of perch *(Perca fluviatilis)* after capture and during subsequent recovery. *Mar. Environ. Res.*, 15:77–95.

Health A.G. (1987). *Water Pollution and Fish Physiology*. CRC Press, Inc., Boca Raton.

Hodson P.V., McWhirter M., Ralph K. et al. (1992). Effects of bleached kraft mill effluent on fish in the St. Maurice River, Quebec. *Environ. Toxicol. Chem.*, 11:1635–1651.

Hontela A., Daniel C. and Ricard A.C. (1996). Effects of acute and subacute exposures to cadmium on the interrenal and thyroid function in rainbow trout, *Oncorhynchus mykiss. Aquatic Toxicol.*, 35:171–182.

Hontela A. and Lederis K. (1995). Diel variations in arginine vasotocin content of goldfish brain and pituitary: effects of photoperiod and pinealectomy. *Gen. Comp. Endocrinol.*, 57:397–404.

Hontela A. and Stacey N.E. (1990). In: Regulation of Reproductive Cycles in Cyprinidae. Munro A.D. et al. (eds.). CRC Press Inc., Boca Raton, Florida, pp. 53–77.

Hontela A., Roy Y., Van Coillie R. et al. (1989). Differential effects of low pH and aluminium on the caudal neurosecretory system of the brook trout, *Salvelinus fontinalis. J. Fish. Biol.*, 35:265–273.

Hontela A., Rasmussen J.B., Ko D. et al. (1991). Arginine vasotocin, an osmoregulatory hormone, as a potential indicator of acid stress in fish. *Can. J. Fish. Aquat. Sci.*, 48:256–263.

Hontela A., Rasmussen J.B., Audet C. and Chevalier G. (1992). Impaired cortisol stress response in fish from environments polluted by PAHs, PCBs, and mercury. *Arch. Environ. Contam. Toxicol.*, 22:278–283.

Hontela A., Rasmussen J.B. and Chevalier G. (1993a). Endocrine responses as indicators of sublethal toxic stress in fish from polluted environments. *Water Poll. Res. J. Can.*, 28:767–780.

Hontela A., Rasmussen J.B., Lederies K. et al. (1993b). Elevated levels of arginine vasotocin in the brain of brook trout *(Salvelinus fontinalis)* from acid lakes: a field test of a potential biomarker for acid stress. *Can. J. Fish. Aquat. Sci.*, 50:1717–1727.

Hontela A., Dumont P., Duclos D. and Fortin R. (1995). Endocrine and metabolic dysfunction in yellow perch, *Perca flavescens*, exposed to organic contaminants and heavy metals in the St. Lawrence river. *Environ. Toxicol. Chem.*, 14:725–731.

Hontela A. (1998). Interrenal dysfunction in fish from contaminated sites: *in vivo* and *in vitro* assessment. *Environ. Toxicol. Chem.*, 17:44–48.

Hontela A., Daniel C. and Rasmussen J.B. (1997). Structural and functional impairment of the hypothalamo-pituitary-interrenal axis in fish exposed to bleached kraft mill effluent in the St. Maurice River, Quebec. *Ecotoxicology*, 6:1–12.

Ilan Z. and Yaron Z. (1983). Interference of o,p DDD with interrenal function and cortisol metabolism in *Sarotherodon aureus* (Steindachner). *J. Fish. Biol.*, 22:657–669.

Jönsson C.J., Lund B.O. and Brandt I. (1993). Adrenolytic DDT-metabolites: studies in mink, *Mustela vison* and otter, *Lutra lutra. Ecotoxicology*, 2:41–53.

Karas P., Neuman E. and Sandström O. (1991). Effects of a pulp mill effluent on the population dynamics of perch, *Perca fluviatilis. Can. J. Fish. Aquat. Sci.*, 48:28–34.

Kendall R.J. and Lacher T.E. (1991). Ecological modeling, population ecology, and wildlife toxicology: a team approach to environmental toxicology. *Environ. Toxicol. Chem.*, 10:297–299.

Kirubagaran R. and Joy K.P. (1989). Toxic effects of mercurials on thyroid function of the catfish, *Clarias batrachus* L. *Ecotoxicol. Environ. Saf.*, 17:265–271.

Kirubagaran R. and Joy K.P. (1991). Changes in adrenocortical-pituitary activity in the catfish, *Clarias batrachus* (L.). after mercury treatment. *Ecotoxicol. Environ. Saf.,* 22:36–44.

Lahlou B. and Giordan A. (1970). Le contrôle hormonal des échanges et de la balance de l'eau chez le Téléostéen d'eau douce *Carassius auratus* intact et hypophysectomisé. *Gen. Comp. Endocrinol.,* 14:491–509.

Lockhart W.L., Uthe J.F., Kenny A.R. and Mehrle P.M. (1972). Methylmercury in northern pike *(Esox lucius):* Distribution, elimination, and some biochemical characteristics of contaminated fish. *J. Fish. Res. Board Can.,* 29:1519–1523.

Maule A.G., Tripp R.A., Kaattari S.A. and Schreck C.B. (1989). Stress alters immune function and disease resistance in chinook salmon *(Oncorhynchus tshawytscha). J. Endocrinol.,* 120:135-142.

Mayer, F.L., Versteeg D.J., McKee M.J. et al. (1992). Physiological and non-specific biomarkers. In: Biomarkers—Biochemical, Physiological, and Histological Markers of Anthropogenic Stress. Huggett R.J., Kimerle R.A., Mehrle P.M. and Bergman H.L. (eds.). Lewis Publishers, Boca Raton, pp. 5-85.

Mazeaud M.M. and Mazeaud F. (1981). Adrenergic responses to stress in fish. In: Stress and Fish. Pickering A.D. (ed.). Academic Press, London, New York, pp. 49–75.

McMaster, M.E., Van Der Kraak G.J., Portt C.B. et al. (1992). Changes in hepatic mixed-function oxygenase (MFO) activity, plasma steroid levels and age at maturity of a white sucker *(Catostomus commersoni)* population exposed to bleached kraft pulp mill effluent. *Aquat. Toxicol.,* 21:199–218.

Menendez R. (1976). Chronic effects of reduced pH on brook trout *(Salvelinus fontinalis). J. Fish. Res. Board Can.,* 33:118–123.

Munkittrick K.R., Port C.B., Van Der Kraak G.J. et al. (1991). Impact of bleached kraft mill effluent on population characteristics, liver MFO activity, and serum steroid levels of a Lake Superior white sucker *(Catostomus commersoni)* population. *Can. J. Fish. Aquat. Sci.,* 48:1371–1380.

Perrot M.N., Carrick S. and Balment R.J. (1991). Pituitary and plasma arginine vasotocin levels in teleost fish. *Gen. Comp. Endocrinol.,* 83:68–74.

Peter R.E., Hontela A., Cook A.F. and Paulencu C.R. (1978). Daily cycles in serum cortisol levels in the goldfish: Effects of photoperiod, temperature and sexual condition. *Can. J. Zool.,* 56:2443–2448.

Pickering A.D. (1993). Endocrine pathology in stressed salmonid fish. *Fish. Res.,* 17:35–50.

Ram R.N., Joy K.P. and Sathyanesan A.G. (1989). Cythion-induced histophysiological changes in thyroid and thyrotrophs of the teleost fish, *Channa punctatus* (Bloch). *Ecotoxicol. Environ. Saf.,* 17:272–278.

Rankin J.C. and Jensen F.B. (1993). Fish Ecophysiology. Chapman and Hall, London.

Ricard A.C., Daniel C., Anderson P. and Hontela A. (1998). Effects of subchronic exposure to cadmium chloride on endocrine and metabolic functions in rainbow trout, *Oncorhynchus mykiss. Arch. Environ. Contam. Toxicol.,* 34:377–381.

Ruby S.M., Idler D.R. and So Y.P. (1987). Changes in plasma, liver, and ovary vitellogenin in landlocked Atlantic salmon following exposure to sublethal cyanide. *Arch. Environ. Contam. Toxicol.,* 16:507–510.

Saito K., Kaneko H., Sato K. et al. (1991). Hepatic UDP-glucuronyltransferase(s) activity toward thyroid hormones in rats: induction and effects on serum thyroid hormone levels following treatment with various enzyme inducers. *Toxicol. Appl. Pharmacol.,* 111:99–106.

Sandström O. and Thoresson G. (1988). Mortality in perch populations in a Baltic pulp mill effluent area. *Mar. Pollut. Bull.,* 19:564–567.

Sangalang G.B. and Freeman H.C. (1974). Effects of cadmium on maturation and testosterone and 11-ketotestosterone production *in vitro* in brook trout. *Biol. Reprod.,* 11:429–435.

Schreck C.B. (1990). Physiological, behavioural, and performance indicators of stress. *Am. Fish. Soc. Symposium,* 8:29–37.

Sheridan M.A. (1986). Effects of thyroxin, cortisol, growth hormone, and prolactin on lipid metabolism of coho salmon, *Oncorhynchus kisutch*, during smoltification. *Gen. Comp. Endocrinol.,* 64:220–238.

Sherwood G.D., Rasmussen J.B., Rowan D.J., Brodeur J. and Hontela A. (2000). Bioenergetic costs of heavy metal exposure in yellow perch (*Perca flarvescens*): *in situ* estimates with a radiotracer (^{137}Cs) technique. *Can. J. Fish. Aquat. Sci.,* 57: 441–450.

Singh S. and Singh T.P. (1987): Impact of malathion and hexachlorocyclohexane on plasma profiles of three sex hormones during different phases of the reproductive cycle in *Clarias batrachus. Pestic. Biochem. Physiol.,* 27:301–308.

Sinha N., Lal B. and Singh T.P. (1991). Carbaryl-induced thyroid dysfunction in the freshwater catfish *Clarias batrachus. Ecotoxicol. Environ. Saf.,* 21:240–247.

Thomas P. (1989). Effects of Aroclor 1254 and cadmium on reproductive endocrine function and ovarian growth in Atlantic croaker. *Mar. Environ. Res.,* 28:499–503.

Thomas P. (1990). Molecular and biochemical responses of fish to stressors and their potential use in environmental monitoring. *Am. Fish. Soc. Symp.,* 8:9–28

Toner E.D. and Lawler G.H. (1969) Synopsis of biological data on the pike *Esox lucius. FAO Fisheries Synopsis* 30.

Van Den Berg K.J., Van Raaij J.A.G.M., Bragt P.C. and Notten W.R.F. (1991). Interactions of halogenated industrial chemicals with transthyretin and effects on thyroid hormone levels *in vivo. Arch. Toxicol.,* 65:15–19.

Van Der Kraak G.J., Munkittrick K.R., McMaster M.E. et al. (1992). Exposure to bleached kraft pulp mill effluent disrupts the pituitary-gonadal axis of white sucker at multiple sites. *Toxicol. Appl. Pharmacol.,* 115:224–233.

Vijayan M.M. and Moon T.W. (1992). Acute handling stress alters hepatic glycogen metabolism in food-deprived rainbow trout (*Oncorhynchus mykiss*). *Can. J. Fish. Aquat. Sci.,* 49:2260–2266.

Wendelaar Bonga S.E., Balm P.H.M. and Flik G. (1988). Control of prolactin secretion in the teleost *Oreochromis mossambicus:* effects of water acidification. *Gen. Comp. Endocrinol.,* 72:1–12.

Wood C.M. (1987). The physiological problems of fish in acid waters. In: Acid Toxicity in Aquatic Animals. Morris R., Brown D.J.A., Taylor E.W. and Brown J.A. (eds.). Soc. for Exp. Biol. Series, Cambridge University Press, 280 pp.

Yadav A.K. and Singh T.P. (1987). Pesticide-induced changes in peripheral thyroid hormone levels during different reproductive phases in *Heteropneustes fossilis. Ecotoxicol. Environ. Saf.,* 13:97–103.

Young G. and Lin R.J. (1988). Responses of the interrenal tissue to adrenocorticotropic hormone after short-term thyroxine treatment of coho salmon (*Oncorhynchus kisutch*). *J. Exp. Zool.,* 245:53–58.

10

Use of Retinoids as Biomarkers

P.A. Spear and D.H. Bourbonnais

INTRODUCTION

The term 'retinoids' refers to a group of compounds that present a molecular structures similar to that of retinol. Many of these compounds are capable of vitamin A activity in biological systems. Even though there are few data on the comparative biochemistry of retinoids, the available information suggests that the basic metabolic mechanisms of retinoids have been preserved during evolution, and that the presence of retinoids is essential to the proper function of all physiological systems of vertebrates that have been studied to date. Both vitamin A deficiency and excess may be harmful. To understand the effects of contamination during laboratory exposures and interpret data from field studies, we must rely, essentially, on our present state of knowledge of retinoids.

1. BIOCHEMISTRY AND BIOLOGICAL ROLE OF RETINOIDS

1.1 Absorption, Storage, and Metabolism

Studies on the biochemistry and biological role of vitamin A are well documented (see, for example, Sporn et al., 1984, 1994; Ganguly, 1989; Saurat, 1991).

Retinoids and their precursors (carotenoids) are not synthesized *de novo* in vertebrates, but are of plant origin and are assimilated by animals through food, in which their concentrations can vary according to geographic or seasonal fluctuations. Although they cannot synthesize vitamin A, animals can store relatively high quantities of inactive retinoids in organs such as the liver, which allows them to compensate for fluctuations in nutrient availability. The vitamin A thus stored can be mobilized to respond to certain physiological demands (e.g., increased requirements for egg formation) or to compensate for an insufficient nutritional input (e.g., during hibernation or migration in birds).

The process of retinoid storage in internal organs seems to go on during the entire lifetime of an animal. On the basis of the studies presently available, it seems that quantities of vitamin A stored are minimal in young animals and that they increase slightly during the phase of rapid growth. Although this tendency is based on information available for only a few species, it should be taken into account during implementation of studies using this type of biomarker.

Figure 1 presents a general outline of vitamin A metabolism. Table 1 presents examples of enzymes implicated in vitamin A metabolism for which adverse effects of xenobiotics have been observed. Food sources can contain substantial levels of plant-derived carotenoids, including provitamins that can be converted into retinoids. According to Underwood (1984), the most important sources of retinoids for animals are isoforms α, β, and γ of carotene, as well as cryptoxanthine. In the intestinal mucus, 15,15'-dioxygenase can convert β-carotene into retinaldehyde, which is then reduced to retinol under the reversible action of retinaldehyde reductase. In most animals, several natural xanthophylls (lycopene, zeaxanthine, and luteine, for example) cannot be metabolized by the 15,15'-dioxygenase. Unlike other vertebrates, fishes, in addition to producing substantial quantities of dehydroretinol from luteine and anhydroluteine, have the capacity to generate retinol from oxygenated carotenoids such as astaxanthine, cantaxanthine, and isozeanthine (Gross and Budowski, 1966; Barua, 1978; Olson, 1983, 1989). Lakshmanan et al. (1972, in Goodman and Blaner, 1984) remarked that the activity of 15,15'-dioxygenase increases as a function of the proportion of plant organic matter in the feeding regime (herbivore to omnivore to carnivore).

The esters of retinol, such as retinyl palmitate, are hydrolysed into retinol in the intestine. Following absorption in mucous cells, the retinol is re-esterified by lecithin retinolacyltransferase (LRAT) or by acyl-coenzyme A retinolacyltransferase (ARAT). The esters of retinol thus formed are incorporated in the chylomicrons that form in mucous cells, and are thus transported by the lymphatic system towards the liver, where they are stored in the hepatocytes and, to a greater extent, in the stellate cells. Retinol is also stored in these cell types and the interconversion of these forms is the result of hydrolysis of esters of retinol and retinolacyltransferase activities.

In fish, apart from retinol and esters of retinol, dehydroretinol as well as esters derived from dehydroretinol are also stored in the liver, the pyloric caecum, and the anterior part of the intestine (Ganguly, 1989; Ndayibagira et al., 1995). This is due to their specialized metabolism of

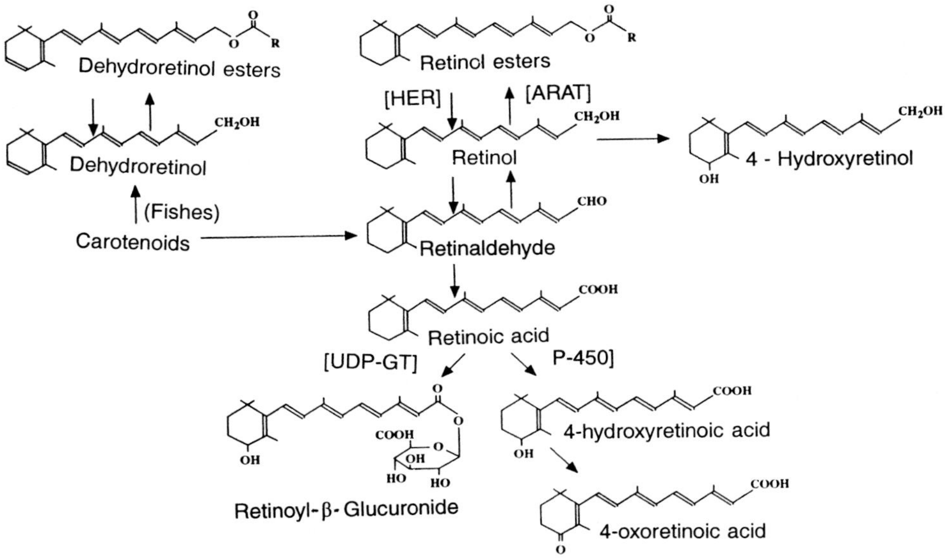

Fig. 1. Metabolic processes of vitamin A and enzymes that can be affected by exposure to xenobiotics. ARAT, acyl-coenzyme A retinolacyltransferase; HER, hydrolase of retinol esters; UDP-GT, uridine diphosphate-glucuronyltransferase; P-450, cytochromes P450.

Table 1. Examples of enzymes involved in the metabolism of vitamin A for which an effect of certain xenobiotics has been observed.

Enzymes	Xenobiotics	References
ARAT (acyl-coenzyme A retinol-acetyltransferase)	Coplanar biphenyls	Jensen et al., 1987; Narbonne et al., 1990
HER (hydrolase of retinol esters)	Coplanar biphenyls	Jensen et al., 1987; Powers et al., 1987; Mercier et al., 1990
UDP-GT (uridine diphosphate-glucuronyltransferase)	Coplanar biphenyls	Spear et al., 1988
	Polycyclic aromatic hydrocarbons	Roberts et al., 1979
	Dioxins	Bank et al., 1990
Cytochrome P450	Coplanar biphenyls	Spear et al., 1988; Gilbert et al., 1995
	Polycyclic aromatic hydrocarbons	Roberts et al., 1979; Koch, 1991

luteine, anhydroluteine, and probably other carotenoids. Nutritional studies have enabled us to show that there are considerable interspecific variations in the metabolic processes of retinoid biosynthesis. For example, after absorption of β-carotene-rich foods, it is the retinoids of the retinol type that are predominant in the sunset platy, *Xiphophorus variatus*, whereas in the guppy, *Lebistes reticulatus*, dehydroretinoids are the most abundant (Gross and Budowski, 1966). At present, it is impossible to confirm whether these interspecific differences depend on a genetically determined biochemical capacity or arise from the carotenoid composition of food resources, which could bring about a predominant biosynthesis of retinol or dehydroretinol.

Hepatic retinol is associated with a transport protein (retinol-binding protein or RBP) to form the complex retinol-RBP. This complex is then liberated in the blood, where it bonds with transthyretine (TTR). Retinol is generally the main form in the blood (0.05 to 0.6 µg/ml according to the species). However, in certain fish species, dehydroretinol is the predominant retinoid in plasma (P.A. Spear, unpublished data).

When it has reached a target site, the retinol dissociates from transport proteins, enters the cell, and binds with the cytosolic retinol-binding protein (CRBP). It can be transformed intracellularly into retinaldehyde and then into retinoic acid, which is linked to a cytosolic transport protein (cytosolic retinoic acid-binding protein or CRABP). The intracellular production of retinoic acid confirms the hypothesis that regulation of vitamin A activity is determined partly by metabolic regulation of this very active form in the specific target cells. Similarly, it seems logical that a less active form such as retinol (or dehydroretinol) should be transported at relatively high levels in the circulatory system.

Retinoic acid is also detected in the blood, but its role in vitamin A activity is not very well understood as compared with intracellular retinoic acid. In the intestinal mucus, a relatively low fraction of retinaldehyde is oxidized irreversibly by aldehyde dehydrogenases to form retinoic acid. Thus, retinoic acid is easily synthesized in fish and has been detected just a few hours after administration of β-carotene in specimens of *Saccobranchus fossilis* previously deficient in vitamin A (Barua and Goswami, 1977). Once it is produced by the intestinal mucous cells, retinoic acid can be transported by the hepatic vein and subjected to further metabolization in the liver. On the other hand, newly absorbed retinoic acid can also be transported in small quantities (of the order of ng/ml) by the circulatory system presumably as a complex with albumin.

Retinoic acid can be transformed into 4-hydroxyretinoic acid by several isoforms of cytochrome P450 and can also be oxidized into 4-oxoretinoic acid by a NAD-dependent dehydrogenase. Moreover, retinoic acid is combined by UDP-GT into retinoyl-β-glucuronide. These metabolites of retinoic acid seem to be among the major forms secreted in the urine and bile. Several other metabolic means of transformation of retinoic acid have been demonstrated, among them decarboxylation, epoxidation, isomerization, and esterification.

1.2 Effects of Vitamin A Deficiency and Excess

A brief summary of nutritional studies will enable us to indicate the significant role of retinoids in various aspects of the physiology of vertebrates and to consider the possible effects of a disturbance in storage of retinoids by environmental contaminants.

1.2.1 Effects of Vitamin A Deficiency

One of the first effects observed is the development of infections facilitated by the existence of an immunodepression. The precise relationship between dysfunction of the immune system and the cellular action of retinoids is not known (Underwood, 1984; Ross and Hammerling, 1994), but it is interesting to recall here that secondary or opportunistic infections have been associated with widespread mortality in several populations of aquatic mammals (beluga whales, seals).

The most significant symptoms of vitamin A deficiency in mammals and birds occur in vision, bone growth, reproduction, and maintenance of epithelial tissues (Thompson, 1976). According to Thompson, several of the effects caused by a deficiency would be the result of epithelial changes. Certain effects of vitamin A deficiency can be explained by the metaplasic keratinization of the epithelium of respiratory, alimentary, and urogenital canals, and saline, lachrymal, and other glands. Underwood (1984) considers that the preliminary signs of deficiency in mammals (with the exception of sensitivity to infection) are loss of appetite, increase in pressure of cerebrorachidial fluid, and decrease in the growth rate.

An inadequate nutritional input of vitamin A affects several aspects of reproduction in birds, including certain secondary sexual characteristics such as testicle weight, spermatogenesis, egg laying, egg size, survival of embryo, and duration of incubation and hatching. The effects on reproduction of mammals comprise inhibition of spermatogenesis, inability to maintain implantation, which leads to spontaneous abortion, resorption of the foetus, and vaginal keratinization accompanied by an increased susceptibility to infection. Acute vitamin A deficiency is accompanied by inhibition of the oestrous cycle. When the deficient embryo or foetus develops to the stage of organogenesis, congenital malformations can be seen, especially craniofacial and urogenital malformations, as well as cardial anomalies. However, embryo mortality may prevent the manifestation of these malformations.

In mammals with a food regime that does not contain vitamin A, a retinoic acid supplement can prevent all the symptoms of deficiency, except changes in vision and reproduction. In birds deficient in vitamin A, a retinoic acid supplement can also lead to reversibility of testicular degeneration (Thompson, 1976). These observations are consistent with the notion that retinoic acid and closely related isomers are the active forms of vitamin A in most tissues.

A parallel has often been established between vitamin A deficiency and toxic effects due to dioxins, furans, and polychlorobiphenyls (PCB) (Vos, 1978; Spear, 1986). In humans, Yusho and Yucheng diseases have been caused by exposure to rice oil contaminated by PCB, dibenzofurans, and other organochlorine products. The most obvious clinical symptoms of these diseases are chloracne, hyperkeratosis, and cutaneous lesions that look like acne eruptions, all typical of vitamin A deficiency, as well as oedema of the eyelid and inflammations around the eyes (Kuratsune, 1980; Rogan, 1989).

Epidemiological investigations in humans enabled the association of a low nutritional input of vitamin A with sensitivity to various cancers (Moon and Itri, 1984; Hong and Itri, 1994). Among the most significant carcinogens in cigarette smoke are polycyclic aromatic hydrocarbons (PAH), which affect storage of vitamin A in laboratory mammals. In parallel, several studies have shown that treatment and prevention of a range of cancers has been possible with the help of certain natural and synthetic retinoids.

Despite many consistencies between vitamin A deficiency and the effects of certain xenobiotics, there are exceptions. For example, the permanent loss of vision is a well-known symptom of vitamin A deficiency in fishes, birds, humans, and other mammals. However, this symptom is not associated with imbalances in retinoids caused by toxic chemicals. The specialized metabolic processes of the eye and its capacity to store retinoids, independently of significant storage sites such as the liver, may explain the apparent insensitivity of the eye to compounds that adversely affect retinoids in other organs and tissues.

Only a few publications address the essential role of active forms of vitamin A in teleosts. According to Kitamura et al. (1967), Poston et al. (1977), and Taveekijakarn et al. (1994), symptoms of vitamin A deficiency in freshwater fish are inhibition of growth, low hepatosomatic index, anemia, dermal depigmentation, hemorrhages in the eyes and at the base of fins, truncated snout, malformations of the opercula, peritoneal oedema, and ocular and histopathological lesions, the ultimate consequence frequently being the death of the individual. The addition of retinyl palmitate or β-carotene in the feed regime prevents the appearance of these symptoms or leads to their reversibility. On the other hand, neither the effects on reproduction, nor the teratogenic effects associated with deficiency in birds and mammals, have been observed in fish.

1.2.2 Effects of Vitamin A Excess

An excess of vitamin A, beyond what can be physiologically tolerated, is responsible for hypervitaminosis A in mammals, birds, and fish. The characteristic signs are bone fragility, lesions of the skin, and congenital malformations that are manifest at the same time as general effects such as weight loss and growth inhibition. Thompson (1976) and Kamm et al. (1984) reviewed studies on nutritional hypervitaminosis A due to the administration of specific retinoids and on its toxic effects.

1.3 Molecular Basis of Vitamin A Activity

In mammals, birds, and amphibians, a series of receptors for retinoic acid have been identified (Chambon et al., 1991). They are divided into two groups, RAR (retinoic acid receptor) and RXR (retinoid X receptor), each having several isoforms. These receptors belong to the family of receptors of steroid and thyroidal hormones (Evans, 1988) and are capable of high-affinity binding with retinoic acid and other ligands. The receptor-ligand complex takes the form of an RAR-RXR heterodimer that has the capacity to bind with DNA and modify the rates of transcription of specific genes. This series of molecular interactions is the basis of vitamin A activity.

The molecular basis of vitamin A activity can explain how retinoic acid helps to prevent certain forms of cancer. It can also explain the profound influence of retinoids on the differentiation and proliferation of different types of cells. For example, the β-RAR receptor expresses itself in the embryonic limb bud of a developing chicken, a retinoic acid gradient being generated from the zone of polarization activity (Thaller et al., 1991; Scott et al., 1994). The differentiation and proliferation of cells located along this gradient is under genetic control in response to the retinoic acid concentration. Thus, the characteristic pattern of digits is produced. If the retinoic acid gradient is reversed or if a second gradient is superimposed on it, that causes respectively an inversion or duplication of the pattern of digits. Retinoic acid is the only embryonic morphogenetic chemical known to date.

These studies demonstrate that abnormal intracellular concentrations of retinoic acid or analogous retinoids lead to changes in genetic transcription that could be linked to teratogenesis. It is therefore logical to suppose that environmental contaminants that alter retinoid metabolism could modify the intracellular concentration of retinoic acid and thereby have an effect on the quantity of this ligand available to bind the receptors.

2. EFFECTS OF ENVIRONMENTAL CONTAMINANTS ON RETINOID STORAGE AND USE OF RETINOIDS AS BIOMARKERS

To date, three types of retinoid storage have been studied in the context of evaluation of the effects of environmental contaminants on vertebrates. These are yolk retinoids in bird eggs, hepatic and intestinal retinoids, and blood retinoids.

2.1 Yolk Retinoids of Bird Eggs

2.1.1 Laboratory Studies

The possibility of using retinoids contained in eggs as biomarkers came to light following an experiment attempting to determine the effects of a coplanar chlorobiphenyl, 3,3',4,4'-tetrachlorobiphenyl (3,3',4,4'-TCB), on reproduction of ring doves, *Streptopelia risoria* (Spear et al., 1989). After this substance was injected into adult doves, significant mortality was observed in embryos from two successive couplings of doves (31% and 38% of embryos respectively) during the first half of the incubation period (i.e. mortality before the 8th day of incubation). The concentrations of retinol and retinyl palmitate in egg yolk decreased significantly between the 3rd and 8th days of incubation. However, the most convincing effect of the intoxication was the increase in retinol/retinyl palmitate ratio in the yolk (Spear et al., 1989).

2.1.2 Field Studies

During the 1960s, it became evident that the reproduction of birds living in the North American Great Lakes region was threatened, the high rate of malformations in juveniles being one of the most obvious manifestations (Hoffman et al., 1987; Gilbertson et al., 1991). This threat still persists for sensitive piscivorous species.

To evaluate the *in situ* influence of environmental contaminants on the retinoid levels of bird eggs, eggs of the herring gull, *Larus argentatus*, were sampled from six island colonies situated in the Great Lakes region. The retinoids contained in the egg yolks, where the embryo is found in the first half of the incubation period (that is, between 2 and 12 days of growth), were quantified (Spear et al., 1990). The retinol and retinyl palmitate concentrations varied significantly from one colony to another. A significant correlation was indicated between the molar ratio of retinol to retinyl palmitate and the toxic equivalents of 2,3,7,8-tetrachlorodibenzo-p-dioxin (TCDD), calculated from congeners of polychlorodibenzodioxins (PCDD) and polychlorodibenzofurans (PCDF) detected in the eggs (Fig. 2; Spear et al., 1990; Peakall, 1992). This relationship could be the result of effects on the process of egg formation or on retinoid metabolism *in ovo*; moreover, such effects could lead to a retinoid imbalance in the embryo during its development.

As part of a monitoring study of the St. Lawrence River, eggs of the great blue heron, *Ardea herodias*, were sampled in several rookeries along the river, as well as in reference colonies (Boily et al., 1994). However, several factors prevented the synchronization of egg sampling: (1) the dimensions of the ecosystem studied, (2) near-simultaneous activation of egg-laying in the different rookeries, which were far from one another, and (3) the difficulty of sampling from nests located in tall trees. Even though earlier studies on herring gulls had demonstrated the advantage of determining the molar ratio of retinol to retinyl palmitate, the results obtained for the great blue heron showed that this parameter was too dependent on the stage of development of the bird to be used in monitoring this species. On the other hand, retinyl palmitate concentration proved to be a reliable parameter because it allowed the discrimination of colonies without being significantly influenced by the age of the embryo. In addition, retinyl palmitate concentration in the yolk was negatively correlated with toxic equivalents of TCDD (Fig. 3) as well as with total concentrations of the three congeners of mono-*ortho*-biphenyl (Fig. 4).

2.1.3 Sources of Variation: Sampling Strategies

Retinoid concentrations in the yolk can vary according to embryonic stage. During the first half of the incubation period, the levels are generally constant. They tend to diminish during the second half of the incubation period, when lipids are transferred to the embryo

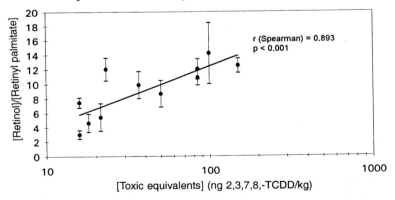

Fig. 2. Correlation between the molar ratio of retinol to retinyl palmitate and toxic equivalents of TCDD (ng/kg) in eggs of herring gulls (*Larus argentatus*) from breeding colonies in the North American Great Lakes (Spear et al., 1990, modified)

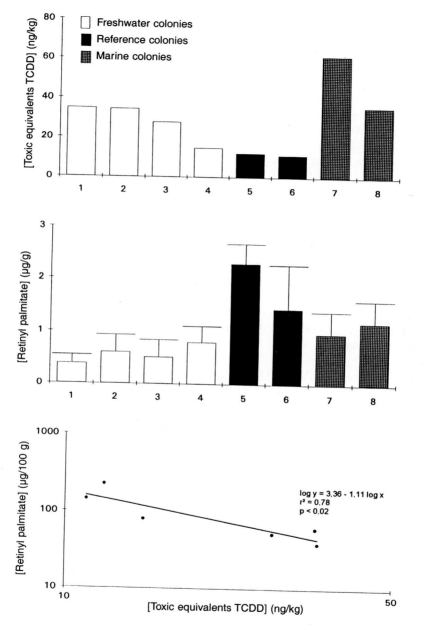

Fig. 3. Correlation between toxic equivalents of TCDD (ng/kg) and retinyl palmitate concentrations (µg/100 g) in egg yolks of the great blue heron (*Ardea herodias*) from colonies of the St. Lawrence River region. Concentrations of PCB 105 and 118 were transformed into toxic equivalents of TCDD (Boily et al., 1994, modified).

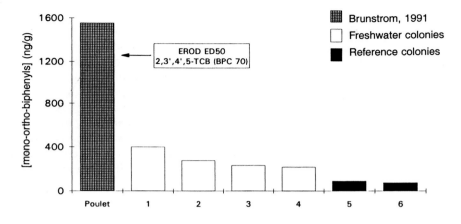

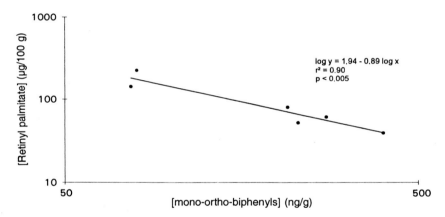

Fig. 4. Correlation between the total concentrations of three congeners of mono-*ortho*-biphenyls (ng/g) and retinyl palmitate concentration (μg/100 g) in egg yolks of the great blue heron (*Ardea herodias*) from colonies of the St. Lawrence River region (Boily et al., 1994, modified)

(Parris et al., 1951; Joshi et al., 1973; Noble, 1987; Spear et al., 1990). In theory, it is preferable to sample eggs during the first half of the incubation period in order to minimize variations associated with the stage of embryo growth. However, when field conditions do not allow synchronization of sampling, it is always possible to submit the data to a statistical treatment to evaluate the influence of embryo age.

The sampling of eggs at the time of hatching ('pipping' stage, that is, from the moment the beak tooth pierces the shell) is one technique that is becoming widespread. Furthermore, bird eggs can be collected in the field earlier in their development and incubated artificially until hatching. These methods of sampling are useful in obtaining embryos at the same stage of development and several tissues and organs (e.g., yolk, plasma, or liver) can be analysed to measure a variety of biomarkers. In theory, samples of this type are poorly adapted for measurement of retinoids because (1) mobilization of yolk retinoids during the second part of the incubation period is intensive, (2) there is a massive transfer of retinoids to the liver of the embryo during the process of yolk resorption, and (3) in certain cases the concentrations of yolk retinoids at hatching are lower than the limits of analytic detection. In spite of these disadvantages, a study based on chicks of the common tern, *Sterna hirundo*, sampled at the time of hatching in eight colonies in the Netherlands, enabled researchers to obtain results comparable to those described earlier for samples of the great blue heron in the St. Lawrence region. In particular, a negative relation was proved between dehydroretinyl palmitate concentration in the yolk sac of chicks at hatching, and concentration of mono-*ortho*-biphenyl congeners (Murk et al., 1994a).

Peakall (1994) pointed out several arguments in favour of the use of biomarkers in bird eggs, including the sensitivity of embryos to contaminants and the ease with which eggs can be sampled. One of the major advantages of analysis of biomarkers in eggs of wild birds is the minimal impact of sampling on the populations concerned. When the nests are easily accessible, eggs can be quickly collected and the adult birds can return to their nests and continue incubating the remaining eggs. Moreover, the subtraction of one egg from a nest does not have a significant effect on annual recruitment in the population.

2.2 Hepatic and Intestinal Retinoids

2.2.1 Birds

2.2.1.1 Laboratory Studies
Laboratory experiments have demonstrated that DDT can lower levels of vitamin A in liver of rock doves (*Columba livia*) (Jeffries and French, 1971). Intoxication by DDT and a mixture of PCB led to diminution of vitamin A in liver of Japanese quail, *Coturnix coturnix japonica* (Cecil et al., 1973). 3,3',4,4'-TCB caused a diminution of retinol concentrations in liver of adult ring doves (Spear et al., 1989) and retinyl palmitate concentrations in eider ducklings, *Somateria mollissima* (Murk et al., 1994b).

2.2.1.2 Field Studies
The existence of a negative relation between the levels of environmental contaminants and hepatic retinoid levels in species of wild fauna was proved during a study on birds nesting on the Great Lakes (Spear

et al., 1986). Adult herring gulls were sampled from several colonies exposed to different levels of contaminants (characterized by concentrations in gull eggs of a representative contaminant, 2,3,7,8-TCDD): Lake Ontario (high exposure), Lakes Superior and Michigan (medium to low exposure), and New Brunswick (low exposure, considered as a reference site). The hepatic concentrations of retinol and retinyl palmitate tended to develop as an inverse function of exposure level (Table 2), the differences between the sampling sites being significant for the two substances.

Following this study, the Canadian Wildlife Service included the measurement of hepatic retinoid concentrations in herring gull, a sentinel species, in its monitoring programme for the Great Lakes. The results obtained in the framework of this programme constituted an independent validation of the use of hepatic retinoids as biomarkers of the effects of organochlorine contamination in the Great Lakes region (Peakall, 1992).

During field studies that were conducted over several years, it appeared that, in herring gulls nesting in the islands of Lake Ontario, the hepatic levels of vitamin A were consistently low (Spear et al., 1992). During the year in which the concentrations were lowest, retinol concentrations in the liver were not higher than 20 µg/g and a negative correlation was proved with hepatic microsomal activity of aryl hydrocarbon hydroxylase (AHH; Spear et al., 1992). Similarly, exposure in laboratory conditions of ring dove, *S. risoria*, to 3,3',4,4'-TCB resulted in lowering of hepatic retinol concentration and a negative correlation was proved between the hepatic retinol concentration and induction of AHH activity in hepatic microsomes (Spear et al., 1986). These two results suggest that metabolic imbalances involving the induction of enzymes could be linked to the apparent incapacity of gulls to store sufficient quantities of retinoids in the liver. The substances capable of inducing

Table 2. Concentrations of retinol (µg/g) and retinyl palmitate (µg/g) in livers of adult herring gulls (*Larus argentatus*) from various North American colonies and concentrations of 2,3,7,8-TCDD (TCDD; pg/g) in eggs of herring gulls from the same colonies (Spear et al., 1986)

| | Retinoids stored in liver | | Exposure to |
| | retinol | retinyl palmitate | TCDD |
Sampling sites	(µg/g)	(µg/g)	(pg/g egg)
New Brunswick	864	1737	3
Lake Superior	382	562	10
Lake Michigan	289	377	13
Lake Ontario	131	231	90
Differences among all sites	$p < 0.005$	$p < 0.005$	
Differences among Great Lakes sites	$p < 0.025$	$p < 0.05$	

cytochrome P-4501A could therefore be implicated in the problems of retinoid storage.

2.2.2 Fish

In order to study the effects that could arise from environmental contaminants on retinoid levels in fish, two species of freshwater fish were sampled close to their spawning sites in Quebec (Spear et al., 1992; Branchaud, 1993; Branchaud et al., 1995). White sucker, *Catostomus commersoni*, and lake sturgeon, *Acipenser fulvescens*, were captured from a contaminated site (Rivière des Prairies, near Montreal) and a reference site located around 1000 km upstream (near the source of the Ottawa River, in La Verendry Park) (Branchaud et al., 1995). Even though the two rivers are part of the same hydrographic system, a series of dams prevents migration between the two study sites. The white sucker of the Rivière des Prairies had lower hepatic concentrations of retinol and retinyl palmitate than those measured in the fish from the reference site. Moreover, the difference between the two groups was accentuated in older fish (Branchaud et al., 1995). Similarly, concentrations of dehydroretinyl palmitate measured in the livers of male and female lake sturgeon were significantly higher in individuals from the reference site than in those from the Rivière des Prairies (Fig. 5).

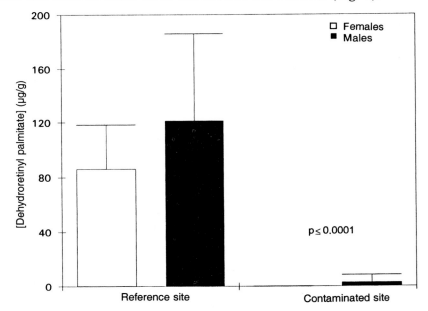

Fig. 5. Concentrations of dehydroretinyl palmitate (µg/g) in livers of male and female lake sturgeon (*Acipenser fulvescens*) captured in the Ottawa River (reference site) and the Rivière des Prairies (contaminated site) (Branchard, 1993)

Since fish generally store vitamin A in their digestive system, the retinoid concentrations were determined in portions of the intestine of lake sturgeon (Ndayibagira et al., 1995). The individuals captured near Montreal showed intestinal retinoid concentrations significantly lower than those measured in sturgeon captured in the Ottawa River (Fig. 6).

Fish from the Montreal site were exposed to numerous types of contaminants, including high levels of coplanar PCBs (P.A. Spear, A. Branchaud and B. Bush, unpublished data). It was possible to demonstrate in laboratory studies that coplanar PCBs affect the storage of retinoids in fish. Thus, exposure to 3,3′,4,4′-PCB dramatically reduced the levels of hepatic vitamin A in lake char, *Salvelinus namaycush* (Palace and Brown, 1994), and reduced concentrations of retinyl palmitate and dehydroretinyl palmitate in the intestine of adult male brook trout (*S. fontinalis*) (Ndayibagira et al., 1995).

In the course of studies on lake sturgeon and white sucker cited above, parameters other than retinoids were evaluated. It was proved that the rates of embryo mortality and teratogenesis were higher in the contaminated site, and that the induction of EROD activity was higher. In lake sturgeon, histopathological investigations demonstrated the existence of lesions in hepatic tissue (accumulation of lipids, proliferation

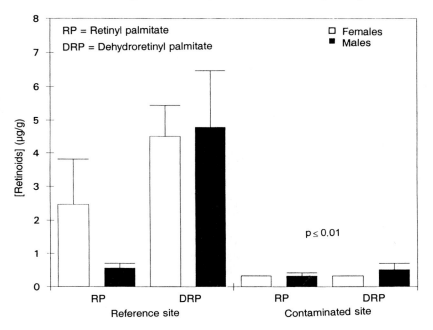

Fig. 6. Concentrations of retinoids (μg/g) measured in intestines of lake sturgeon (*Acipenser fulvescens*) captured in the Ottawa River (reference site) and the Riviere des Prairies (contaminated site) (Ndayibagira et al., 1995, modified)

of cells of bile ducts, periportal fibrosis and inflammation), which were significantly higher in fish captured in the contaminated site (Rousseaux et al., 1995). Despite various alterations detected in lake sturgeon in the Rivière des Prairies, no effect on the dynamics of the population concerned has yet been proved (Fortin et al., 1993). However, it is interesting to note that several fish species of the St. Lawrence, including the American eel, tomcod, and shad (Y. Mailhot, personal communication; Scott and Crossman, 1974; Hodson et al., 1993), have suffered population declines.

2.2.3 Mammals

2.2.3.1 Laboratory Studies

Early laboratory studies on the action of PAHs revealed that benzanthracene, 1,2-dibenzanthracene, 3-methylcholanthrene, and benzo(a)pyrene cause a decline in vitamin A levels in the liver of mammals (Goerner, 1939; Bauman et al., 1941), as do persistent environmental contaminants such as DDT, PCB, and TCDD (Phillips, 1963; Bitman et al., 1972; Thunberg et al., 1979). Animals subjected to food regimes deficient in vitamin A and exposed to dibenzanthracene, phenobarbital, or PCB, alone or mixed, developed symptoms of vitamin A deficiency that were not visible in animals subjected to the same food regime but not exposed to contaminants (Bauman et al., 1941; Backes et al., 1979; Darjono et al., 1983). Furthermore, Innami et al. (1974) showed that a feed supplement of vitamin A provides partial protection against the effects of a PCB mixture.

In the laboratory, consumption of feed contaminated by a commercial mixture of PCB (Clophen A50) for a period of 13–14 weeks by adult female mink (*Mustela vison*) led to significant reduction in their hepatic retinoid levels (Hakansson et al., 1992). The same result was obtained with a non-*ortho* mixture of PCB in proportions equal to those of the commercial mixture, which was not the case with other mixtures of PCB. The reduction in hepatic retinoid levels was thus attributed to non-*ortho* coplanar congeners (Hakansson et al., 1992).

Laboratory studies conducted on mammals show that several organochlorine compounds and PAHs can affect levels of retinoids and that there is an interaction between the effects due to these contaminants and the symptoms of vitamin A deficiency. It can therefore be difficult to precisely identify the contaminant or contaminants responsible for these adverse effects in the field.

2.2.3.2 Field Studies

The present population of beluga whales, *Delphinapterus leucas*, in the St. Lawrence river is only 10% of the pre-industrial population.

Immunodepression and alteration in reproduction are among the possible causes of the increased scarcity of this species (Martineau et al., 1987, 1988). Unlike other species of aquatic mammals, this population is confined to a space that is limited and highly contaminated because it does not migrate to the ocean, where exposure would be much lower. Studies conducted on these animals have shown the presence of high levels of PCB, DDT, and other organochlorine compounds in their tissues and the existence of several pathologies, including biliary cancer. Measurement of retinoids (P.A. Spear, unpublished data) in livers from stranded beluga whales revealed that two individuals out of three had very low levels of retinoids. While *post mortem* biochemical analysis of tissue cannot be interpreted with the desired scientific rigour, these preliminary results give the same indications as those obtained in mammals, wild birds, and fish exposed in the laboratory. They also emphasize the primary need to find non-destructive techniques for measuring biomarkers (particularly for large mammals and for endangered species).

2.3 Blood Retinol

The effects of dioxin-type contaminants on blood retinol have been the subject of several studies, the results of which are often contradictory. During laboratory studies on mammals, it was reported that blood retinol concentration increases, decreases, remains unchanged, or even fluctuates during a single experiment. To date, no field study establishing a clear relation between the concentrations of environmental contaminants and of blood retinol has been published. Even in the absence of toxic substances, the interpretation of variations in blood retinol concentration remains problematic (Underwood, 1984). It would therefore be too soon to consider blood retinol as a suitable biomarker.

During a study on the effects of 3,3',4,4'-TCB on reproduction of ring doves, males and females that laid eggs in which the embryos failed to develop showed significantly higher concentrations of serum retinol than the controls (Spear et al., 1989). Levels of plasma retinol in eider ducklings were not affected by 3,3',4,4'-TCB or by a PCB mixture (Clophen A50; Murk et al., 1994b). In chicks (*Gallus gallus*) subjected to a regime low in vitamin A, the serum retinol concentration was not affected significantly by 3,3',4,4'-TCB, whereas, in chicks subjected to iodine deficiency plus vitamin A deficiency, a decrease in serum retinol was detected (Spear and Moon, 1986). The plasma retinol concentrations measured in common tern chicks captured at different colony sites showed differences depending on the colony of origin, without any relationship with the level of contamination by dioxin-type compounds (Murk et al., 1994a).

During an experiment conducted under semi-controlled conditions, female common seals, *Phoca vitulina*, were fed for two years, either with fish from the Wadden Sea containing high concentrations of organochlorine compounds, or with less contaminated or uncontaminated fish from the northeast Atlantic. The reproductive success of the seals fed with the contaminated fish was significantly altered (Reijnders, 1986). A decrease in plasma retinol was observed in gestating and nongestating females fed with fish from the Wadden Sea (Brouwer et al., 1989). According to Brouwer and van den Berg (1986), PCB metabolites can bind with transthyretine, which would explain the diminution of plasma retinol concentrations. These results are significant in light of the critical role of vitamin A in reproduction and the reported decline of the common seal population in the Wadden Sea (Reijnders, 1980).

3. INTERPRETATION OF FLUCTUATIONS IN RETINOID LEVELS

In field studies done to date, relationships between retinoid concentrations and those of various organochlorine contaminants were demonstrated (coplanar PCBs, PCDD, and PCDF; Spear et al., 1986, 1990; Boily et al., 1994; Murk et al., 1994a). The toxic action of contaminants of this type is based on a sequence of intracellular events: high-affinity binding with the *Ah* receptor, a covalent binding of the xenobiotic-*Ah* receptor complex with DNA, alteration of genetic transcription and an increase in the synthesis of cytochromes P-4501A, and also the induction of enzymes such as UDP-GT (Poland et al., 1979; Safe, 1994). These inducible enzymes can metabolize retinol and retinoic acid (Fig. 1), the form P4501A2 presenting one of the highest rates of hydroxylation of retinoic acid and retinol in the reconstituted enzyme system (Roberts et al., 1992).

The rates of retinoid metabolism are substantially increased in mammals by dioxin-type contaminants such as coplanar biphenyls (Spear et al., 1988) and PAHs (Roberts et al., 1979; Koch, 1991). Gilbert et al. (1995) reported that hydroxylation of retinoic acid in the hepatic microsomes of rainbow trout, *Oncorhynchus mykiss*, was increased by a factor of 6 when fish were exposed to 3,3',4,4'-TCB. Exposure to such contaminants caused not only a diminution of retinoid concentrations in the liver, but also an increase in urinary and faecal excretion of retinoids (Hakansson and Ahlborg, 1985; Bank et al., 1989). It should be noted that metabolism of retinoids by cytochrome P-450 is not limited to the liver, but also occurs in the intestine, which could help explain the diminution of retinoid levels in the intestines of fish.

Such imbalances in metabolism of retinoic acid could alter the availability of this ligand for nuclear receptors, affecting the vitamin A activity in specific cells. Alternatively, the increased production of retinoic acid metabolites may be associated with adverse effects. Metabolites of retinoic acid were considered to be relatively inactive products. However, 4-oxo-retinoic acid shows a high affinity for the β-RAR receptor of amphibian embryo and is teratogenic in embryos of mice and amphibians (Creech Kraft et al., 1989; Pijnappel et al., 1993). These studies suggest that the enzymes induced by certain contaminants could increase the intracellular production of retinoids that are potentially teratogenic. In this context, it is interesting to recall that embryo mortality and teratogenic effects are associated with low levels of retinoids measured in fish and birds that live in highly contaminated habitats (Spear et al., 1986, 1992; Branchaud et al., 1995).

Apart from hydroxylation and glucuronidation of retinoic acid and retinol, some contaminants can affect interconversion between esters of retinol and retinol (Fig. 1). The effect of coplanar biphenyls on the activity of hydrolase of retinol ester remains poorly understood. Some studies have shown that some coplanar biphenyls are non-competitive inhibitors of retinyl palmitate hydrolase in rats (Jensen et al., 1987; Powers et al., 1987; Mercier et al., 1990), whereas other studies have described contrary effects. Esterification of retinol by ARAT is inhibited by certain coplanar biphenyls (Jensen et al., 1987; Narbonne et al., 1990).

4. COMPARATIVE SENSITIVITY OF RETINOIDS AND OTHER BIOMARKERS

Few studies have compared the sensitivity of retinoid biomarkers with that of other biomarkers. The most extensive study in this regard consisted of the sub-chronic exposure of rats to a coplanar biphenyl, 3,3',4,4',5-pentachlorobiphenyl (3,3',4,4',5-PCB). The most sensitive effect was the induction of EROD activity, which was demonstrated at a dose of 0.1 ng/g of 3,3',4,4',5-PCB in the food (Chu et al., 1994). The concentration of hepatic vitamin A proved slightly less sensitive; a significant diminution of this parameter was detected in female rats at the next higher dose level of 1.0 ng/g of 3,3',4,4',5-PCB in the food. No other significant change appeared for other biochemical, hematological, or histological parameters at this dose, although minimal multifocal changes were observed in the thymus of 3 male rats among the 10 individuals contaminated (Chu et al., 1994). The level of vitamin A in mammals therefore seems highly sensitive to this type of environmental contaminant.

5. CONCLUSION

In all the vertebrates studied to date, it has been demonstrated that retinoids are essential to life, and that retinoid imbalances can affect several aspects of growth, reproduction, and development.

Very low retinoid levels have been detected in fish and birds from populations that live in relatively contaminated habitats. The inability of animals to store sufficient quantities of vitamin A can be explained by changes induced by toxic chemicals in the metabolism of retinoids. Field studies have demonstrated that retinoid concentrations in bird eggs showed a high correlation with tissue levels of dioxins, coplanar PCB, and related organochlorine compounds.

The levels of stored retinoids can therefore be affected chronically by a certain number of environmental contaminants, which could lead to adverse effects in populations of wild vertebrates. Field validation studies demonstrated that the use of stored retinoids as biomarkers has several advantages, including (1) their potential for integration of effects of exposure over time and (2) their mechanistic relation to many harmful effects (e.g., growth, reproduction, development, immunocompetence, susceptibility to cancers) and thus effects on the populations concerned.

REFERENCES

Backes W.L., Krause R.F. and Canady W.J. (1979). Observations relating phenobarbital induction of cytochrome P-450 to the obligatory destruction or release of vitamin A. *Nutr. Rep. Internat.*, 119:9–14.

Bank P.A., Salyers K.L. and Zile M.H. (1989). Effect of tetrachlorodibenzo-*p*-dioxin (TCDD) on the glucuronidation of retinoic acid in the rat. *Biochim. Biophys. Acta*, 993:1–6.

Barua A.B. (1978). Biosynthesis of dehydroretinol. *Wld. Rev. Nutr. Diet.*, 31:89–94.

Barua A.B. and Goswami U.C. (1977). Formation of vitamin A in a freshwater fish: isolation of retinoic acid. *Biochem. J.*, 164:133–136.

Bauman C.A., Foster E.G. and Lavik P.S. (1941). The effect of certain carcinogens on vitamin A in the liver. *J. Nutr.*, 21:431–444.

Bitman J., Cecil H.C, and Harris S.J. (1972). Biological effects of polychlorinated biphenyls in rats and quail. *Environ. Health Perspect.*, I:145–149.

Boily M.H., Champoux L., Bourbonnais et al. (1994). β-Carotene and retinoids in eggs of Great Blue Herons (*Ardea herodias*) in relation to St. Lawrence River contamination. *Ecotoxicology*, 3:271–286.

Branchaud A. (1993). Étude écotoxicologique de la vitamine A et des malformations chez le meunier noir. Dissertation in biology, University of Quebec at Montreal, Montreal.

Branchaud A., Gendron A., Fortin R. et al. (1995). Vitamin A stores teratogenesis and EROD activity in white sucker, *Catostomus commersoni*, from Rivière des Prairies near Montreal and a reference site. *Can. J. Fish. Aquat. Sci.*, 52:1703–1713.

Brouwer A. and Van Den Berg K.J. (1986). Binding of a metabolite of 3,4,3',4'-tetrachlorobiphenyl to transthyretin reduces serum vitamin A transport by inhibiting the formation of the protein complex carrying both retinol and thyroxin. *Toxicol. Appl. Pharmacol.*, 85:301–312.

Brouwer A., Reijnders P.J.H. and Koeman J.H. (1989). Polychlorinated biphenyl (PCB)-contaminated fish induce vitamin A and thyroid hormone deficiency in the common seal (*Phoca vitulina*). *Aquat. Toxicol.*, 15:99–106.

Cecil H.C., Harris S.J., Bitman J. and Fries G.F. (1973). Polychlorinated biphenyl-induced decrease in liver vitamin A in Japanese quail and rats. *Bull. Environ. Contam. Toxicol.*, 9:179–185.

Chambon P., Zelent A., Petkovich M. et al. (1991). The family of retinoic acid nuclear receptors. In: Retinoids 10 years on. Saurat J.H. (ed.). Karger, Basel, pp. 10–27.

Chu I., Villeneuve D.C., Yagmanis A. et al. (1994). Subchronic toxicity of 3,3′4,4′,5-pentachlorobiphenyl in the rat. I. Clinical, biochemical, hematological, and histological changes. *Fundam. Appl. Toxicol.*, 22:457–468.

Creech Kraft J., Lofberg B., Chahoud I. et al. (1989). Teratogenicity and placental transfer of all-*trans*, 13-*cis*, 4-oxo-all-*trans*, and 4-oxo-13-*cis*-retinoic acid after administration of a low oral dose during organogenesis in mice. *Toxicol. Appl. Pharmacol.*, 100:162–176.

Darjono S., Sleight S.D., Stowe H.D. and Aust S.D. (1983). Vitamin A status, brominated biphenyl (PBB) toxicosis, and common bile duct hyperplasia in rats. *Toxicol. Appl. Pharmacol.*, 71:184–193.

Evans M.E. (1988). The steroid and thyroid hormone receptor superfamily. *Science*, 240:889–895.

Fortin R., Mongeau J.R., Desjardins G. and Dumont P. (1993). Movements and biological statistics of lake sturgeon (*Acipenser fulvescens*) populations from the St. Lawrence and Ottawa river system, Quebec. *Can. J. Zool.*, 71:638–650.

Ganguly J. (1989). *Biochemistry of vitamin A*. CRC Press, Boca Raton.

Glibert N.L., Cloutier M.J. and Spear P.A. (1995). Retinoic acid hydroxylation in rainbow trout (*Oncorhynchus mykiss*) and the effect of a coplanar PCB, 3,3′,4,4′-tetrachlorobiphenyl. *Aquat. Toxicol.*, 32:177–187.

Gilbertson M., Kubiak T., Ludwig J. and Fox G. (1991). Great Lakes embryo mortality, edema, and deformities syndrome (GLEMEDS) in colonial fish-eating birds: similarity to chick edema disease. *J. Toxicol. Environ. Health.* 33:455–520.

Goerner A. (1939). The influence of a carcinogenic compound on the hepatic storage of vitamins. *J. Nutr.*, 18:441–446.

Goodman D.S. and Blaner W.S. (1984). Biosynthesis, absorption and hepatic metabolism of retinol. In: The Retinoids, Vol. 2. Sporn M.B., Roberts A.B. and Goodman D.S. (eds.). Academic Press, Toronto, pp. 2–39.

Gross J. and Budowski P. (1966). Conversion of carotenoids into vitamins A1 and A2 in two species of freshwater fish. *Biochem. J.*, 101:747–754.

Håkansson H. and Ahlborg U.G. (1985). The effect of 2,3,7,8-tetrachlorodibenzo-*p*-dioxin (TCDD) on the uptake, distribution and excretion of a single oral dose of [11, 12 –^{3}H] retinyl acetate and on the vitamin A status in the rat. *J. Nutr.*, 115:759–771.

Håkansson H., Manzoor E. and Ahlborg U.G. (1992). Effects of technical PCB preparations and fractions thereof on vitamin A levels in the mink (*Mustela vison*). *Ambio*, 21:588–590.

Hoffman D.J., Rattner B.A., Sileo L. et al. (1987). Embryotoxicity, teratogenicity, and aryl hydrocarbon hydroxylase activity in Forster's terns on Green Bay, Lake Michigan. *Environ. Res.*, 42:176–184.

Hodson P.V., Castonguay M., Couillard C.M. et al. (1993). Spatial and temporal variations in chemical contamination of American eels, *Anguilla rostrata*, captured in the estuary of the St. Lawrence River (Quebec, Canada). *Can. J. Fish. Aquat. Sci.* 51:464–478.

Hong W.K. and Itri L.M. (1994). Retinoids and human cancer. In: The Retinoids: Biology, Chemistry and Medicine, 2nd ed. Sporn M.B., Roberts A.B. and Goodman D.S. (eds.). Raven Press, New York, pp. 597–630.

Innami S., Nakamura A. and Nagayama S. (1974). Polychlorobiphenyl toxicity and nutrition: II. PCB toxicity and vitamin A$_2$. *J. Nutr. Sci. Vitaminol.*, 20:363–370.

Jeffries D.J. and French M.C. (1971). Hyper and hypothyroidism in pigeons fed DDT: an explanation for the thin egg shell phenomenon. *Environ. Pollut.*, 1:235–242.

Jensen R.K., Cullum M.E., Deyo J. and Zile M.H. (1987). Vitamin A metabolism in rats chronically treated with 3,3',4,4',5,5'-hexabromobiphenyl. *Biochim. Biophys. Acta.*, 926:310–320.

Joshi P.S., Mathur S.N., Murthy S.K. and Ganguly J. (1973). Vitamin A economy of the developing chick embryo and of the freshly hatched chick. *Biochem. J.*, 136:757–761.

Kamm J.J., Ashenfelder K.O. and Ehmann C.W. (1984). Preclinical and clinical toxicology of selected retinoids. In: The Retinoids, Vol. 2. Spron M.B., Roberts A.B. and Goodman D.S. (eds.). Academic Press, Montreal, pp. 288–326.

Kitamura S., Suwa T., Ohara S. and Nakagawa K. (1967). Studies of vitamin requirements of rainbow trout—III. Requirement for vitamin A and deficiency symptoms. *Bull. Japan. Soc. Sci. Fish.*, 33:1126–1130.

Koch B. (1991). Effets des xénobiotiques sur le transport, la répartition et le métabolisme oxydatif de la vitamine A chez le rat. Doctoral thesis, University of Bordeaux, France.

Kuratsune M. (1980). Yusho. In: Halogenated Biphenyls, Terphenyls, Naphthalenes, Dibenzodioxins and Related Products. Kimbrough R.D. (ed.). Elsevier/North-Holland Biomed. Press, Amsterdam, pp. 287–302.

Martineau D., Béland P., Desjardins C. and Lagacé A. (1987). Levels of organochlorine chemicals in tissues of beluga whales (*Delphinapterus leucas*) from the St. Lawrence Estuary, Quebec, Canada. *Arch. Environ. Contam. Toxicol.*, 16:137–147.

Martineau D., Lagacé A., Béland P. et al. (1988). Pathology of stranded beluga whales (*Delphinapterus leucas*) from the St. Lawrence estuary, Quebec. *Can. J. Comp. Pathol.*, 98:287–311.

Mercier M., Pascal G. and Azais-Braesco V. (1990). Retinyl ester hydrolase and vitamin A status in rats treated with 3,3',4,4'-tetrachlorobiphenyl. *Biochim. Biophys. Acta*, 1047:70–76.

Moon R.C. and Itri L.M. (1984). Retinoids and cancer. In: The Retinoids, Vol. 2. Sporn M.B., Roberts A.B. and Goodman D.S. (eds.). Academic Press, Toronto, pp. 327–371.

Murk A.J., Bosveld A.T.C., Van Den Berg M. and Brouwer A. (1994a). Effects of polyhalogenated aromatic hydrocarbons (PHAHS) on biochemical parameters in chicks of the common tern (*Sterna hirundo*). *Aquat. Toxicol.*, 30:91–115.

Murk A.J., Van Den Berg J.H.J., Fellinger M. et al. (1994b). Toxic and biochemical effects of 3,3',4,4'-tetrachlorobiphenyl (CB-77) and Clophen A50 on eider ducklings (*Somateria mollissima*) in a semifield experiment. *Environ. Pollut.*, 80:21–30.

Narbonne J.F., Grolier P., Albrecht R. et al. (1990). A time course investigation of vitamin A level and lipid composition of the liver endoplasmic reticulum in rats following treatment with congeneric polychlorobiphenyls. *Toxicology*, 60:253–261.

Ndayibagira A., Cloutier M.J., Anderson P.D. and Spear P.A. (1995). Effects of 3,3',4,4'-tetrachlorobiphenyl (TCBP) on the dynamics of vitamin A in brook trout (*Salvelinus fontinalis*) and intestinal retinoid concentrations in two populations of lake sturgeon (*Acipenser fulvescens*). *Can. J. Fish. Aquat. Sci.*, 52:512–520.

Noble R.C. (1987). Lipid metabolism in the chick embryo: some recent ideas. *J. Exp. Zool. Suppl.*, 1:65–73.

Olson J.A. (1983). Formation and function of vitamin A. In: Biosynthesis of Isoprenoid Compounds. Porter J.W. and Spurgeon S.L. (eds.). John Wiley and Sons, New York, pp. 371–412.

Olson J.A. (1989). Provitamin A function of carotenoids: The conversion of β-carotene into vitamin A. *J. Nutr.*, 119:105–108.

Palace V.P. and Brown S.B. (1994). HPLC determination of tocopherol, retinol, dehydroretinol and retinylpalmitate in tissues of lake char (*Salvelinus namaycush*) exposed to coplanar 3,3',4,4'5-pentachlorobiphenyl. *Environ. Toxicol. Chem.*, 13:473–476.

Parrish D.B., Williams R.N. and Sanford P.E. (1951). The state of vitamin A in livers and unabsorbed yolks of embryonic and newly hatched chicks. *Arch. Biochem. Biophys.*, 34:64–66.

Peakall D.B. (1992). *Animal Biomarkers as Pollution Indicators.* Chapman and Hall, New York.

Peakall D.B. (1994). Biomarkers in egg samples. In: Nondestructive Biomarkers in Vertebrates. Fossi M.C. and Leonzio C. (eds.). Lewis Publishers, London, pp. 201–216.

Phillips W.E.J. (1963). DDT and the metabolism of vitamin A and carotene in the rat. *Can. J. Biochem. Physiol.*, 41:1793–1802.

Pijnappel W.W.M., Hendriks H.F.J., Folkers G.E. at al. (1993). The retinoid ligand 4-oxo-retinoic acid is a highly active modulator of positional specification. *Nature*, 366:340–344.

Poland A., Greenlee W.F. and Kende A.S. (1979). Studies of the mechanism of action of the chlorinated dibenzo-*p*-dioxins and related compounds. In: Health effects of halogenated hydrocarbons. Nicholson W.J. and Moore J.A. (eds.). *Ann. N. Y. Acad. Sci.*, 320:214–230.

Poston H.A., Riis R.C., Rumsey G.L. and Ketola H.G. (1977). The effect of supplemental dietary amino acids, minerals and vitamins on salmonids fed cataractogenic diets. *Cornell Vet.*, 67:472–509.

Powers R.H., Gilbert L.C. and Aust S.D. (1987). The effect of 3,4,3',4'-tetrachlorobiphenyl on plasma retinol and hepatic retinyl palmitate hydrolase activity in female Sprague Dawley rats. *Toxicol. Appl. Pharmacol.*, 89:370–377.

Reijnders P.J.H. (1980). Organochlorine and heavy metal residues in harbour seals from the Wadden Sea and their possible effects on reproduction. *Neth. J. Sea Res.*, 14:30–65.

Reijnders P.J.H. (1986). Reproductive failure in common seals feeding on fish from polluted coastal waters. *Nature*, 324:456–457.

Roberts A.B., Nichols M.D., Newton D.L. and Sporn M.B. (1979). *In vitro* metabolism of retinoic acid in hamster intestine and liver. *J. Biol. Chem.*, 254:6296–6302.

Roberts E.S., Vaz A.D. and Coon M.J. (1992). Role of iozymes of rabbit microsomal cytochromes P-450 in the metabolism of retinoic acid, retinol, and retinyl. *Mol. Pharmacol.*, 41:427–433.

Rogan W. (1989). Yu-Cheng. In: Halogenated Biphenyls, Terphenyls, Naphthalenes, Dibenzodioxins and Related Products, 2nd ed. Kimbrough R.D. and Jensen A.A. (eds.). Elsevier, Amsterdam, pp. 401–415.

Ross A.C. and Hammerling U.G. (1994). Retinoids and the immune system. In: The Retinoids: Biology, Chemistry and Medicine, 2nd ed. Sporn M.B., Roberts A.B. and Goodman D.S. (eds). Raven Press, New York, pp. 521–544.

Rousseaux C.G., Branchaud A. and Spear P.A. (1995). Evaluation of liver histopathology and EROD activity in St. Lawrence lake sturgeon (*Acipenser fulvescens*) in comparison with a reference population. *Environ. Toxicol. Chem.*, 14:843–849.

Safe S.H. (1994). Polychlorinated biphenyls (PCBs): Environmental impact, biochemical and toxic responses, and implications for risk assessment. *Crit. Rev. Toxicol.*, 24:87–149.

Saurat J.H. (1991). *Retinoids 10 years on.* Karger, Basel.

Scott W.B. and Crossman E.J. (1974). *Poissons d'eau douce du Canada.* Ministère de l'environnement, Ottawa, Canada, Rep. 184

Scott W.J. Jr., Walter R., Tzimas G. et al. (1994). Endogenous status of retinoids and their cytosolic binding proteins in limb buds of chick vs mouse embryos. *Develop. Biol.*, 165:397–409.

Spear P.A. (1986). The action of a dioxin-like compound upon avian thyroid and vitamin A homeostasis. PhD. thesis, University of Ottawa, Canada.

Spear P.A. and Moon T.W. (1986). Thyroid-vitamin A interactions in chicks exposed to 3,4,3',4'-tetrachlorobiphenyl: influence of low dietary vitamin A and iodine. *Environ. Res.*, 40:188–198.

Spear P.A., Bilodeau A.Y. and Branchaud A. (1992). Retinoids: from metabolism to environmental monitoring. *Chemosphere*, 25:1733–1738.

Spear P.A., Garcin H. and Narbonne J.F. (1988). Increased retinoic metabolism following 3,3',4,4',5,5'-hexabromobiphenyl injection. *Can. J. Physiol. Pharmacol.*, 66:1181–1186.

Spear P.A., Moon T.W. and Peakall D.B. (1986). Liver retinoid concentrations in natural populations of herring gulls (*Larus argentatus*) contaminated with 2,3,7,8-tetrachlorodibenzo-*p*-dioxin and in ring doves (*Streptopelia risoria*) injected with a dioxin analogue. *Can. J. Zool.*, 64:204–208.

Spear P.A., Bourbonnais D.H., Norstrom R.J. and Moon T.W. (1990). Yolk retinoids (vitamin A) in eggs of the herring gull and correlations with polychlorinated dibenzo-*p*-dioxins and dibenzofurans. *Environ. Toxicol. Chem.*, 9:1053–1061.

Spear P.A., Bourbonnais D.H., Peakall D.P. and Moon T.W. (1989). Dove reproduction and retinoid (vitamin A) dynamics in adult females and their eggs following exposure to 3,3',4,4'-tetrachlorobiphenyl. *Can. J. Zool.*, 67:908–913.

Sporn M.B., Roberts A.B. and Goodman D.S. (1984). *The Retinoids, Vol. 2*. Academic Press, Toronto.

Sporn M.B., Roberts A.B. and Goodman D.S. (1994). *The Retinoids: Biology, Chemistry and Medicine*, 2nd ed. Raven Press, New York.

Taveekijakarn P., Miyazaki T., Matsumoto M. and Arai S. (1994). Vitamin A deficiency in cherry salmon. *J. Aquat. Anim. Health*, 6:251–259.

Thaller C., Smith S.M. and Eichle G. (1991). Retinoids in vertebrate development: Pattern formation in limbs and in the central nervous system. In: Retinoids 10 years on. Saurat J.H. (ed.). Karger, Basel, pp. 89–108.

Thompson J.N. (1976). Fat-soluble vitamins. *Comp. Anim. Nutr.*, 1:99–135.

Thunberg T., Ahlborg U.G. and Johnson H. (1979). Vitamin A (retinol) status in the rat after a single oral dose of 2,3,7,8-tetrachlorodibenzo-*p*-dioxin. *Arch. Toxicol.*, 42:265–274.

Underwood B.A. (1984). Vitamin A in animal and human nutrition. In: The Retinoids, Vol. 1. Sporn M.B., Roberts A.B. and Goodman D.S. (eds.). Academic Press, Montreal, pp. 281–392.

Vos J.G. (1978). 2,3,7,8-tetrachlorodibenzo-*p*-dioxin: effects and mechanisms. In: Chlorinated phenoxy acids and their dioxins, Vol. 27. Ramel C. (ed.). Ecol. Bull., Stockholm, pp. 165–176.

11

Lysosomal Fragility as Cytological Biomarker

J. Pellerin-Massicotte and R. Tremblay

INTRODUCTION

Lysosomes are organelles that respond to numerous stress factors, in mammals as well as in other vertebrates or in invertebrates. Lysosomal fragility can be considered, like other cellular, physiological, or biochemical biomarkers, a significant indicator of response to stress factors. Destabilization of the lysosomal membrane has chiefly been studied experimentally in the laboratory to determine the causes of certain pathologies in mammals. These studies concerned:

- the relation between lysosomal stability and the resistance of mucous cells to lesions (Waldron-Edward and Greenberg, 1980);
- the role of lysosomal stability in the first hours following coronary blockage (Ricciutti, 1972);
- its implication in formation of stomach ulcers (Porta et al., 1986), hepatic lesions due to alcohol (Wilson et al., 1990, 1992), during severe cardio-respiratory stress (Bonin, 1972), or following exposure to ionizing radiations (Abok et al., 1984).

Lysosomal fragility is a biomarker at the cellular level. It has the peculiarity of integrating the effects of multiple stress factors, and it can also be influenced by hormonal factors (Decker, 1977; Philip and Kurup, 1978).

In this chapter, the term *stress* is used to designate the consequence of the action of abiotic or other factors at the individual level or at the population level, that surpass the limits of the adaptive capacity of the organisms or that disturb their physiological functioning to such a point that their chances of survival are reduced. The term *stress factor* designates here any condition or situation that causes resource mobilization in the organisms and that increases their energy expenditure, the stress therefore being the response of the biological systems to the stress factor through an expenditure of energy.

The level of lysosomal activity is very important for proper physiological functioning, and lysosomes are present in all organisms and in all tissues except red blood cells (Holtzman, 1989). This characteristic is one of the potential advantages of lysosomal fragility in terms of its use as a biomarker of stress. However, the abundance of lysosomes is not the same in all organs. The cells of the gastroderm of Coelentera and the digestive cells of bivalves and gastropods present high number of lysosomes, which explains why the first studies on the effects of environmental stress were done on these organisms (Moore, 1976; Moore and Stebbing, 1976).

1. LYSOSOMAL ACTIVITY

Lysosomal activity is a normal physiological phenomenon found in the processes of intracellular digestion such as autophagy, heterophagy, and other related processes that occur in the cell cytoplasm. A sound understanding of these processes is essential to any possible use of lysosomal fragility as a biomarker.

Heterophagy (Fig. 1) begins with *phagocytosis* (Fig. 2), during which exogenous material is captured by the plasma membrane directly from the extracellular space. After vesicles or vacuoles are formed, they fuse with the lysosomal membrane and lysosomal hydrolases begin intracellular digestion.

Autophagy, defined as degradation of part of its own cytoplasm by the lysosomes in a given cell (Fig. 3), is a normal phenomenon, but can be amplified by stress due to a lesion, lack of nutrients, excessive accumulation of xenobiotics, or the presence of a virus (Bayne et al., 1985).

Autophagy is also considered a physiological mechanism of survival in stress conditions where there is resorption or autodigestion of cellular components. Four types of autophagous processes have been observed:

- classical autophagy, in which the cell isolates and degrades a region of its cytoplasm within a vacuole bounded by a membrane;
- lysosomal digestion of part of the cell membrane after the latter has been internalized;
- crinophagy, in which the intracellular secretory structures fuse with the lysosomes rather than with the plasma membrane;
- degradation of endogenous reserves.

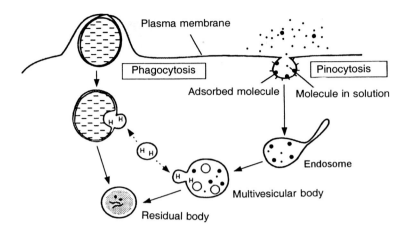

Fig. 1. Heterophagy. The cells can integrate xenobiotics present in solid form (dust particles or PAH, for example) or in solution (micropollutants in solution) by an invagination of the cell membrane, which forms a phagocytous vacuole (at left in the figure) or a pinocytous vacuole (at right in the figure). In the two cases, the lysosomes fuse with these vacuoles and discharge into them their hydrolytic enzymes, the fusion yielding multivesicular bodies. After the foreign components are digested, there remains an organelle called the residual body, which isolates foreign matter that cannot be digested.

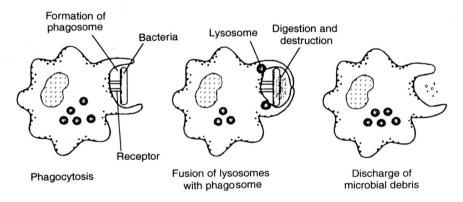

Fig. 2. Phagocytosis. The lysosomal activity can be enhanced by the presence of bacteria or viruses. This peculiarity can be used as a biomarker of water quality in oyster or mussel farms, by measurement of lysosomal activity of the hemocytes, the phagocytous cells in molluscs. The choice of tissue or cells in which the lysosomal activities will be measured depends on the objective of the study (modified from Holtzman, 1989).

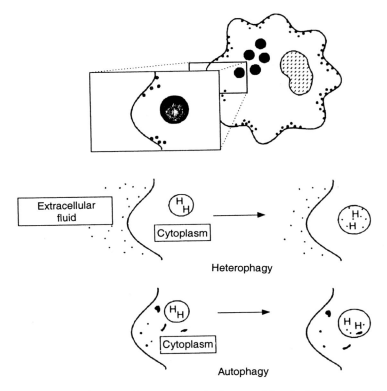

Fig. 3. Autophagy. The lysosomal activity can be increased by the integration of foreign particles (heterophagy) and/or by the degradation of cell fragments (autophagy). This last process can be hampered by a state of physiological stress, or during a normal renewal of macromolecules. Cytochemical measurements involve the estimation of the basal level of lysosomal activity on control glass slides that have not been subjected to a destabilization caused by acid pH, and subtraction of these values from those found during evaluation of the lysosomal latency period. For details on measurement methods, Bayne et al. (1985) give all the parameters of technical protocols (modified from Holtzman, 1989).

2. FRAGILIZATION OF THE LYSOSOME MEMBRANE: MECHANISMS AND IMPLICATIONS IN RESEARCH ON CELLULAR BIOMARKERS

2.1 Mechanisms of Action of Stress Factors on the Lysosome Membrane

The hydrolytic enzymes present in the lysosomes have a latency period that is explained by the natural impermeability of the lysosome membrane to certain substrates and by the linkage of enzymes with the internal membrane of lysosomes, which makes them inactive.

The lysosomal membrane can, however, become destabilized or more fragile in certain physiological or pathological conditions (Fig. 4).

Lysosomes contain about thirty different hydrolytic enzymes and can isolate several contaminants. Generally, they are implicated in many physiological functions such as intracellular digestion, conservation, digestion, resorption, cellular proliferation, and immune mechanisms. Complexes that are liposoluble, or at least only slightly polar, are isolated and accumulated in the lysosomes, where they can precipitate in the form of insoluble phosphate crystals. Several heavy metals, such as zinc, iron, cadmium, or uranium, thus form cytoplasmic inclusions in bivalves (George, 1983). Various environmental stress factors, biotic (density) or abiotic (salinity, temperature), can also affect the integrity of the lysosome membrane by destabilizing the lipoprotein matrix, which leads to the release of hydrolytic enzymes in the cytoplasm, where energy reserves are then degraded (Moore et al., 1980). According to some authors, this phenomenon explains the natural variability in growth of organisms that live in the estuarine environment. In fact, long periods of emersion in an intertidal habitat lead to an increase in autophagy and resulting proteolysis, exposure to air being the stress factor responsible (Tremblay and Pellerin-Massicotte, 1996). This example demonstrates, if necessary, that it is essential to take abiotic factors into account while

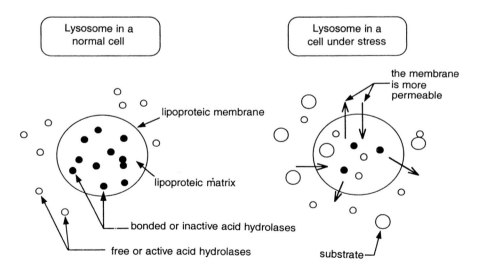

Fig. 4. Effect of stress on lysosomes (modified from Bayne et al., 1985). The lysosomes in non-stressed cells are mostly impermeable to substrates. In stress conditions, the lysosome membrane is destabilized, which leads to the penetration of substrates and the activation of hydrolytic enzymes that were previously inactive. Hydrolases that cause cytolytic damage are then released in the cytoplasm.

implementing a general biomarker of stress such as lysosomal fragility in the field.

The increased fragility of the lysosome membrane observed in the presence of micropollutants is explained by increase in detoxification enzymatic activities that can lead to bioactivation of the pollutant and overproduction of free radicals that can damage lysosomes (Fig. 5).

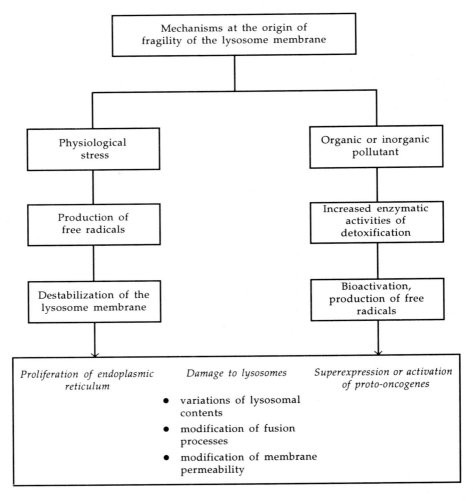

Fig. 5. Mechanisms of action of stress factors on the lysosome membrane. Sequence of steps leading to increase in lysosomal fragility in the presence of a physiological stress or a contaminant.

2.2 Research on Cellular Biomarkers

As we will see, the system of lysosome membranes is activated during numerous physiological processes. It is therefore complex and several parameters can be used as biomarkers of stress (Moore, 1990). Frequently used parameters are (Moore et al., 1994):

- changes in volume of lysosomes (Fig. 6);
- accumulations of lipids and lipofuschin, a stress- and ageing-specific pigment;
- destabilization of the lysosome membrane, which disorganizes cellular functioning;
- inhibition of the functioning of the lysosome proton pump.

The lysosomal system enables the degradation of macromolecules ingested by cells by means of a battery of hydrolytic enzymes functioning at low pH in the lysosomal compartment. This pH is maintained by ATP-dependent proton pumps located in the membrane.

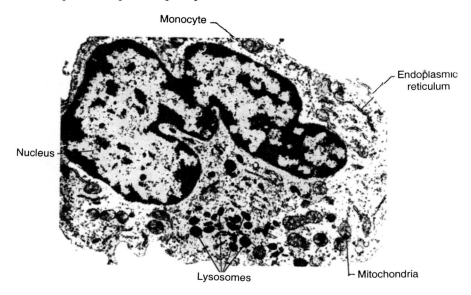

Fig. 6. Cellular location of lysosomes in a monocyte (modified from Ham, 1974). The lysosomes newly formed from a Golgi body are called primary lysosomes. These, after having fused and reacted with foreign or intracellular substances, are then called secondary lysosomes, phagocytous or pinocytous vacuoles. If there are several fusions, the secondary lysosomes are then called multivesicular bodies. Finally, what remains after digestion in the secondary lysosome is recognized as a residual body. One can see in this figure that the lysosomes are generally grouped in the cell. They are not visible under light microscope, which is why cytochemical measurements are most often used to detect lysosomal activity.

2.3 Study Methods

Most studies attempting to estimate lysosomal fragility were performed on cells of the digestive gland of bivalve molluscs, the lysosomal system of these cells being considered a potential target of toxic effects of several xenobiotics (Moore, 1991). The criteria chosen to indicate this adverse action include:

- cytochemical tests, which enable estimation of lysosome membrane stability;
- measurement of percentage of latency period, based on the revelation of hydrolase activity (Fig. 7);
- estimation of lysosomal levels of lipofuschin and neutral lipids (Bitensky et al., 1973; Moore, 1988).

All these measurements can be made with the help of cytochemical techniques, with great experimental rigour and in a very short time. The digestive gland must be excised in the field just after sampling, cut into equal pieces, and immediately frozen at −70°C (Pellerin-Massicotte et al., 1993). The histochemical reactions can then be done later in the

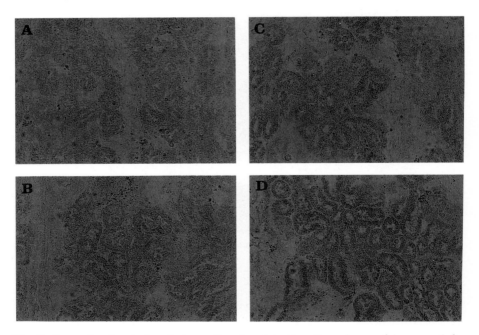

Fig. 7. Example of cytochemical reaction for detecting the release of lysosomal β-N-acetyl-hexosaminidase. A, negative control; B, slight reaction after 18 min. incubation; C, moderate reaction after 10 min. incubation; D, strong reaction after 2 min. incubation.

laboratory. The experimental protocols are described in detail in Bayne et al. (1985) and Krishnakumar et al. (1994).

The activity of various lysosomal enzymes in the tissues can also be measured by using purified fractions of cytoplasmic organelles, and choosing enzymes with a well-known physiological activity. Examples are β-N-acetyl-hexosaminidase, β-N-acetyl-D-glucosaminidase, acid phosphatases, and arylsulphatase (Patel and Patel, 1985; Cajaraville et al., 1995), present in the primary and tertiary lysosomes and in the phagocytous vacuoles. It has been demonstrated that the lysosomal activity and latency period for these enzymes in the digestive gland of *Anadara granosa* (Mollusca, Bivalva) are varied and seem to be influenced by physicochemical characteristics of the habitat of the individuals studied. Arylsulphatase activity is lower than that measured in hepatic cells of mammals, but its lysosomal latency period is comparable to that evaluated in other aquatic organisms such as lobster (Stauber et al., 1975) or certain fish such as *Tilapia mossambica* (Warrier et al., 1972).

The activity of β-N-acetyl-D-glucosaminidase in tissues in which energy macromolecules are stored reflects the lysosomal activity in these cells and indicates the degree of mobilization of these reserves by autophagy (Peek and Gabbot, 1990). Measurement of this lysosomal enzyme activity has also been used to characterize the condition of reproductive functions in bivalves. The production of gametes requires a significant nutrient input, which could come from metabolic reserves contained in the tissues (Pipe, 1987). In the cold estuarine environment, the egg-laying period of *Mytilus edulis* (Mollusca, Bivalva) and *Mya arenaria* (Mollusca, Bivalva) occurs toward the end of spring, in May and June. Gametogenesis resumes at the end of summer in *M. edulis*, perhaps because of thermal stress, and in autumn in *M. arenaria*. This pattern is found in measurements of lysosomal β-N-acetyl-D-glucosaminidase activity made on the two species. Measurement of the activity of this enzyme will therefore be useful as a biomarker of the effect of an abiotic and/or biotic stress and gives indications even about the proper functioning of a precise physiological function if the lysosomes of different organs are used.

Laboratory experiments gave the opportunity to evaluate the sensitivity of lysosomal responses, whether for toxic substances or for essential elements. However, few studies have shown the relevance of the use of lysosomal responses as biomarker of exposure to environmental xenobiotics in indigenous populations of aquatic bivalves, or indicated the existence of a link with growth retardation in contaminated environments (Krishnakumar et al., 1994).

3. LYSOSOMAL RESPONSES TO CONTAMINANTS IN INDIGENOUS POPULATIONS OF INVERTEBRATES AND VERTEBRATES

The lysosomal response may vary according to the organism chosen, invertebrate or fish, and according to the age class considered. In Bivalvia, for example, hydrolytic enzymes are released in the cytoplasm after exposure to hydrocarbons, to modify the fluidity of lysosome membranes, which causes an increase in the fusion of lysosomes and an increase in their volume (Lowe et al., 1981; Moore and Clarke, 1982). In fish, fragilization of the lysosome membrane precedes the release of hydrolytic enzymes. Differences depending on age of organisms in the recovery of membrane integrity after a stress have been observed in mussels, the adaptability being inferior in the oldest organisms (Hole et al., 1988). In the winkle *Littorina littorea* (Mollusca, Gasteropoda, Prosobranchea), the progress of the response is different after exposure in a laboratory to sublethal concentrations of α-naphthol (Cajaraville et al., 1989) and cadmium (Marigomez et al., 1989). A ballooning of the lysosomes is observed and precedes their fusion. It is later followed by destabilization of the lysosome membrane. However, in the case of cadmium, fragilization of the lysosome membrane appears only when storage capacity of lysosomes has been exceeded, which restricts the use of this biomarker in winkle from sites with high, but not lethal, cadmium contamination.

3.1 Lysosomal Responses to Stress in Mollusca

One of the most interesting examples of the use of lysosome membrane fragility *in situ* is provided by the comparative study of responses of *Mytilus edulis* L. (Mollusca, Bivalva) sampled in 9 sites on a pollution gradient (Krishnakumar et al., 1994). Variations in lysosomal response were estimated in several reference and contaminated sites. The study of mussels from several reference sites is a useful element, because it gives the opportunity to minimize the influence of abiotic and biotic factors by characterizing the range of normal response of considered parameters, which enables better estimation of the specific responses to the contaminated site. On the other hand, we must be aware that, in mussel, sectioning of byssus causes a significant stress, mobilizing energy reserves for its reconstruction. Therefore, some care must be given during sampling to leave the byssus intact to prevent modification and reduction of the lysosomal response reading (Pellerin-Massicotte et al., 1993).

Krishnakumar et al. (1994) have moreover attempted to correlate the time required to destabilize the lysosomal membrane, the percentage of latency of destabilization processes, N-acetyl β-hexosaminidase (NAH) activity, and quantities of neutral lipids and lipofuschin with exposure to anthropic factors (Figs. 8 and 9). They state that the lipofuschin content increases and that less time is required to destabilize the lysosomal membrane when the mussels have bioaccumulated polycyclic aromatic hydrocarbons (PAH). A significant relation with a decline in the body mass has also been observed.

A close relation between *in situ* exposure to heavy metals and an increase in hepatic contents of lipofuschin associated with a lysosomal fragilization has also been observed in *Mytilus galloprovincialis* (Mollusca, Bivalvia) (Regoli, 1992). The experimental plan implemented comprises a study of mussels in polluted environments and transfer of mussels to a less polluted environment to distinguish the responses due

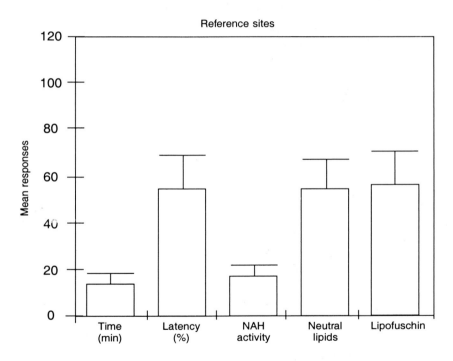

Fig. 8. Response of lysosomal markers in reference sites (according to data from Krishnakumar et al., 1994). This figure illustrates the different lysosomal responses observed *in situ* in *Mytilus edulis* in several reference sites from the Californian coast. The mean time of destabilization, percentage of latency of destabilization processes, N-acetyl β-hexosaminidase (NAH) activity, and concentration of neutral lipids and lipofuschin are shown. It may be noted that there is a relatively large variability between individuals and sites for the percentage of latency and concentration of lipids and lipofuschin.

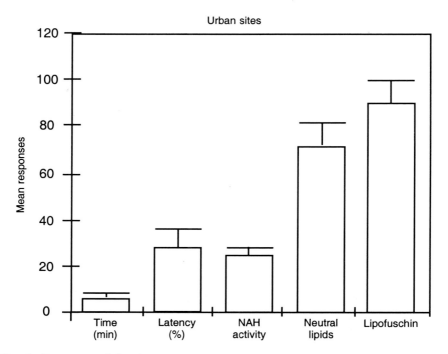

Fig. 9. Response of lysosomal markers in urban sites (according to data from Krishnakumar et al., 1994). This graph illustrates different mean lysosomal responses observed *in situ* in *Mytilus edulis* exposed at different levels of pollution on the Californian coast. The mean time of destabilization, percentage of latency of destabilization processes, N-acetyl β-hexosaminidase (NAH) activity, and concentration of neutral lipids and lipofuschin are shown. A significant reduction of mean time of lysosomal destabilization and latency percentage can be seen. NAH activity has slightly increased, while the concentrations of neutral lipids and lipofuschin show significant increases. If one compares these data to those in Figure 8, one observes that the most marked differences for these lysosomal biomarkers have been obtained for latency percentage, time of destabilization, and lipofuschin content.

to natural fluctuations from those due to exposure to metals. The results obtained are convincing and show the sensitivity of lysosomes to metallic pollutants present in the environment. The same protocol was used with mussel and soft-shell clams (*Mya arenaria*) in a cold estuarine environment. The results concerning the destabilization of the lysosomal membrane of the digestive gland (Fig. 10) showed the existence of a direct relation with sublethal levels of heavy metals such as mercury and zinc (Pellerin-Massicotte et al., 1993) and with the level of energy reserves in the gonads (Figs. 11 and 12) and in reserve tissues such as the mantle.

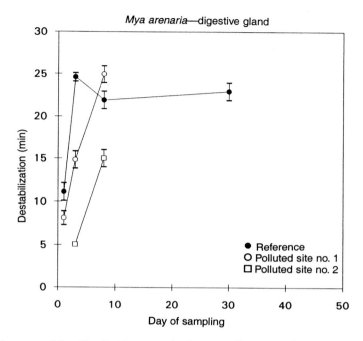

Fig. 10. Lysosomal fragility in *Mya arenaria* along a pollution gradient. These results are taken from an experimental transfer of *Mya arenaria* from a reference site to the same site and to two sites in Baie des Anglais (Baie Comeau, Quebec, Canada) that represent a pollution gradient. The transfer produced a stress characterized by the degree of fragility of the lysosome membrane on the day following the transfer, in the reference site. In this reference site, on day 3 the physiological condition of individuals returned to normal within 23 min. At the least polluted station (site 1), the stress lasted a few days more with a moderate state of stress perceptible at day 3, and a return to normal after 8 days, which was not the case in the other polluted site (site 2), where the stress was significant on day 3 and moderate on day 8.

The lysosomal response in target organs such as gonads is slightly different and more complex to interpret. In fact, in *M. galloprovincialis* exposed to crude oil in the laboratory, β-glucuronidase activity of lysosomes of ovocytes was high, whereas the lysosomes were small and numerous. It seems, therefore, that the ovocytes develop a response different from that observed in the somatic cells such as digestive cells, for example. The budding of lysosomes leads to fission rather than to fusion after exposure to petroleum hydrocarbons (Cajaraville et al., 1991). The use of this biomarker for ovocytes is therefore not recommended before thorough studies can be undertaken, given the uncertainties about the mechanisms at work and the interpretation that one can make of the responses observed.

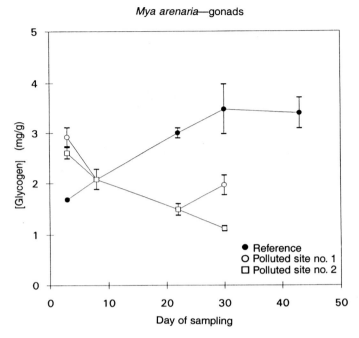

Fig. 11. Variations in glycogen concentrations in *Mya arenaria* along a pollution gradient (modified from Pellerin-Massicotte et al., 1993). The experimental conditions are the same as for the results presented in Fig. 10. The glycogen concentration in gonads of *Mya arenaria* could be related to the level of destabilization of the lysosome membrane in the reference site and site 2. In site 1, environmental factors other than anthropic ones could be the cause of the decline in glycogen concentration.

3.2 Lysosomal Responses to Stress in Fishes

The lysosomal system of cells of aquatic organisms is particularly sensitive to pollutants and several tests have been developed on fish to determine which part of the lysosomal function is affected by xenobiotics. A non-specific decrease in endocytosis was observed in dab (*Limanda limanda*) from highly contaminated zones (Moore et al., 1994). Hepatocytes from individuals living along a contamination gradient in the plume of the Elbe in the North Sea have been used *in vitro*. In *Platichthys flesus*, a fish from sites representing a contamination gradient of the Elbe, pronounced and severe histopathological hepatic lesions were closely correlated to concentrations of contaminants and with fragilization of the lysosome membrane of hepatic cells, suggesting a lysosomal response specific to the presence of contaminants (Kohler, 1989).

Lysosomes in fish respond to inorganic as well as organic contaminants and the response develops in two stages. The first stage is

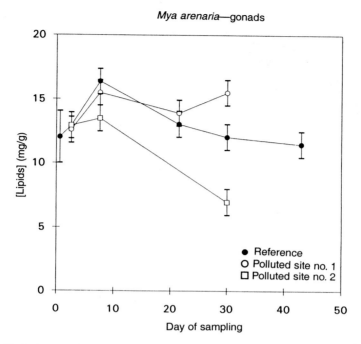

Fig. 12. Variations in lipid concentrations in *Mya arenaria* along a pollution gradient (modified from Pellerin-Massicotte et al., 1993). The experimental conditions are the same as for the results presented in Fig. 10. The lipid concentration in gonads of *Mya arenaria* could be related to the level of destabilization of the lysosome membrane, for all the three sites.

characterized by an increase of number and size of lysosomes, associated with the accumulation of pigments and intralysosomal lipids, which constitutes one protective response. The fragilization of the lysosome membrane constitutes the second stage, with the release of hydrolytic enzymes in the cytoplasm and nucleoplasm, leading to the formation of hepatic lesions (Kohler, 1991). It has been suggested that the membrane integrity of liver lysosomes in fish may be correlated with the adaptive capacity of hepatic cells, and fragilization of the lysosome membrane may be associated with dysfunction of the detoxifying capacity of the liver and irreversible pathological modifications (Kohler, 1989, 1991).

4. APPLICABILITY AT THE ECOSYSTEM LEVEL

The fragility of the lysosomal membrane has significant potential as a biomarker that can be used in the field because it integrates the effects of various stress factors. It meets all the criteria of a good biomarker. In particular, it is relatively easy to quantify the fragility of the lysosomal membrane in several individuals in response to various levels of exposure to the stress agent (Pellerin-Massicotte et al., 1993). Measurement of the fragility of the lysosome membrane is biologically useful because it is linked to biological processes with demoecological significance such as reproduction and growth (Tremblay and Pellerin-Massicotte, 1996; Kohler, 1989, 1991).

As with all biomarkers, the variability of the response due to biotic and abiotic factors must be well controlled and understood. This variability, which is due to a natural genetic heterogeneity, is itself useful for population studies. The diversity of age classes and size classes as well as the different exposure to chemical and environmental stress factors (dissolved oxygen, suspended solids, predation, food quality, sampling and handling) can also generate a variability of responses. This variability can be controlled by:

- acquiring a thorough knowledge of field conditions, at a small, medium, or large spatio-temporal scale;
- having a very good understanding of the physiology of the organism chosen.
- studying the responses of autochthonous organisms and comparing them to those of organisms transferred or caged in reference sites and polluted sites (Pellerin-Massicotte et al., 1993; Pellerin-Massicotte, 1994).
- choosing organisms of the same size and homogenous in terms of their reproductive status;
- linking measurements of energy reserves (lipids and glycogen) in two key tissues, such as the digestive gland and the gonads, and according to a temporal scale appropriate to the organism chosen;
- associating the measurement of lysosomal fragility to other biomarkers that are complementary, which helps in the identification of populations subject to stress (Pellerin-Massicotte et al., 1993).

In our view, the potential for use of this cytological marker is significant, especially because of its availability at all trophic levels of the ecosystem and in every tissue of an individual.

REFERENCES

Abok K., Rundquist I., Forsberg B. and Brunk U. (1984). Dimethylsulfoxide increases the survival and lysosomal stability of mouse peritoneal macrophages exposed to low-LET ionizing radiation and/or ionic iron in culture. *Virchows Archiv. (Cell Pathol.),* 46:307–320.

Bayne B.L., Brown D.A., Burns K. and Dixon D.R. (1985). *The Effects of Stress and Pollution on Marine Animals.* Praeger Publishers, New York.

Bitensky L., Butcher R.S. and Chayen J. (1973). Quantiative cytochemistry in the study of lysosomal function. In: Lysosomes in Biology and Pathology, Vol. 3. Dingle J.T. (ed.). Elsevier, Amsterdam, pp. 465–510.

Bonin M.L. (1972). The relationship of lysosomal stability and plasma enzyme levels to stress. Dissertation submitted to the Faculty of Graduate School, Marquette University, Milwaukee, Wisconsin.

Cajaraville M.P., Marigomez J.A. and Angulo E. (1989). A stereological survey of lysosomal structure alterations in *Littorina littorea* exposed to l-naphthol. *Comp. Biochem. Physiol.*, 93C:231–237.

Cajaraville M.P., Marigomez J.A. and Angulo E. (1991). Automated measurement of lysosomal structure alterations in oocytes of mussels exposed to petroleum hydrocarbons. *Arch. Environ. Contam. Toxicol.*, 21:395–400.

Cajaraville M.P., Robledo Y., Etxeberria M. and Marigomez I. (1995). Cellular biomarkers as useful tools in the biological monitoring of environmental pollution: Molluscan digestive lysosomes. In: Cell Biology in Environmental Toxicology. Cajaraville M.P. (ed.). University of the Basque Country Press Service, Bilbao, pp. 29–55.

Decker R.S. (1977). Lysosomal properties during thyroxine-induced lateral motor column neurogenesis. *Brain Res.*, 132:407–422.

George S.G. (1983). Heavy metal detoxication in *Mytilus* kidney: an *in vitro* study of Cd- and Zn-binding to isolated tertiary lysosomes. *Comp. Biochem. Physiol.*, 76C:59–65.

Ham A.W. (1974). *Histology*. J.B. Lippincott Company, Philadelphia.

Hole L.M., Moore M.N. and Belfamy D. (1988). Age-related effects of stress on molluscal lysosomal stability. *J. Histochem. Cytochem.*, 36:862.

Holtzman E. (1989). *Lysosomes*. Plenum Press, New York.

Köhler A. (1989). Cellular effects of environmental contamination in fish from River Elbe and the North Sea. *Mar. Environ. Res.*, 28:417–424.

Köhler A. (1991). Lysosomal perturbations in fish liver as indicators for toxic effects of environmental pollution. *Comp. Biochem. Physiol.*, 100C:123–127.

Krishnakumar P.K., Casillas E. and Varanasi U. (1994). Effect of environmental contaminants on the health of *Mytilus edulis* from Puget Sound, Washington, USA. 1. Cytochemical measures of lysosomal responses in the digestive cells using automatic image analysis. *Mar. Ecol. Prog. Ser.*, 106:249–261.

Lowe D.M., Moore M.N. and Clarke K.R. (1981). Effects of oil on digestive cells in mussels: quantitative alterations in cellular and lysosomal structure. *Aquat. Toxicol.*, 1:213–226.

Marigomez J.A., Vega M.M., Cajaraville M.P. and Angulo E. (1989). Quantitative responses of the digestive lysosomal system of winkles to sublethal concentrations of cadmium. *Cell. Mol. Biol.*, 35:555–562.

Moore M.N. (1976). Cytochemical demonstration of latency of lysosomal hydrolases in digestive cells of the common mussel, *Mytilus edulis* and changes induced by thermal stress. *Cell Tiss. Res.*, 175:279–287.

Moore M.N. (1988). Cytochemical responses of the lysosomal system and NADPH-ferrihemoprotein reductase in molluscan digestive cells to environmental and experimental exposure to xenobiotics. *Mar. Ecol. Prog. Ser.*, 46:81–89.

Moore M.N. (1990). Lysosomal cytochemistry in marine environmental monitoring. *Histochem. J.*, 22:187–191.

Moore M.N. (1991). Environmental distress signals: cellular reactions to marine pollution. In: Histo- and cytochemistry as a tool in environmental toxicology. Graumann W. and Drukker J. (eds.). *Prog. Histochem. Cytochem.*, 23:2–19.

Moore M.N. and Clarke R.K. (1982). Use of microstereology and quantitative cytochemistry to determine the effects of crude oil-derived aromatic hydrocarbons on lysosomal structure and function in a marine bivalve mollusc, *Mytilus edulis. Histochem. J.*, 14:713–718.

Moore M.N. and Stebbing A.R.D. (1976). The quantitative cytochemical effects of three metal ions on a lysosomal hydrolase of a hydroid. *J. Mar. Biol. Assoc. U.K.*, 56:995–1005.

Moore M.N., Koehn R.K. and Bayne B.L. (1980). Leucine aminopeptidase (Aminopeptidase l), N-Acetyl-β-hexosaminidase and lysosomes in the mussel, *Mytilus edulis* L., in response to salinity changes. *J. Exp. Biol.*, 214:239–249.

Moore M.N., Köhler A., Lowe D.M. and Simpson M.G. (1994). An integrated approach to cellular biomarkers in fish. In: Non-Destructive Biomarkers in Vertebrates. Fossi M.C. and Leonzio C. (eds.). Lewis Publishers, Boca Raton.

Patel S. and Patel B. (1985). Effect of environmental parameters on lysosomal marker enzymes in the tropical blood clam *Anadara granosa*. *Mar. Biol.*, 85:245–252.

Peek K. and Gabbott P.A. (1990). Seasonal cycle of lysosomal enzyme activities in the mantle tissue and isolated cells from the mussel *Mytilus edulis*. *Mar. Biol.*, 104:403–412.

Pellerin-Massicotte, J. (1994). Oxidative processes as indicators of chemical stress in bivalves. *J. Aquat. Ecosys. Health.*, 3:101–111.

Pellerin-Massicotte J., Vincent B. and Pelletier E. (1993). Évaluation écotoxicologique de la qualité de la baie des Anglais (Québec). *Wat. Poll. Res. J. Can.*, 28:665–689.

Philip B. and Kurup P.A. (1977). Cortisol and lysosomal stability in normal and atheromatous rats. *Atherosclerosis*, 27:129–139.

Pipe R.K. (1987). Ultrastructural and cytochemical study on interactions between nutrient storage-cells and gametogenesis in the mussel *Mytilus edulis*. *Mar. Biol.*, 96:519–528.

Porta R., Niada R., Pescador R. et al. (1986). Gastroprotection and lysosomal membrane stabilization by sulglicotide. *Arzneim.-Forsch./Drug Res.*, 36:1079–1082.

Regoli F. (1992). Lysosomal responses as a sensitive stress index in biomonitoring heavy metal pollution. *Mar. Ecol. Prog. Ser.*, 84:63–69.

Ricciutti M.A. (1972). Myocardial lysosome stability in the early stages of acute ischemic injury. *Am. J. Cardiol.*, 30:492–497.

Stauber W.T., Canonico P.G., Milanesi A.A. and Bird J.W.C. (1975). Lysosomal enzymes in aquatic species IV. *Comp. Biochem. Physiol.*, 50B:379–384.

Tremblay R. (1992). Variabilité à court terme de paramètres physiologiques chez *Mya arenaria*. Dissertation presented at the University of Quebec at Rimouski.

Tremblay R. and Pellerin-Massicotte J. (1996). Effect of the tidal cycle on lysosomal membrane stability in the digestive gland of *Mya arenaria* and *Mytilus edulis* L. *Comp. Biochem. Physiol.*, 117A:99–104.

Waldron-Edward D. and Greenberg L. (1980). Lysosomal stability and mucosal resistance to injury. *Int. Congress Ser.*, 537:87–100.

Warrier S.B.K., Ninjoor V., Sawant P.L. et al. (1972). Differential release of latent lysosomal hydrolases in muscle of *Tilapia mossambica* by whole body gamma irradiation. *J. Biochem. Biophys.*, 9:278–279.

Wilson J.S., Korsten M.A., Apte M.V. et al. (1990). Both ethanol consumption and protein deficiency increase the fragility of pancreatic lysosomes. *J. Lab. Clin. Med.*, 115:749–755.

Wilson J.S., Apte M.V., Thomas M.C. et al. (1992). Effects of ethanol, acetaldehyde and cholesteryl esters on pancreatic lysosomes. *Gut*, 33:1099–1104.

12

Molecular, Genetic and Population Bases of Insect Resistance to Insecticides

M. Amichot, J.-B. Bergé, A. Cuany, N. Pasteur, D. Pauron and M. Raymond

INTRODUCTION

All living beings are constantly subjected to physical, chemical, or biological stresses. Many of these stresses are of 'natural' origin (e.g., food toxins); others are due to human activities (e.g., industry, agriculture). Adaptation to these stresses is a necessary condition for the survival of living species, and resistance, which is a hereditary adaptation, is one of its consequences (Sawicki and Denholm, 1984).

This chapter addresses specifically the resistance of invertebrates, particularly of insects, to insecticides. At present, it is difficult to know whether the resistance genes existed before treatments or whether they are the result of genome modifications (mutations), but it is clear that their selection by insecticides profoundly alters the genetic characteristics of populations. Most individuals that, before the beginning of treatments, have alleles of sensitivity are replaced, after a few treatment cycles, by individuals with resistant alleles. In this sense, resistance is an ecotoxicological manifestation of the use of pesticides. Can it, as such, be used as a biomarker? Conventionally, resistance is considered a biomarker of individual sensitivity, expressing a reduction in the capacity of individuals to respond to exposure to toxicants (Koeman et al., 1993; Timbrell et al., 1994). The experimental results that are presented and discussed in this chapter show that, if considered as a variable that integrates the 'toxicological history' of the population, resistance can be used as a biomarker of long-term effect of exposure of natural populations to toxins in their environment.

The numerous researches on biochemical mechanisms of resistance and the genes that control them (see the works of Green et al., 1990; Roush and Tabashnick, 1990; Mullin and Scott, 1992; and the syntheses of Poirie and Pasteur, 1991 and Berge et al., 1996) are the source of much of our biochemical and molecular knowledge of enzyme or protein systems affected by exposure to pesticides: induction of biotransformation enzymes (Terriere, 1983, 1984; Cohen, 1986; Capua et al., 1991; Lagadic, 1991; Lagadic et al., 1993; Scott et al., 1996; Amichot et al., 1998) and inhibition of target enzymes (Yamamoto et al., 1983; Devonshire and Moores, 1984; Raymond et al., 1985; Hall and Spierer, 1986; Fournier et al., 1993). These systems interact directly with xenobiotics and most can be characterized at the individual level (Pasteur and Georghiou, 1989; Dary et al., 1990, 1991; De Sousa et al., 1995), which is essential if they are to be used as biomarkers of the current or past presence of toxicants.

1. BIOCHEMICAL AND GENETIC MECHANISMS OF RESISTANCE OF INSECTS TO INSECTICIDES

There are several types of mechanisms of insect resistance to insecticides. Behavior-based resistance is very little understood (Sparks et al., 1989), but certainly merits closer study within the framework of the characterization of biomarkers of resistance. The most frequent mechanisms of resistance are of 'physiological' and mostly 'biochemical' origin (Fig. 1). They result from a modification of the molecular target of insecticides and/or an increase in the efficacy of protein systems involved in the detoxification of these products (Brown, 1990; Soderlund and Bloomquist, 1990; Poirie and Pasteur, 1991; Berge et al., 1996).

1.1 Targets of Insecticides

Insecticides can be classified as a function of the protein target on which they act. The very large majority of them disturb the functioning of the nervous system (Fig. 2). Organophosphorus compounds and carbamates are anticholinesterasic. They act by inhibiting the activity of AChE (Bocquene et al., 1997), an enzyme essential to the transmission of nerve impulses in the cholinergic synapses located in the central nervous system. The pyrethroids and certain organochlorine compounds such as DDT act on voltage-dependent sodium channels or CNaV (Narahashi,

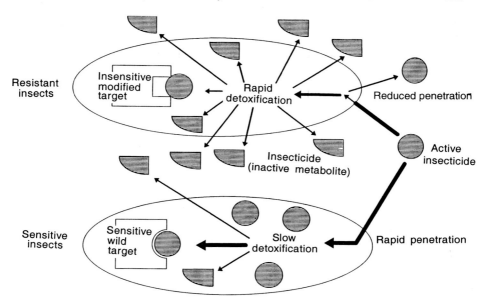

Fig. 1. The most common physiological and biochemical mechanisms of resistance of insects to insecticides: slow penetration of the insecticide into the insect, detoxification by an enhancement of the metabolism of the insecticide, and modification of the molecular target of the insecticides.

1986; Sattelle and Yamamoto, 1988; Soderlund and Bloomquist, 1989; Narahashi et al., 1992; Bloomquist, 1993), the functional integrity of which is essential to the depolarization of nervous membranes during the passage of the action potential. Pyrethroids delay the closing of this channel and disturb the functioning of the entire nervous system. The last class of insecticides with a well-known mode of action is that of organochlorine compounds of the dieldrin type. These molecules bond with GABA receptor (GABAr) and inhibit the functioning of the associated chlorine channel (Eldefrawi and Eldefrawi, 1988; Lunt et al., 1988; Sattelle, 1990; ffrench-Constant, 1991; Lee et al., 1993). The opening of this channel induces a hyperpolarization of the nervous membrane and its inactivation, if prolonged, disturbs the overall function of the nervous system.

The sensitivity of targets is a function of their affinity for the pesticide. It depends on the structure of the binding site, itself a function of the sequence of amino acids of the protein. This explains why (or why not) occasional mutations that diminish the affinity of the insecticide for the target protein result in the latter's insensitivity to the toxin and finally in the resistance of the animal. For example, a mutation transforming a serine into alanine is responsible for the resistance of GABAr to dieldrin in all dipteran or coleopteran insects studied to date

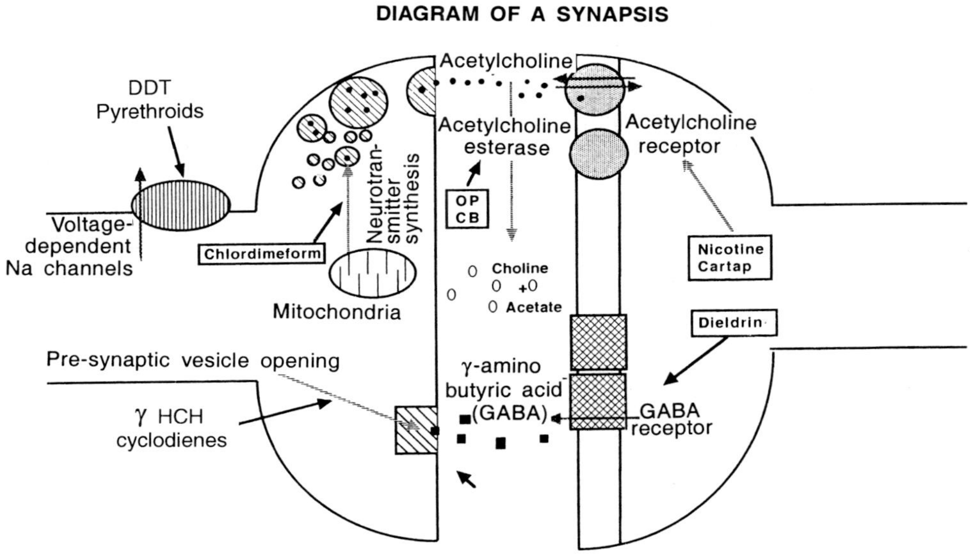

DIAGRAM OF A SYNAPSIS

Fig. 2. Principal modes of action of insecticides presently used on target proteins, the majority of which are found in synapses of the nervous system. The organophosphorus compounds (OP) and carbamates (CB) act by inhibiting AChE activity. The pyrethroids and DDT act by keeping open the voltage-dependent sodium channel. Dieldrin acts by hampering the opening of the chlorine channel associated with GABA receptor. Cartap and nicotine act on acetylcholine receptors. γ-HCH (lindane) and cyclodianes act on the opening of presynaptic vesicles that release neurotransmitters in the intersynaptic space.

(ffrench-Constant et al., 1993a, 1994). Similarly, AChEs become highly resistant when several mutations, present at the same time, modify the structure of the protein (Mutero et al., 1994). The high resistance of the sodium channel to pyrethrinoids and to DDT is also associated with a low number of mutations (Taylor et al., 1993; Williamson et al., 1996), although atypical mutations have been identified in certain cases of low resistance (Amichot et al., 1992).

Because their function may be disturbed following interaction with insecticides, the targets of these molecules can serve as biomarkers of present or past contamination of organisms. It is relatively easy to study functional modifications in targets when the target is an enzyme whose activity or inhibition can be measured. In the case of AChE, for example, inhibition may be used to indicate an exposure to anticholinesterasic substances (Bocquene et al., 1997). Such parameters, which rapidly or even instantaneously respond to exposure, are mostly used in the aquatic environment, especially on fishes and lamellibranch molluscs (Bocquene et al., 1997; see chapter 6). However, this approach is useful only if the sensitivity of AChEs to pesticides is known. Tests enabling measurement of the characteristics of AChE inhibition have been set up on insects to detect resistant AChEs. They may be used as ecotoxicological tests (Byrne and Devonshire, 1991) in order to indicate the effects of possible past exposures to xenobiotics, though it may not always be possible to identify with certainty the nature of products that are the origin of the modification of AChE.

Measurement of alteration in the functioning of ionic channels is more tricky because it requires the use of electrophysiological techniques. However, if the mutation or mutations responsible for the resistance correspond to a site of action of a DNA restriction enzyme, the resistance can be detected by relatively simple RFLP techniques. This is the case, in many resistant insects, with the GABA receptor, which can therefore be considered a potential biomarker (ffrench-Constant et al., 1993a; Bloomquist, 1994; Cousteau and ffrench-Constant, 1995).

1.2 Detoxication Enzymes

Any molecule that penetrates an organism, whether that organism is sensitive or resistant, binds with proteins and undergoes metabolic transformations. Insecticides are no exception to this rule. Numerous enzymes participate in the detoxification of insecticides, but three types of enzymes are essential for resistance: the esterases, glutathione S-transferases, and monooxygenases associated with cytochrome P450 (Table 1). The challenge posed to the researcher for the past ten years has been to understand why these activities are higher in resistant insects.

Table 1. Characteristics of the principal biotransformation enzymes known to be responsible for insecticide resistance cases

| Enzyme | Catalysed reaction | Resistance mechanism | | | Examples |
		Enzyme overproduction	Gene amplification	Occasional mutation	
Esterases (EST)	Hydrolysis of insecticide	No	No	Probable	Specific resistance to malathion of *Trilobium castaneum* and *Lucilia cuprina*
	Fixing of insecticides	Yes	Yes	?	Esterases A2, A4, A5, B1, B2, B4, B5, B7 of mosquitoes Esterases E4 and FE4 of aphids
		Yes	No	?	Esterase A1 of *Culex*
Monooxygenase cytochrome P450 (MFO)	Oxidation	Yes	No	Probable	Resistance of drosophila and housefly
Glutathione S-transferases (GST)	Conjugation of insecticide with glutathione	Yes	No	?	Resistance of drosophila and housefly

1.2.1 Esterases

The molecules of the principal groups of insecticides (organophosphorus compounds, carbamates, pyrethroids) have an ester function, essential to the expression of their toxicity. There are two types of esterases involved in the metabolism of insecticides. The first hydrolyse carboxylic ester bonds. For example, malathion contains two ethyl ester bonds that are hydrolysed by a malathion-carboxylesterase (or MCE) that releases one or two carboxylic functions, the negative charge of which prevents the molecule from positioning itself on the anionic site of AChE, which is also negatively charged. Certain insects resist malathion because of an increase in MCE activity. Purification of this esterase in the coleopteran *Tribolium castaneum* enabled the demonstration that the difference between sensitive and resistant individuals is due to an increase in its specific activity, the origin of which is probably linked to one or several occasional mutations modifying the sequence of amino acids of the enzyme (Haubruge, 1995).

The second type of esterase is characterized by a high activity on chromogenous substrates (e.g., naphthylacetate) in resistant individuals, while this activity is very low in sensitive animals. In fact, in aphids (Devonshire and Moores, 1982) and mosquitoes (Fournier et al., 1987; Mouches et al., 1987), the increase in activity is due to an overproduction of proteins with naphthylacetate-hydrolase activity, itself due, in most cases, to an amplification of corresponding genes (Mouches et al., 1986; Field et al., 1988; Devonshire and Field, 1991). This amplification can reach significant levels. Some resistant mosquitoes have 250 times more DNA coding for these esterases than sensitive ones (Mouches et al., 1987). In mosquitoes, there are at least two loci (*est*) that code for these proteins responsible for resistance to organophosphorus compounds. The locus *est*-3 produces EST-A, while the locus *est*-2 produces EST-B. Each of these loci presents several allelic forms that can be differentiated by electrophoresis (EST-A1, EST-A2, etc.; EST-B1, EST-B2, etc.). The resistant insects are often characterized by the association of the overproduction of particular alleles of these two loci (for example, EST-A2 is overproduced with EST-B2, EST-A4 with EST-B4, etc.). In this case, the genes coding for these two types of esterases are coamplified (Rooker et al., 1996). The catalytic activity of these esterases on insecticides is relatively low but they effectively fixed them (Cuany et al., 1993), playing thus the role of fixing-proteins. In these conditions, it is logical that individuals having a high quantity of fixing-esterases have a high level of resistance. Theoretically, the level of resistance is correlated with the degree of overproduction of the fixing-protein, itself dependent on the level of amplification of the gene that codes for it.

1.2.2 Cytochrome P450 Monooxygenases

Monooxygenases correspond to a polyproteic enzymatic system that comprises cytochrome P450 (P450) and electron transfer proteins (P450-reductase, cytochrome b5 and its reductase). The P450 contains the site on which the substrate is fixed. It is therefore responsible for the specificity of the substrate and the type of transformation it will undergo. In most cases, it is also the limiting element for the rapidity of the reaction. P450 are present in all organisms, from bacteria up to plants, invertebrates, and higher mammals. They can metabolize most organic substrates. Some of these substrates are of endogenous origin (steroid hormones, fatty acids, etc.), and others are xenobiotic (pesticides, toxins in foods) (Estabrook, 1996). This lack of specificity is only apparent as there are, in each organism, a large number of P450 isoforms. More than 800 genes coding for P450s have been sequenced (Nelson, website). This great diversity has necessitated the establishment of a general nomenclature of P450, based on the homologies of sequences between the proteins of this superfamily (cytochrome P450). When two P450 have more than 40% homology, they belong to the same family, designated by an arabic numeral. If the homology is higher than 60%, the P450 belong to the same subfamily designated by a capital letter. If the homology is higher than 80%, they are considered identical genes, again designated by an arabic numeral. Thus, CYP1A1 is the first protein of P450 sequenced in the subfamily A and in the family 1. The DNA coding for the P450 is designated by the same code but in italics (the gene *CYP1A1* codes for the protein CYP1A1). The first insect P450 was cloned in the housefly (Feyereisen et al., 1989). Since then, several other DNA have been isolated in many families. One of them, the family *CYP4*, is relatively ubiquitous and comprises numerous genes (Snyder and Davidson, 1983; Bradfield et al., 1991; Gandhi et al., 1992; Frolov and Alatorsev, 1994; Amichot et al., 1994; Scott et al., 1994; Dunkov et al., 1996). Other families, such as *CYP6* (Feyereisen et al., 1989; Berenbaum et al., 1992; Carino et al., 1992, 1994; Waters et al., 1992; Cohen and Feyereisen, 1995; Hung et al., 1995; Maitra et al., 1995; Tomita and Scott, 1995; Xiao-Ping and Hobbs, 1995; Scott et al., 1996), *CYP9* (Hodgson et al., 1995; Dunkov et al., 1996) and *CYP18* (Dunkov et al., 1996), are more specific to insects.

Cytochrome P450s are essential for the adaptation of organisms to environmental variations, whether 'natural' or associated with human activities. For example, some insects can feed on plants that are naturally toxic to other organisms because they possess P450 capable of degrading the toxic substances (especially phytotoxins) synthesized by the plants to defend themselves from this type of aggression (Cohen et al., 1989; Lee and Berenbaum, 1989). The P450s are often compared to compounds of immune systems because, like the latter, they have a wide variability and are implicated in defence reactions. This comparison is supported by the fact that the biosynthesis of many P450s is induced by the presence of the molecules they metabolize. This is the phenomenon of induction (Ahmad et al., 1986; Yu, 1986; Brattsten, 1988; Mullin, 1988; Snyder et al., 1993).

The understanding of P450 induction is less developed in invertebrates than in vertebrates. In insects, particularly drosophila, many molecules are known to induce a transitory increase in enzymatic activities associated with P450: EROD, ECOD, hydroxylations of various sites of testosterone, etc. (Amichot et al., 1998). While an increase in activity on some of these can be attributed specifically to a given P450 in vertebrates (e.g., EROD activity is representative of CYP1A1), such relations have not been clearly established in invertebrates. To know the specific level of a P450, immunological measurements must be used (western blot) or the level of mRNA must be quantified (northern blot). It has been shown that, in drosophila, certain products, such as phenobarbital, are 'general' inducers, causing the synthesis of mRNA of P450 of the CYP6A and CYP4 type, while others, such as prochloraze, are more specific (Amichot et al., 1994; Brun et al., 1996). Antibodies are available to measure the levels of CYP6A2 in drosophila, and nucleic acid probes to measure the levels of the mRNA corresponding to several genes of families CYP6 and CYP4 have also been elaborated. We are presently developing other probes to measure the expression level of a larger number of P450 and enable, at some point, a more precise identification of their inducers.

Apart from induction, which is an 'instantaneous' adaptation of organisms to survive a toxic xenobiotic (Brattsten, 1988), some individuals have genotypes adapted permanently to the presence of toxins. This is the phenomenon of resistance, which occurs in all organisms (e.g., resistance of plants to herbicides, resistance of fungi to fungicides, resistance of insects to insecticides) (Ford et al., 1987). In most cases, the resistance of insects to insecticides linked to the intervention of P450 is characterized by a constitutional increase in the level of an enzyme or by a modification of the specific activity of the enzyme on insecticides (Brun, 1996; Scott, 1996). In nature, individuals resistant to insecticides by P450-dependent metabolism can be detected by qualitative methods (analysis of specific activity) or quantitative methods (analysis of level of CYP proteins or P450 mRNA).

P450s can be used as biomarkers because of their inducibility and/ or the role they may play in resistance to pesticides (Lagadic et al., 1994). The use of P450 induction as biomarker of exposure to pollutants has been well described in several organisms, vertebrates and invertebrates (Monod, 1997; Narbonne and Michel, 1997; see chapters 3 and 4).

In vertebrates, the use of P450 induction as biomarker of exposure to toxins is frequently limited by ignorance about reference levels, because, at present, very little information is available on the biological characteristics (enzymatic polymorphism, level of adaptation of detoxification systems, etc.) of species in which this phenomenon could be studied. This knowledge is, however, essential for the use of sentinel species. For example, since P450 is constitutionally overexpressed in strains of insect resistant to insecticides and it is no longer inducible, it seems evident that these strains cannot be used to study P450-type biomarkers. Conversely, when resistance derives from a modification of a specific P450 activity, that activity can possibly be used as biomarker of the effect of repeated exposures to toxins. Given the manner in which resistance develops (see later), P450-dependent activities of resistant individuals may reconstruct part of the 'toxicological history' of the population, as we have mentioned at the beginning of this chapter.

In such a context, one of the advantages of using invertebrates is the possibility of using laboratory models that are well known genetically, such as drosophila (for the aerial environment) (Morton, 1993) or the nematode *Caenorhabditis elegans* (for the edaphic environment). These models can be used either by caging laboratory colonies in the field, or by collecting adequate information directly on autochthonous populations, because these organisms are relatively ubiquitous.

1.2.3 Glutathione *S*-transferases

Glutathione *S*-transferases (GSTs) represent another type of biotransformation enzymes that could be implicated in resistance of insects to insecticides (Clark, 1989, 1990; Vos and Van Bladeren, 1990; Yu, 1996). Unfortunately, little information has been acquired on the genetic mechanisms at the origin of GST-based resistance in insects. The few data available on the housefly (Fournier et al., 1992) suggest that it must be a mechanism similar to that of P450, i.e. an increase in GST quantity without genic amplification and enhancement of a particular GST activity.

2. MECHANISMS OF DISPERSAL OF GENES OF RESISTANCE TO INSECTICIDES

Generally, resistance to pesticides can have very important consequences for ecological balance. A well-known case of resistance in arthropods is that of populations of secondary pests of certain crops, following the appearance of resistance in these species but not in their natural enemies (Croft and Stricker, 1983; Sawicki et al., 1989). There are, however, few thorough studies of these phenomena. One of the avenues of present research concerns the study of the consequence of an insect's resistance on its parasites, particularly on the immunological responses of the host (Delpuech et al., 1996).

2.1 Dynamics of the Development of Resistance

Research that attempts to identify and quantify the factors that influence the dynamics of resistance in natural populations is, by contrast, well developed. Phenomenologically, spatial evolution of resistance is always the same. Resistance appears in a geographically limited area and, if the selection pressure continues, it extends gradually to increasingly larger areas. The resistance factor (RF), defined as the ratio of resistant individuals to sensitive individuals, increases in parallel with the geographic extension of resistance. For example, the first case of resistance in the mosquito *Culex pipiens* in the south of France was identified in 1972 near Lunel in Herault. In 1978, resistant mosquitoes were present in the coastal area between the Rhone valley and the Spanish border, and the RF reached 80. In 1992, the resistance had invaded the entire French coast and also existed in Spain, northern Italy, and Corsica with an RF value of 200.

2.2 Intra- and Interspecific Variability

While certain populations of target insects rapidly develop resistance to most of the insecticides to which they are successively exposed, many other populations of the same species or closely related species remain relatively sensitive, even though they may be subjected to similar patterns of treatment with the same products (Roush and McKenzie, 1987). Variations in structure and size of the populations probably contribute to these differences.

The development of resistance depends mainly on interactions among four factors:

- mutations, which are the origin of resistance alleles (or genes);
- migration, which enables these genes to spread outside their original geographical area;
- selection, which retains the genes best adapted to environmental conditions;
- genetic drift, a random phenomenon associated with small populations.

The nature of mutations at the origin of resistance alleles have already been described in the earlier sections, therefore they will not be discussed here again. However, the question of their frequency is a key issue in the study of resistance. The case of *C. pipiens* is one of the best known from this point of view, owing to the study of amplified esterases with a widely varying geographical distribution. The coamplified genes *est*A4 and *est*B4, as well as *est*A5 and *est*B5, exist only around the Mediterranean Sea. The amplified gene *est*B1 is relatively widely distributed, in North and South America, the Caribbean region, and China. However, the greatest geographic distribution remains that of the coamplified genes *est*A2 and *est*B2, which have been found practically throughout the area of distribution of the species. The variability of the amplicon containing these genes has been analysed in mosquitoes of widely varying geographic origins, by two independent methods. First, the patterns of restriction of DNA hybridizing with a clone of the gene *est*B2 were studied with 13 restriction enzymes (Raymond et al., 1991). Then, an intron of gene *est*A2 was sequenced (Guillemaud et al., 1996). These analyses have revealed no variation, either in the coding DNA or in the non-coding DNA. In the first case, it could be because of a selection pressure in favour of the presence of these esterases, but it is more difficult to use the same argument for non-coding DNA, which is not subject to any known selection pressure. Unfortunately, no non-amplified haplotype alleles coding *est*A2 and *est*B2 have ever been found. However, the comparison of this result with those obtained in mosquitoes, of which the esterases A and B were not amplified, provides useful information. All the studies (Qiao and Raymond, 1995; Guillemaud et al., 1996) reveal high variability both within a given locality and between different regions. The probability that an amplicon carrying the genes coding *est*A2 and *est*B2 could be amplified two times is less than 10^{-10}. It is therefore probable that this amplification produced only once, and that its present worldwide distribution is the result of gene migration. In the south of France, where genes of resistance have been regularly mapped since 1974, these esterases appeared in 1986 near Marseilles (probably at Marignane). In eight years, they crossed the Rhone valley up to Lyon and have extended towards the east and west up to Montpellier (Rivet et al., 1993). These results are compatible with studies that have shown that gene flows are very significant in the south of France (Chevillon et al., 1995). The passive migration associated with human activities of marine and air transport may explain the presence of these esterases on different continents and in isolated islands (Pasteur et al., 1995).

2.3 Genes of Resistance and Environmental Factors

It no longer needs to be demonstrated that resistance genes are selected in the presence of insecticides and gradually become more frequent. On the other hand, the nature of the forces of selection that act on these genes when the treatments are stopped is still not well understood.

The degree to which environmental variations modify the phenotypic expression of a genotype is known as phenotypic plasticity (Hoffmann and Parsons, 1991). A phenotypically plastic genotype is considered capable of maintaining relatively high levels of fitness (adaptability) in diverse environmental conditions. However, high levels of phenotypic plasticity are associated with low levels of stress resistance. Hoffmann and Parsons (1991) suggest that, if the plasticity of response is dependent on the same mechanisms as genetic variation in the presence of a stress, increase in stress resistance can reduce the level of plasticity. Increase in stress tolerance can also lead to a decrease in fitness when stress stops. It is well established that genotypes that present very good fitness in given environmental conditions can often have low fitness in different conditions (Mueller et al., 1991). If the capacity of a population to adapt itself to a change in the environment depends on a certain genetic variability, reduction of genetic variability under the effect of pollutants can lead to an increase in resistance to specific pollutants, but could also lead to reduction in the capacity of the population to support other types of pollutants or stress of natural origin (ffrench-Constant et al., 1993b). Evaluation of the sensitivity of each 'physiotype' after exposure to toxins, coupled with the determination of relative proportions of each 'physiotype' in the natural populations exposed *in situ*, can provide information in the manner in which effects develop in the population (Depledge, 1990).

Our studies on the mosquito *C. pipiens* on an 800 km transect, between Valence (Spain) and the French-Italian border, showed high homogeneity of polymorphism of several genes that were not selected by insecticides (genes that are neutral with respect to selection), which shows that genetic exchanges between mosquitoes from different points of this zone are significant (Chevillon et al., 1995). These results indicate that any new gene appearing in one place very quickly invades the entire zone. The study of resistance genes showed a very high heterogeneity. Their frequency was extremely high in populations of treated regions and much lower in the untreated regions. This suggests that, in untreated regions, the resistance genes are cross-selected. Research is underway to measure the intensity of cross-selection acting on resistance genes. For example, the study of variations in frequency of resistance genes in hibernating females proves to be particularly useful since there is no anti-mosquito treatment during the winter. It has thus been shown that, though the frequency of genes coding for the overproduced esterases (*est*A4/*est*B4) does not undergo any modification in individuals, that of the resistant gene coding for AChE decreases from 25% to 9%, which indicates a mortality rate of carriers of this allele of about 1% a day (Chevillon, 1994).

3. ECOTOXICOLOGICAL SIGNIFICANCE OF RESISTANCE

In a book addressing the use of biomarkers in ecotoxicology, the resistance of organisms to pesticides may appear extraneous. We consider, however, that this phenomenon is one of the first 'durable' ecotoxicological manifestations of pesticide use. Indeed, resistance to pollutants significantly modifies the genomes of organisms and genetic equilibrium in the network of specific isolates in the populations.

At least theoretically, the development of resistance is faster in small, subdivided populations than in panmictic, large populations. This is particularly true when the resistance is associated with recessive characters. However, if the immigration of sensitive individuals is significant, the frequency of resistance may increase more slowly than the intensity of pesticide treatment would allow us to suppose (Tabashnik, 1990). At the same time that the size of population is reduced by the impact of treatments, the number of alleles in its gene pool diminishes. Effects linked to inbreeding may appear when the population becomes homozygous for recessive lethal genes (Maynard Smith, 1989). Such 'founder effect', which is produced when the population size is found to be considerably reduced, may accelerate speciation. This could explain why populations composed of opportunistic species having converging genetic characteristics are found in highly contaminated environments (Gray, 1989).

3.1 Resistance and Ecological Fitness

Genome modifications associated with resistance can be considerable, as for example in the case of genic amplification of esterases. Before treatment, the resistance alleles, if they exist, are in a minority. After treatment, they are the most numerous. This genome modification may pertain to the selection of alleles, which in themselves must not perceptibly disturb the physiology of the individual, even though we have shown that certain alleles can have a considerable effect on the ecological fitness of the population because of the genetic cost they entail (Chevillon, 1994).

The numerous studies performed on this subject (see the syntheses of Roush and McKenzie, 1987 and Roush and Daly, 1990) revealed interesting relations between reduction in fitness and mechanisms of resistance. In addition to the fact that most serious disadvantages are generally associated with esterase-linked resistance, probably because these enzymes represent a relatively large proportion of the proteins of resistant individuals (Devonshire and Moores, 1982; Mouches et al., 1987), resistant colonies of *Myzus persicae* can be more exposed to predation than sensitive aphids because of reduced response of the alarm pheromone (Dawson et al., 1983). However, resistance does not seem to affect fitness significantly when it is linked to an increase in monooxygenase activity. Thus, the reproductive efficacy of resistant colonies of *Metaseiulus occidentalis* (Roush and Hoy, 1981; Roush and Plapp, 1982a), *Musca domestica* (Roush and Plapp, 1982b), or *Helicoverpa armigera* (Daly et al., 1988) is only slightly or not affected, confirming the fact that intervention of microsomal monooxygenases does not cause a significant expenditure of energy (Neal, 1987; Brattsten, 1988). In the same manner, little or no reproductive disadvantage could be related to mechanisms of *kdr* type in *Musca domestica* and *Boophilus microplus*, with mechanisms of alteration of AChE in *Tetranychus urticae*, or even with MCE in *Anopheles arabiensis* and *Tribolium castaneum* (Roush and Daly, 1990).

Nevertheless it should be noted that these observations come from laboratory studies that, even when they are done with resistant colonies taken directly from the field, are not necessarily representative of real effects of the resistance on the fitness of individuals within natural populations. Massive declines in fertility, viability, and other aspects of fitness can be produced in natural populations in which genetic diversity is reduced following stress. This type of effect can, to some extent, give insight into the state of the ecosystem as a whole (Rapport, 1989).

3.2 Perspectives

Chemical pollutants present in the environment pose a selection pressure on living organisms and such a selection may lead to modifications of the frequency of genotypes within populations and to a subsequent increase in the tolerance of populations (Forbes and Forbes, 1994).

In future, a more careful consideration of resistance in the framework of evaluation of ecotoxicological effects of pesticides will depend on intensification of research on two aspects: (1) the effect of pesticide treatments on non-target organisms and (2) the repercussions of selection of resistant individuals on characteristics of populations other than genetic characteristics alone, and on the communities to which these populations belong. The first point needs to be addressed soon with regard to arthropod natural enemies for the purpose of reinforcement of regulations in development of new pesticides. For other non-target organisms, it seems that more thorough studies can be undertaken only on species that have particular characteristics, related especially to how easily they can be studied, and on the available knowledge on ecology and genetic characteristics. They may be chosen even because they are characteristic of certain communities. One species of choice for these studies is *D. melanogaster*, because of its very wide geographical distribution and above all the advanced state of present knowledge of its genome (Morton, 1993). Moreover, the fact that parasites of this species are well studied identifies it as an excellent model for analysing the impact of resistance on the immune defences insects use against parasites. Other insects, such as the housefly (*Musca domestica*) or cotton pests (*Spodoptera* spp.), in which phenomena of induction and resistance are also well known, could be models more particularly adapted to the evaluation of ecotoxicological risks in agricultural areas.

Studies on resistance of invertebrates to pesticides have enabled characterization of the protein systems implicated in phenomena of intoxication and detoxification. In particular, they have demonstrated the need to acquire precise knowledge of systems in order to elaborate reliable biochemical tests to detect the resistance or to use them as biomarkers. It is not sufficient, in discriminating between these phenotypes, to observe that there is a difference in activity of the enzymatic system considered (AChE, P450, GST,...) between sensitive and resistant individuals or between exposed and unexposed individuals. For example, P450-dependent activity would be the same in a sensitive individual after induction as in a resistant individual never exposed to an inducer (Lagadic and Cresteil, 1993; Amichot et al., 1998). The same observation has been made for levels of mRNA and P450 specific proteins (Brun et al., 1996). Similarly, if the biological model has AChEs that are naturally mildly or not sensitive to a pesticide, no phenomenon of inhibition will be observed, even if this pesticide is present in high concentration in the environment.

4. CONCLUSION

Resistance of insects to pesticides is clearly the result of an ecotoxicological problem. Given the manner in which it develops within populations, resistance can be considered as a biomarker of long-term effect of pesticides, offering an integrated picture of the ecotoxicological history of a given site. In addition, when they are induced by pesticides, the enzymes that are the basis of this resistance can be used as biomarkers indicating the instantaneous effect of exposure to toxicants. Whatever the significance given to resistance as a biomarker, its *in situ* validity will depend on fundamental knowledge acquired on the repercussions of its response at the population level.

REFERENCES

Ahmad S., Brattsten L.B., Mullin C.A. and Yu S.J. (1986). Enzymes involved in the metabolism of plant allelochemicals. In: Molecular Aspects of Insect-Plant Associations. Brattsten L.B. and Ahmad S. (eds). Plenum Press, New York, pp. 73–151.

Amichot M., Castella C., Cuany A. et al. (1992). Target modification as a molecular mechanism of pyrethroid resistance in *Drosophila melanogaster*. *Pestic. Biochem. Physiol.*, 44:183–190.

Amichot M., Brun A., Cuany A. et al. (1994). Expression study of CYP genes in *Drosophila* strains resistant or sensitive to insecticides. In: Cytochrome P-450. 8th. International Conference. Lechner M.C. (ed). John Libbey Eurotext, Paris, pp. 689–692.

Amichot M., Brun A., Cuany A. et al. (1998). Induction of cytochrome P450 activities in *Drosopohila melanogaster* strains susceptible or resistant to insecticides. *Comp. Biochem. Physiol., Part C*, 121:311–319.

Berenbaum M.R., Cohen M.B. and Schuler M.A. (1992). Cytochrome P450 monooxygenase genes in oligophagous lepidoptera. In: Molecular Basis of Insecticide Resistance: Diversity Among Insects. Mullin C.J. and Scott J.G. (eds). American Chemical Society Symposium Series 505, ACS, Washington DC, pp. 114–124.

Bergé J.B., Chevillon C., Raymond M. and Pasteur N. (1996). Résistance des insectes aux insecticides: Mécanismes moléculaires et épidémiologie. *C.R. Soc. Biol.*, 190:445–454.

Bloomquist J.R. (1993). Neuroreceptor mechanisms in pyrethroid mode of action and resistance. *Rev. Pestic. Toxicol.*, 2:185–230.

Bloomquist J.R. (1994). Cyclodiene resistance at the insect GABA receptor/chloride channel complex confers broad cross resistance to convulsants and experimental phenylpyrazole insecticides. *Arch. Insect. Biochem. Physiol.*, 26:69–79.

Bocquené G., Galgani F. and Walker C.H. (1997). Les cholinestérases, biomarqueurs de neurotoxicité. In: Biomarqueurs en écotoxicologie: Aspects fondamentaux. Lagadic L., Caquet Th., Amiard J.C. and Ramade F. (eds). Masson, Paris, pp. 209–239.

Bradfield J.Y., Lee Y.H. and Keeley L.L. (1991). Cytochrome P450 family 4 in a cockroach: Molecular cloning and regulation by hypertrehalosemic hormone. *Proc. Natl. Acad. Sci. USA*, 88:4558–4562.

Brattsten L.B. (1988). Potential role of plant allelochemicals in the development of insecticide resistance. In: Novel Aspects of Insect-Plant Interactions. Barbosa P. and Letourneau D.K. (eds). John Wiley, New York, pp. 313–348.

Brown T.M. (1990). Biochemical and genetic mechanisms of insecticide resistance. In: Managing Resistance to Agrochemicals. From Fundamental Research to Practical Strategies. Reen M.B., Lebaron H.M. and Moberg W.K. (eds). American Chemical Society Symposium Series No. 421. ACS, Washington DC, pp. 61–76.

Brun A. (1996). *Étude de l'expression des cytochromes P450: Analyse de la résistance au DDT d'une souche de Drosophila melanogaster*. Thesis, Université de Nice.

Brun A., Cuany A., Le Mouël T., Bergé J.B. and Amichot M. (1996). Inducibility of the *Drosophila melanogaster* cytochrome P450 gene, CYP6A2, by phenobarbital in insecticide susceptible or resistant strains. *Insect Biochem. Molec. Biol.*, 26:697–703.

Byrne F.J. and Devonshire A.L. (1991). *In vivo* inhibition of esterase and acetylcholinesterase activities by profenofos treatments in the tobacco whitefly *Bemisia tabaci* (Genn): Implications for routine biochemical monitoring of these enzymes. *Pestic. Biochem. Physiol.*, 40:198–204.

Capua S., Cohen E. and Gerson U. (1991). Induction of aldrin epoxidation and glutathione S-transferase in the mite *Rhizoglyphus robini*. *Entomol. Exp. Appl.*, 59:43–50.

Cariño F., Koener J.F., Plapp F.W. Jr and Feyereisen R. (1992). Expression of the cytochrome P450 gene CYP6A1 in the house fly, *Musca domestica*. In: Molecular Basis of Insecticide Resistance: Diversity among Insects. Mullin C.J. and Scott J.G. (eds). American Chemical Society Symposium Series 505, ACS, Washington DC., pp. 31–40.

Cariño F., Koener J.F., Plapp F.W. Jr and Feyereisen R. (1994). Constitutive overexpression of the cytochrome P450 gene *CYP6A1* in a house fly strain with metabolic resistance to insecticides. *Insect Biochem. Mol. Biol.*, 24:411–418.

Chevillon C., Pasteur N., Marquine M. et al. (1995). Population structure and dynamics of selected genes in the mosquito *Culex pipiens*. *Evolution*, 49:997–1007.

Chevillon C. (1994). *Évolution de mécanismes adaptatifs: flux géniques, sélection et contre-sélection. Cas de la résistance de Culex pipiens aux insecticides organophosphorés.* Thesis, Université de Montpellier II.

Clark A.G. (1989). The comparative enzymology of the glutathione *S*-transferases from non-vertebrate organisms. *Comp. Biochem. Physiol.*, 92B:419–446.

Clark A.G. (1990). The glutathione *S*-transferases and resistance to insecticides. In: Glutathione *S*-transferases and Drug Resistance. Hayes J.D., Pickett C.B. and Mantle T.J. (eds). Taylor and Francis, London, pp. 369–378.

Cohen E. (1986). Glutathione *S*-transferase activity and its induction in several strains of *Tribolium castaneum*. *Entomol. Exp. Appl.*, 41:39–44.

Cohen M.B., Berenbaum M.R. and Schuller M.A. (1989). Induction of cytochrome P-450-mediated metabolism of xanthotoxin in the black swallowtail. *J. Chem. Ecol.*, 15:2347–2355.

Cohen M.B. and Feyereisen R. (1995). A cluster of cytochrome P450 genes of the *CYP6* family in the housefly. *DNA Cell Biol.*, 14:73–82.

Cousteau C. and ffrench-Constant R.H. (1995). Detection of cyclodiene insecticide resistance-associated mutations by a single stranded conformational polymorphism analysis. *Pestic. Sci.*, 43:267–271.

Croft B.A. and Strickler K. (1983). Natural enemy resistance to pesticides: documentation, characterization, theory and applications. In: Pest Resistance to Pesticides. Georghiou G.P. and Saito T. (eds). Plenum Press, New York, pp. 669–702.

Cuany A., Handani J., Bergé J.B. et al. (1993). Action of esterase B1 on chlorpyrifos in organophosphate-resistant *Culex* mosquitoes. *Pestic. Biochem. Physiol.*, 45:1–6.

Daly J.C., Fisk J.H. and Forrester N.W. (1988). Selective mortality in field trials between strains of *Heliothis armigera* (Lepidoptera: Noctuidae) resistant and susceptible to pyrethroids: functional dominance of resistance and age class. *J. Econ. Entomol.*, 81:1000–1007.

Dary O., Georghiou G.P., Parsons E. and Pasteur N. (1990). Microplate adaptation of Gomori's assay for quantitative determination of general esterase activity in single insects. *J. Econ. Entomol.*, 83:2187–2192.

Dary O., Georghiou G.P., Parsons E. and Pasteur N. (1991). Dot-blot test for identification of insecticide-resistant acetylcholinesterase in single insects. *J. Econ. Entomol.*, 84:28–33.

Dawson G.W., Griffiths D.C., Pickett J.A. and Woodcock C.M. (1983). Decreased response to alarm pheromone by insecticide-resistant aphids. *Naturwissenschaften*, 70:254–255.

Delpuech J.M., Frey F. and Carton Y. (1996). Action of insecticides on the cellular immune reaction of *Drosophila melanogaster* against the parasitoid *Leptopilina boulardi*. *Environ. Toxicol. Chem.*, 15:2267–2271.

Depledge M.H. (1990). New approaches in ecotoxicology: Can inter-individual physiological variability be used as a tool to investigate pollution effects? *Ambio*, 19:251–252.

De Sousa G., Cuany A., Amichot M. et al. (1995). A fluorometric method for measuring ECOD activity on individual abdomen of *Drosophila melanogaster*: Application to the study on resistance of insects to insecticides. *Anal. Biochem.*, 229:86–91.

Devonshire A.L. and Field L.M. (1991). Gene amplification and insecticide resistance. *Annu. Rev. Entomol.*, 36:1–23.

Devonshire A. and Moores G.D. (1982). A carboxylesterase with a broad substrate specificity causes organophosphorus, carbamate and pyrethroid resistance in peach-potato aphids (*Myzus persicae*). *Pestic. Biochem. Physiol.*, 18:235–246.

Devonshire A. and Moores G.D. (1984). Different forms of insensitive acetylcholinesterase in insecticide-resistant house flies (*Musca domestica*). *Pestic. Biochem. Physiol.*, 21:336–340.

Dunkov B., Rodrigaiz-Arnaiz R., Pittendrigh B. et al. (1996). Cytochrome P450 gene clusters in *Drosophila melanogaster*. *Mol. Gen. Genetics*, 251:290–297.

Eldefrawi M.E. and Eldefrawi A.T. (1988). Action of toxicants on GABA and glutamate receptors. In: Molecular Basis of Drug and Pesticide Action. Lunt GG (ed). Elsevier Science Publishers BV (Biomedical Division), pp. 207–221.

Estabrook R.W. (1996). The remarkable P450s: A historical overview of these versatile hemeprotein catalysts. *FASEB J.*, 10:202–204.

Feyereisen R., Koener J.F., Farnsworth D.E. and Nebert D.W. (1989). Isolation and sequence of a cDNA encoding a cytochrome P450 from an insecticide-resistant strain of the house fly, *Musca domestica*. *Proc. Natl. Acad. Sci. USA*, 86:1465–1469.

ffrench-Constant R.H., Mortlock D.P., Shaffer C.D. et al. (1991). Molecular cloning and transformation of cyclodiene resistance in *Drosophila*: an invertebrate gamma-aminobutyric acid subtype A receptor locus. *Proc. Natl. Acad. Sci. USA*, 88:7209–7213.

ffrench-Constant R.H., Rocheleau T.A., Steichen J.C. and Chalmers A.E. (1993a). A point mutation in a *Drosophila* GABA receptor confers insecticide resistance. *Nature*, 363:449–451.

ffrench-Constant R.H., Steichen J.C. and Ode P.J. (1993b). Cyclodiene insecticide resistance in *Drosophila melanogaster* (Meigen) is associated with a temperature-sensitive phenotype. *Pestic. Biochem. Physiol.*, 46:73–77.

ffrench-Constant R.H., Steichen J.C. and Brun L.O. (1994). A molecular diagnostic for endosulfan insecticide resistance in the coffee berry borer *Hypothenemus hampei* (Coleoptera: Scolitidae). *Bull. Ent. Res.*, 84:11–16.

Field L.M., Devonshire A.L. and Forde B.G. (1988). Molecular evidence that insecticide resistance in peach-potato aphids (*Myzus persica* Sulz.) results from amplification of an esterase gene. *Biochem. J.*, 251:309–312.

Forbes V.E. and Forbes T.L. (1994). *Ecotoxicology in Theory and Practice*. Ecotoxicology Series 2. Chapman and Hall, London.

Ford M.G., Holloman D.W., Khambay B.P.S. and Sawicki R.M. (1987). *Combating Resistance to Xenobiotics. Biological and Chemical Approaches*. Ellis Horwood Series in Biomedicine. Ellis Horwood Ltd., Chichester.

Fournier D., Bride J.M., Mouchès C. et al. (1987). Biochemical characterization of the esterases A1 and B1 associated with organophosphate resistance in the *Culex pipiens* complex. *Pestic. Biochem. Physiol.*, 27:211–217.

Fournier D., Bride J.M., Poirié M. et al. (1992). Insect glutathione-transferases. Biochemical characteristics of the major form from houseflies susceptible and resistant to insecticides. *J. Biol. Chem.*, 267:1840–1845.

Fournier D., Mutero A., Pralavorio M. and Bride J.M. (1993). *Drosophila* acetylcholinesterase: mechanism of resistance to organophosphates. *Chem. Biol. Interactions*, 87:233–238.

Frolov M.V. and Alatorsev V.E. (1994). Cluster of cytochrome P450 genes on the X chromosome of *Drosophila melanogaster*. *DNA Cell Biol.*, 13:663–668.

Gandhi R., Varak E. and Goldberg M.L. (1992). Molecular analysis of cytochrome P450 gene of family 4 on the *Drosophila* X chromosome. *DNA Cell Biol.*, 11:394–404.

Gray J.S. (1989). Effects of environmental stress on species rich assemblages. *Biol. J. Linn. Soc.*, 37:19–32.

Green M.B., Lebaron H.M. and Moberg W.K. (1990). *Managing Resistance to Agrochemicals. From Fundamental Research to Practical Strategies*. American Chemical Society Symposium Series No. 421. ACS, Washington, DC.

Guillemand T., Rooker S., Pasteur N. and Raymond N. (1996). Testing the unique amplification event and the worldwide migration hypothesis of insecticide resistance genes with sequence data. *Heredity*, 77:535–543.

Hall L.M.C. and Spierer P. (1986). The *Ace* locus of *Drosophila melanogaster*: Structural gene for acetylcholinesterase with an unusual 5' leader. *EMBO J.*, 5:2949–2954.

Haubruge E. (1995). *Étude des phénomènes responsables de la résistance spécifique au malathion chez Tribolium castaneum HERBST (Col., Tenebrionidae)*. Thesis, Faculty of Agricultural Sciences, Gembloux.

Hodgson E., Rose R.L., Thompson D.M. et al. (1995). Expression of cytochrome P450 in insects. *9th Int. Conference on Cytochrome P450: Biochemistry, Biophysics and Molecular Biology, Zurich.* Abstr. SL-32, p. 259.

Hoffmann A.A. and Parsons P.A. (1991). *Evolutionary genetics and environmental stress.* Oxford Science Publications, Oxford.

Hung C.F., Harrison T.L., Berenbaum M.R. and Schuler M.A. (1995). CYP6B3, a second furanocoumarin-inducible cytochrome P450 expressed in *Papilio polyxenes*. *Insect Biochem. Mol. Biol.*, 25:149–160.

Koeman J.H., Köhler-Günther A., Kurelec et al. (1993). Applications and objectives of biomarkers research. In: Biomarkers. Research and Application in the Assessment of environmental health. Peakall D.B. and Shugart L.R. (eds). NATO Advanced Science Institutes Series, Vol. H 68. Springer Verlag, Berlin, Heidelberg, pp. 1–13.

Lagadic L. (1991). *Induction, par le lindane, des enzymes de biotransformation chez Spodoptera littoralis (Boisd.) (Lepidoptera: Noctuidae): Conséquences sur la toxicité et la métabolisation in vivo de la cyfluthrine*. Thesis, Doctorate in Sciences, Université Paris XI, Orsay.

Lagadic L. and Cresteil T. (1993). Enhanced *in vitro* metabolism of testosterone by microsomes from insecticide-resistant *Spodoptera littoralis* larvae. *Insect. Biochem. Molec. Biol.*, 23:475–480.

Lagadic L., Caquet Th. and Ramade F. (1994). The role of biomarkers in environmental assessment (5). Invertebrate populations and communities. *Ecotoxicology*, 3:193–208.

Lagadic L., Cuany A., Bergé J.B. and Echaubard M. (1993). Purification and partial characterization of glutathione *S*-transferases from insecticide-resistant and lindane-induced susceptible *Spodoptera littoralis* (Boisd.) larvae. *Insect. Biochem. Molec. Biol.*, 23:467–474.

Lee H.J., Rocheleau T., Zhang H.G. et al. (1993). Expression of a *Drosophila* GABA receptor in a baculovirus insect cell system—Functional expression of insecticide susceptible and resistant GABA receptors from the cyclodiene resistance gene rdl. *FEBS Lett.*, 335:315–318.

Lee K. and Berenbaum M.R. (1989). Action of antioxidant enzymes and cytochrome P-450 monooxygenases in the cabbage looper in response to plant phototoxins. *Arch. Insect. Biochem. Physiol.*, 10:151–162.

Lunt G.G., Brown M.C.S., Riley K. and Rutherford D.M. (1988). The biochemical characterization of insect GABA receptors. In: Molecular Basis of Drug and Pesticide Action. Lung G.G. (ed). Elsevier Science Publishers BV (Biochemical Division), pp. 185–192.

Maitra S., Dombrowski S.M., Waters L.C. and Ganguly R. (1995). Isolation and characterization of new family genes and their expression in insecticide resistant and susceptible strains of *Drosophila melanogaster*. *3rd Symposium on Cytochrome P450 Biodiversity, Woods Hole.* Abstr. II–4.

Maynard Smith J. (1989). *Evolutionary Genetics.* Oxford University Press, Oxford.

Monod G. (1997). L'induction du cytochrome P4501A1 chez les poissons. In: Biomarqueurs en eécotoxicologie: Aspects fondamentaux. Lagadic L., Caquet Th., Amiard J.C. and Ramade F. (eds). Masson, Paris, pp. 33–54.

Morton R.A., (1993). Evolution of *Drosophila* insecticide resistance. *Genome*, 36:1–7.

Mouchés C., Pasteur N., Bergé J.B. et al. (1986). Amplification of an esterase gene is responsible for insecticide resistance in a California *Culex* mosquito. *Science*, 233:778–780.

Mouchés C., Magnin M., Bergé J.B. et al. (1987). Overproduction of detoxifying esterases in organophosphate-resistant *Culex* mosquitoes and their presence in other insects. *Proc. Natl. Acad. Sci., USA*, 84:2113–2116.

Mueller L.D., Guo P. and Ayala F.J. (1991). Density-dependent natural selection and trade-offs in life history traits. *Science*, 253:433–435.

Mullin C.A. (1988). Adaptive relationships of epoxide hydrolase in herbivorous arthropods. *J. Chem. Ecol.*, 14:1867-1888.

Mullin C.J. and Scott J.G. (1992). *Molecular Basis of Insecticide Resistance: Diversity among Insects*. In: American Chemical Society Symposium Series 505. ACS, Washington DC, pp. 114–124.

Mutero A., Pralavorio M., Bride J.-M. and Fournier D. (1994). Resistance associated point mutations in insecticide insensitive acetylcholinesterase. *Proc. Natl. Acad. Sci. USA.*, 91:5922–5926.

Narahashi T. (1986). Mechanisms of action of pyrethroids on sodium and calcium channel gating. In: Neuropharmacology and Pesticide Action. Ford M.G., Lunt C.G., Reay R.C. and Usherwood P.N.R. (eds). Ellis Horwood Ltd., Chichester, England, pp. 36–60.

Narahashi T., Frey J.M., Ginsburg K.S. and Roy M.L. (1992). Sodium and GABA-activated channels as the targets of pyrethroids and cyclodienes. *Toxicol. Lett.*, 64/65:429–436.

Narbonne J.F. and Michel X. (1997). Systèmes de biotransformation chez les mollusques. In: Biomarqueurs en écotoxicologie: Aspects fondamentaux. Lagadic L., Caquet Th., Amiard J.C. and Ramade F. (eds). Masson, Paris, pp. 11–31.

Neal J.J. (1987). Metabolic costs of mixed function oxidase induction in *Heliothis zea*. *Entomol. Exp. Appl.*, 43·175–179.

Nelson D.R., Koymans L., Kamataki et al. (1996). P450 superfamily: Updated on new sequences, gene mapping, accession numbers and nomenclature. *Pharmacogenetics*, 6:1–42.

Pasteur N. and Georghiou G.P. (1989). Improved filter paper test for detecting and quantifying increased esterase activity in organophosphate-resistant mosquitoes (Diptera: Culicidae). *J. Econ. Entomol.*, 82:347–353.

Pasteur N., Marquine M., Rousset F. et al. (1995). The role of passive migration in the dispersal of resistance genes in *Culex pipiens quinquefasciatus* within French Polynesia. *Genet. Res.*, 66:139–146.

Poirié M. and Pasteur N. (1991). La résistance des insects aux insecticides. *La Recherche*, 22:874–882.

Qiao C.L. and Raymond M. (1995). The same esterease B1 haplotype is amplified in insecticide resistant mosquitoes of the *Culex pipiens* complex from the Americas and China. *Heredity*, 74:339–345.

Rapport D.J. (1989). Symptoms of pathology in the Gulf of Bothnia (Baltic Sea): Ecosystem response to stress from human activity. *Biol. J. Linn. Soc.*, 37:33–49.

Raymond M., Callaghan A., Fort P. and Pasteur N. (1991). Worldwide migration of amplified insecticide resistance genes in mosquitoes. *Nature*, 350:151–153.

Raymond M., Pasteur N., Fournier D. et al. (1985). Le gène d'une acétylcholinestérase insensible au propoxur détermine la résistance de *Culex pipiens* L. à cet insecticide. *C. R. Acad. Sci. Paris*, 300:509–512.

Rivet Y., Marquine M. and Raymond M. (1993). French mosquito populations invaded by A2-B2 esterases causing insecticide resistance. *Biol. J. Linnean Soc.*, 49:249–255.

Rooker S., Guillemaud T., Bergé J.B. et al. (1996). Coamplification of esterase A and B genes as a single unit in the mosquito *Culex pipiens*. *Heredity*, 77:555–561.

Roush R.T. and Daly J.C. (1990). The role of population genetics in resistance research and management. In: Pesticide Resistance in Arthropods. Roush R.T. and Tabashnik B.E. (eds). Chapman and Hall, New York and London, pp. 97–152.

Roush R.T. and Hoy M.A. (1981). Laboratory, glasshouse, and field studies of artificially selected carbaryl resistance in *Metaseiulus occidentalis*. *J. Econ. Entomol.*, 74:142–147.

Roush R.T. and Mckenzie J.A. (1987). Ecological genetics of insecticide and acaricide resistance. *Annu. Rev. Entomol.*, 32:361–380.

Roush R.T. and Plapp F.W. Jr (1982a). Biochemical genetics of resistance of aryl car-bamate insecticide in the predaceous mite, *Metaseiulus occidentalis J. Econ. Entomol.*, 75:304–307.

Roush R.T. and Plapp F.W. Jr (1982b). Effects of insecticide resistance on biotic potential of the house fly (Diptera: Muscidae). *J. Econ. Entomol.*, 75:708–713.

Roush R.T. and Tabashnik B.E. (1990). *Pesticide Resistance in Arthropods*. Chapman and Hall, New York and London.

Sattelle D.B. (1990). GABA receptors of insects. *Adv. Insect Physiol.*, 22:1–113.

Sattelle D.B. and Yamamoto D. (1988). Molecular targets of pyrethroid insecticides. *Adv. Insect Physiol.*, 20:147–213.

Sawicki R.M. and Denholm I. (1984). Adaptation of insects to insecticides. In: Origins and Development of Adaptation. Evered D. and Collins G.M. (eds). Ciba Founda-tion Symposium Series 102. Pitman Books, London, pp. 152–166.

Sawicki R.M., Denholm I., Forrester N.W. and Kershaw C.D. (1989). Present insecticide-resistance management strategies in cotton. In: Pest Management in Cotton. Green M.B. and Lyon D.J. De B. (eds). Ellis Horwood Series in Agricultural Sciences. Ellis Horwood Ltd., Chichester, pp. 31–43.

Scott J.G. (1996). Cytochrome P450 monooxygenase-mediated resistance to insecticides. *J. Pestic. Sci.*, 21:241–245.

Scott J.G., Collins F.H. and Feyereisen R. (1994). Diversity of cytochrome P450 genes in the mosquito *Anopheles albaminus. Biochem. Biophys. Res. Comm.*, 205:1452–1459.

Scott J.G., Sridhar P. and Liu N. (1996). Adult specific expression and induction of cytochrome P450$_{lpr}$ in house flies. *Arch. Insect Biochem. Physiol.*, 31:313–323.

Snyder M.J. and Davidson N. (1983). Two gene family clustered in a small region of the *Drosophila* genome. *J. Biol. Mol.*, 166:101–118.

Snyder M.J., Hsu E.L. and Feyereisen R. (1993). Induction of cytochrome P-450 activities by nicotine in the tobacco hornworm, *Manduca sexta. J. Chem. Ecol.*, 19:2903–2916.

Soderlund D.M. and Bloomquist J.R. (1989). Neurotoxic actions of pyrethroids insecti-cide. *Annu. Rev. Entomol.*, 34:77–96.

Soderlund D.M. and Bloomquist J.R. (1990). Molecular mechanisms of insecticide resist-ance. In: Pesticide Resistance in Arthropods. Roush R.T. and Tabashnik B.E. (eds). Chapman and Hall, New York and London, pp. 58–96.

Sparks T.C., Lockwood J.A., Byford R.L. et al. (1989). The role of behaviour in insecti-cide resistance. *Pestic. Sci.*, 26:383-399.

Tabashnik B.E. (1990). Modeling and evaluation of resistance management tactics. In: Pesticide Resistance in Arthropods. Roush R.T. and Tabashnik B.E. (eds). Chapman and Hall, New York and London, pp. 153–182.

Taylor M.F.J., Heckel D.G., Brown T.M. et al. (1993). Linkage of pyrethroid insecticide resistance to a sodium channel locus in the tobacco budworm. *Insect Biochem. Mol. Biol.*, 23:763–775.

Terriere L.C. (1983). Enzyme induction, gene amplification and insect resistance to in-secticides. In: Pest Resistance to Pesticides. Georghiou G.P. and Saito T. (eds). Ple-num Press, New York, pp. 265–298.

Terriere L.C. (1984). Induction of detoxication enzymes in insects. *Annu. Rev. Entomol.*, 29:71–88.

Timbrell J.A., Draper R. and Waterfield C.J. (1994). Biomarkers in toxicology: new uses for old molecules. *Toxicol. Ecotoxicol. News*, 1:4–14.

Tomita T. and Scott J.G. (1995). cDNA and deduced protein sequence of *CYP6D1*: The putative gene for a cytochrome P450 responsible for pyrethroid resistance in house fly. *Insect Biochem. Mol. Biol.*, 25:275–283.

Vos R.M.E. and Van Bladeren P.J.V. (1990). Glutathione S-transferases in relation to their role in the biotransformation of xenobiotics. *Chem. Biol. Interact.*, 75:241–265.

Waters L.C., Zelhof A.C., Shaw B.J. and Ch'ang L.Y. (1992). Possible involvement of the long terminal repeat of transposable element 17.6 in regulating expression of an insecticide resistance-associated P450 gene in *Drosophila. Proc. Natl. Acad. Sci. USA*, 89:4855–4859.

Williamson M.S., Martinez-Torres D., Bell C.A. and Devonshire A.L. (1996). Identification of mutations in the housefly para-type sodium channel gene associated with knock-down resistance (kdr) to pyrethroid insecticides. *Molec. Genet.*, 252:51–60.

Xiao-Ping W. and Hobbs A.A. (1995). Isolation and sequence analysis of a cDNA clone for a pyrethroid inducible cytochrome P450 from *Helicoverpa armigera*. *Insect Biochem. Mol. Biol.*, 25:1001–1009.

Yamamoto I., Takahashi Y. and Kyomura N. (1983). Suppression of altered acetylcholinesterase of the green rice leafhopper by N-propyl and N-methyl carbamate combinations. In: Pest Resistance to Pesticides. Georghiou G.P. and Saito T. (eds). Plenum Press, New York, pp. 579–594.

Yu S.J. (1986). Consequences of induction of foreign compound-metabolizing enzymes in insects. In: Molecular Aspects of Insect-Plant Associations. Brattsten L.B. and Ahmad S. (eds). Plenum Press. New York, pp. 153–174.

Yu S.J. (1996). Insect glutathione S-transferases. *Zool. Stud.*, 35:9–19.

13

Consequences of Individual-Level Alterations on Population Dynamics and Community Structure and Function

Th. Caquet and L. Lagadic

INTRODUCTION

The ecotoxicological effects induced by the introduction of a pollutant in an ecosystem may appear at various levels of biological organization, from populations and communities up to the ecosystem itself (Fig. 1).

Biomarkers may be identified at different levels within the individual (molecular, biochemical, and cellular biomarkers) or at the level of the whole individual (physiological or behavioural biomarkers). Their response generally corresponds either to direct interaction between toxicants and a biological target (e.g., acetylcholinesterases, cytochrome P450-dependent monooxygenases, metallothioneins), or to metabolic effects (e.g., energy metabolism, hormonal metabolism), physiological effects (e.g., growth, fecundity), or behavioural effects (locomotion, response to stimuli) that may result from this interaction. The ecotoxicological validation of some of these biomarkers should consist in establishing a mechanical or causal relationship between changes in the considered parameter in the exposed individuals and more or less durable effects measured on the corresponding population and, in a second step, on the community to which this population belongs.

In this chapter we attempt to identify possible biomarkers that may be used to predict ecotoxicological effects of environmental contaminants. After we briefly recall the main ecological mechanisms that

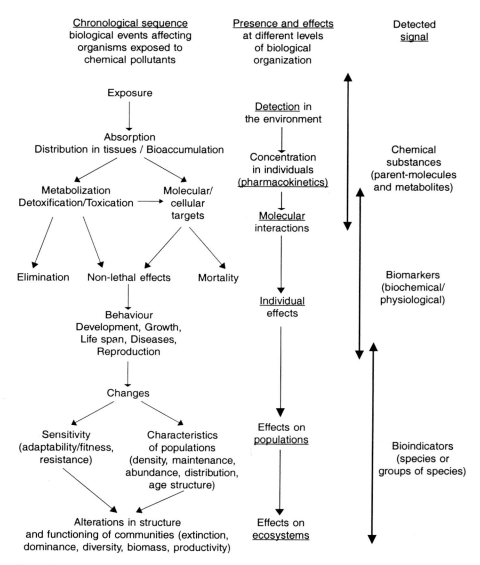

Fig. 1. Theoretical chronology of effects induced by the introduction of a pollutant in the natural environment (adapted from Lagadic and Caquet, 1996)

regulate the structure and dynamics of populations, communities, and ecosystems, we will discuss some key processes in which biomarkers may be identified. Plants are involved in the processes that govern the structure and function of communities and ecosystems. However, due to the available literature, we will mainly refer to animals in the following.

1. BASIC MECHANISMS OF POPULATION, COMMUNITY AND ECOSYSTEM DYNAMICS

1.1 Populations

In the ecological sense, a population is a set of individuals of the same species that live within a given ecosystem. Populations exhibit five major types of relations with their environment (Barbault, 1992):

- interactions with physical and chemical factors of the environment;
- vertical interactions with prey and predators and/or parasites;
- horizontal interactions with other populations belonging to the same trophic level (competition);
- interactions between individuals within the population;
- positive interactions with other species that are sometimes biologically very different (mutualism or symbiosis).

Retroactions between individual characteristics and environmental factors and the considered population modulate the dynamics and evolution of the latter (Berryman, 1981). The ecotoxicological effects at the population level frequently derive from changes in performances (i.e., embryonic development, growth, reproduction) of at least some of the individuals (Munkittrick and McCarty, 1995). The optimal accomplishment of these phenomena depends essentially on the absence of severe alterations in various processes (Table 1).

When selecting relevant biomarkers for prediction of the effects of pollutants on the dynamics of populations, parameters that allow us to identify possible changes in one or several of these processes should be emphasized. However, even when a considerable proportion of individuals in a population are affected by a contaminant and that impact may be proved by the measurement of certain biomarkers, that does not necessarily mean that this will be the case for the entire population. Indeed, all the individuals within a population often do not completely express their biotic potential (e.g., reproduction), especially because of inter-individual competition (e.g., crowding effect, competition for resources). Thus, adverse effects of a contamination on a fraction of a population may result in an increase in the rate of survival or reproduction of unaffected or less affected individuals (see chapter 12). This may occur when some individuals are exposed to lower levels of contaminants and/or present biological differences (especially physiological or genetic differences). In particular, populations with a high degree of demographic heterogeneity (coexistence of several age classes or development stages) or genetic heterogeneity (high degree of polymorphism) will often be much less affected than more homogeneous ones (for example, laboratory strains).

Table 1. Main processes whose alteration by a pollutant may disturb the dynamics of a population

| Function | Processes that can be disturbed at | |
	individual level	inter-individual level
Reproduction	Orientation Equilibrium Locomotive activity Regulation (hormones) Acquisition of primary and secondary sexual characteristics Gonadogenesis, gametogenesis Embryonic development Parturition Optimal energy use: metabolism, resource allocation	Aggregation Search for sexual partners Mating Emission of gametes Courtship behaviour Nesting Predatory capacities
Development and growth	Foraging Various taxes (e.g., phototaxis) Equilibrium Locomotive activity Learning Regulation: growth factors Optimal energy use: metabolism, resource allocation, fitness	Brooding Feeding of young Teaching of young Protection against environmental factors
Survival	Foraging Various taxes (e.g., phototaxis) Equilibrium Locomotive activity Homeostatic mechanisms Detoxification, capacity for tissue repair, resistance to parasites and diseases	Aggression Protection against environmental factors Protection against predators or parasites

1.2 Communities

A community is a multispecific system of populations that are or could be interconnected, thus constituting a system that has its own structure and function (Barbault, 1992). Interactions among populations may, for example, be competitive, trophic (e.g., predator-prey relationships) or cooperative (e.g., mutualism). Three major types of factors influence the organization and dynamics of communities: the probability of colonization, physical factors, and biotic factors. These factors in turn influence various biological processes, that determine the structure and functioning of the community (Fig. 2).

The mechanisms and general rules that regulate the structure of communities have been the subject of several studies (May, 1973; Cohen, 1978; Martinez and Lawton, 1995). Many studies have focused on the relations between the number of species present in a given community

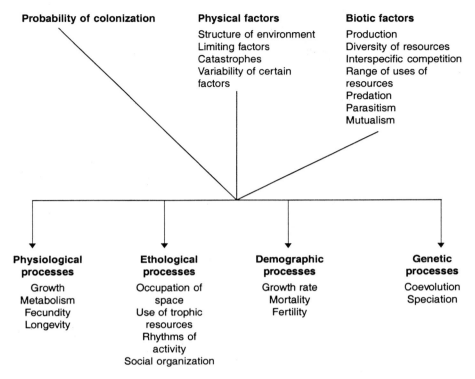

Fig. 2. Relations between the major factors responsible for the organization and dynamics of communities and the resulting processes at the community level (adapted from Barbault, 1992)

and the properties of the corresponding food web: e.g., total number of trophic links, number of trophic links per species, and connectance, i.e., the ratio between the number of observed links and the number of possible links (Havens, 1997). Moreover, several studies in natural or experimental ecosystems (mesocosms) have demonstrated the consequences of interactions between trophic levels on the abundance of organisms that constitute the food webs. This is, for example, the case of cascading trophic interactions (Paine, 1980; Carpenter et al., 1985), frequently called 'trophic cascades' (Carpenter and Kitchell, 1993). It has been shown that the abundance of primary producers may be affected either by consumers (top-down or descending effect) or by nutrient availability (bottom-up or ascending effect). The two types of effect may occur in the same type of ecosystem (see, for example, Carpenter and Kitchell, 1992; DeMelo et al., 1992; Hunter and Price, 1992; Power, 1992; Strong, 1992). Moreover, one trophic level can simultaneously exert a top-down and a bottom-up effect on another trophic level within a single ecosystem. In

certain lakes, for example, planktonophagous fish may favour the multiplication of phytoplankton (Fig. 3) by feeding on herbivorous
zooplankton organisms (top-down effect) and by increasing the concentration of nutrients (essentially phosphorus) in the water column by
various mechanisms (bottom-up effect; see, for example, Andersson et
al., 1978; Shapiro and Wright, 1984; Mazumder et al., 1988; Drenner et
al., 1990; Reinertsen et al., 1990; Lazzaro et al., 1992; Vanni and Layne,
1997; Vanni et al., 1997).

A community may be characterized in numerical and/or functional
terms.

In numerical terms, the structure of a community may be characterized using various descriptors such as species density, richness, or

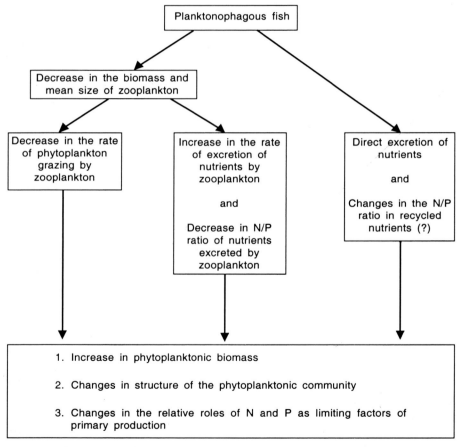

Fig. 3. Mechanisms involved in the control of the phytoplanktonic community in lake
ecosystems by planktonophagous fishes (after Vanni and Layne, 1997) (?: supposed
effect)

diversity. These quantitative descriptors may serve to compare communities or successive stages of the same community. Moreover, this numerical structure may indicate a type of biological organization. However, it indicates nothing in terms of interactions between the different populations.

From a functional point of view, a community may be characterized by defining the relations of competition, predation, or cooperation between its different constitutive populations. This implies that the identification of the place and function of each population in the community (niche) is possible, particularly in relation to other populations (niche overlap). In this perspective, the organization of a community can be characterized in relation to spatial, trophic, and temporal dimensions of the niche (Barbault, 1992).

To date, the evaluation of the impact of pollutants on communities has essentially been performed using numerical descriptors of their structure. It is in this context, for example, that many indices have been developed and/or used (see for example Pascoe and Edwards, 1984; Blandin, 1986; Hellawell, 1986):

- diversity indices
- dominance indices
- distribution of abundance and/or biomass within different species (rank-frequency diagrams);
- relative abundance of species or groups of species (indicator species);
- synthetic indices linking the abundance of species and their potential as indicators.

Diversity and dominance indices as well as models of distribution of species abundances are the more or less direct reflection of the total number of species present in a community and of the distribution of individuals among these different species. These descriptors are purely quantitative and do not involve qualitative biological criteria such as the role of the species in the functioning of the community and ecosystem, or their bioindicator potential. Furthermore, the use of purely quantitative descriptors raises the problem of the definition of reference values.

On the other hand, methods based on the presence or absence and/or abundance of indicator species or groups of indicator species provide an opportunity to obtain information that is ecologically highly relevant. Biological indication has been defined in many ways by various authors, some including even biomarkers, but at the level of a community, it is the notion of indicator species that is the most often taken into account. Indicators species constitute a particular type of ecological indicator, which has been defined by Blandin (1986) as a population or

set of populations that, by its qualitative and/or quantitative characteristics, gives information on the state of an ecological system and that, by
changes in these characteristics, enables the detection of possible modifications of this system.

The use of indicator species and bioindication are outside the scope
of this chapter. However, it should be noted that although biological
indicators are ecologically relevant, they frequently are not specific and
do not generally allow us to accurately identify the causes of the disturbance observed. When they are measured in indicator species, biomarkers
may help to refine the environmental diagnostic by providing indications on the nature of the pollutants. Should the research of biomarkers
therefore be limited to bioindicators or should it be extended to species
that are not necessarily bioindicators but have a definitive structural
and functional role within a particular community ('key-stone' species;
Paine, 1966)?

If the measurement of a biomarker may give an early warning of
alteration of functional relationships within a community, it is advisable to examine mechanisms that influence the extent of the different
dimensions of the niche. The three main axes that define the actual
niche of a species are the *spatial axis*, which covers the various types of
exploitable environments or microenvironments, the *trophic axis*, which
represents the categories of prey (in the wide sense) that can be consumed, and the *temporal axis*, which reflects the modalities of use and
distribution over time of the two preceding resources. A parameter that
may indicate the alteration of at least one of these three axes may be
considered a biomarker. The alteration may concern the resources or
their use. Moreover, changes may occur in trophic ratios between species that may have consequences on the structure of the food web or on
its functioning.

In this perspective, behavioural processes and mechanisms that regulate the growth and reproduction of organisms appear to be promising
areas for the identification of such biomarkers.

1.3 Ecosystems

According to the definition of Tansley (1935), an ecosystem is a spa-
tially delimited ecological entity characterized by a set of abiotic components (biotope) and a set of communities (biocoenosis). Since this
original definition, the concept of ecosystem has considerably evolved
(Willis, 1997), especially through the development of the theory of
ecological systems (see, for example, Odum, 1964, 1968; Patten, 1966;
Van Dyne, 1966), systems modelling (Neel and Olson, 1962; Mauersberger
and Straskraba, 1987; Jorgensen, 1988), and application of thermodynamics to the functioning of ecosystems (Watt, 1968; Jorgensen, 1992).

Within an ecosystem, organisms can be grouped schematically as producers, consumers, and decomposers. From a functional perspective, an ecosystem is characterized by a one-way flow of energy and internal recycling of matter. It should be noted that some of the nutrients can be imported from or exported to other ecosystems.

The response of such a system to any kind of disturbance can be characterized qualitatively by its stability, i.e. its ability to return to its initial state, and quantitatively by the rapidity of that return, which is also called the resilience of the system (Holling, 1973; Webster et al., 1975; Beddington et al., 1976; Harrison, 1979; DeAngelis, 1980,1992; Pimm, 1982, 1984, 1991). Resilience corresponds to the qualitative persistence of relations within the system despite quantitative fluctuations of certain parameters. The nature of the relationship between the characteristics of an ecosystem and its resilience is still a matter of discussion, even though numerous observations indicate that there is generally a positive correlation between species diversity in natural ecosystems and their stability and resilience (Frontier, 1977).

The quantification of properties of an ecosystem is a difficult task, which is frequently performed using models that assume the existence of a stable asymptotic equilibrium (Patten and Witkamp, 1967; Jordan et al., 1972; Pimm and Lawton, 1977; Harte, 1979; DeAngelis, 1980). It is still the subject of new developments, for example to improve the characterization of the intermediate stages of the response of ecosystems to disturbances (Neubert and Caswell, 1997).

The maturity of ecosystems is another parameter linked to their dynamics. In theory, ecosystems tend toward a more or less stable state, the climax, through several successive stages (ecological succession). This evolution is characterized by an increase in their stability. When pollution occurs, two cases may be observed:

– In the case of an accidental (acute) pollution, a return to a preceding stage of the succession is most often accompanied by a reduction in the diversity of communities. The same phenomenon may be observed following a natural catastrophe (fire, for example). The resistance of ecosystems to such disturbances is not necessarily related to their complexity.

– In the case of a chronic pollution, gradual changes of the structure of the ecosystem lead to an 'adaptation' to new environmental conditions. These changes are accompanied by modifications in the structure and functioning of communities (e.g., disappearance of certain species, which are replaced by others, modifications in productivity). It is commonly acknowledged that these phenomena will lead to simplification of the ecosystem structure and favour respiration to the detriment of production (Woodwell, 1970).

In the 1980s, the idea of 'ecosystem health' appeared (Schaeffer et al., 1985; Rapport, 1989; Costanza et al., 1992), according to which it should be possible to establish a diagnostic on the state of an ecosystem from a four-step process, inspired by the process used for elaboration of a diagnostic on humans or animals (Schaeffer, 1996):

- collecting 'historical' data on the ecosystem and making physical, chemical, and biological measurements;
- evaluating the data obtained during the first step;
- comparing the data with reference values, in order to identify the possible disorder(s);
- applying logical reasoning to observations in order to identify the problem.

Even though several studies, seminars, and publications have been devoted to this approach (see, for example, Schaeffer et al., 1985, 1988; Costanza et al., 1992, 1995; Munkittrick and McCarty, 1995; Rapport, 1995; Schaeffer, 1996; Attrill and Depledge, 1997), theoretical developments and refinement of specific tools are required before it can be used on a routine basis (Schaeffer, 1996). Even the meaning of the word 'health' for an ecosystem is still a matter of debate (Calow, 1995; Rapport, 1995). A comprehensive assessment of the health of an ecosystem requires ecological and ecotoxicological information (e.g., biodiversity, health of individuals, diversity and availability of habitats, level of contamination of habitats and organisms), as well as socio-economic data and information on human health (Rapport, 1995). The precise choice of indicators to be used depends on the ecosystem considered (Wilson et al., 1995; Shear, 1996; Yazvenko and Rapport, 1996). In this context, certain biomarkers may play a significant role, especially as indicators of the health of organisms. Nevertheless, this implies necessarily that those biomarkers must be validated *in situ*.

Considering the difficulties of characterizing the properties and the health of an ecosystem, the establishment of simple relations between the responses of biomarkers measured in a group of individuals and the properties of the ecosystem they belong to is unlikely. This is why the following pages are restricted to insights on possible ways to identify biomarkers that can be used to predict potential effects of contaminants on populations and communities.

Behaviour is a key phenomenon in the life of many organisms. It simultaneously requires an optimal functioning of the nervous system and the availability of energy reserves. Moreover, acquisition of energy requires that several behavioural processes occur in an optimal way (e.g., locomotion, foraging). The interactions between environmental contaminants and living organisms may have various consequences, but alterations of the nervous system are undoubtedly a significant source of dysfunction of many processes involved in survival, development, and reproduction (Fig. 4). Apart from direct disturbances of neurohormonal control of reproductive processes, changes in energy allocation are another important phenomenon that may have consequences at the population level. In the following sections, we successively consider behavioural biomarkers, biomarkers related to the control of reproduction, development, and growth, and energy-related biomarkers.

2. BIOMARKERS AND BEHAVIOUR

The decline of a population or reduction of species richness in response to the presence of contaminants may be due to various alterations in the behaviour of intoxicated individuals (Olla et al., 1980; Haynes, 1988; Cohn and MacPhail, 1996). According to some authors, behaviour of organisms itself may be considered as a biomarker (Peakall and Walker, 1996). Other authors consider that behaviour is an integrated biological

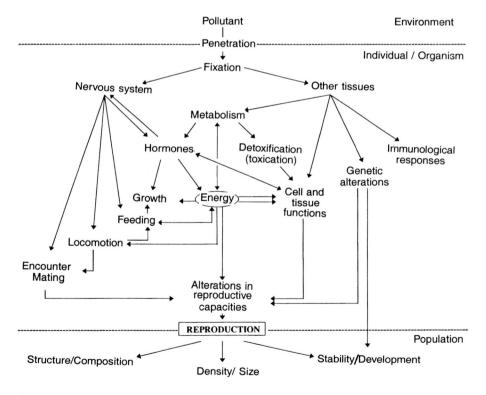

Fig. 4. Potential impact of pollutants on animal reproduction via modifications in the availability of metabolic energy and hormonal control

phenomenon and that behaviour-related biomarkers should be identi-fied at the level of basic processes involved in the elaboration of the behavioural response, such as transmission of sensory or motor nerve impulse (Van Gestel and Van Brummelen, 1996). The two approaches are complementary and it should be possible to consider the identifica-tion of biomarkers at the level of information processing (biomarkers of neurotoxicity, for example) as well as at the level of behavioural se-quences.

2.1 Biomarkers of Neurotoxicity

Exposure to neurotoxic substances may result in behavioural changes that may be evaluated indirectly by the measurement of biomarkers related to the functioning of the nervous system. The activity of cholinesterases, especially acetylcholinesterase (AChE), is a typical biomarker of neurotoxicity (Bocquené et al., 1997; see chapter 6).

Frequently used as biomarker in vertebrates and especially in birds (Greig-Smith, 1991; Peakall, 1992), measurement of AChE activity has rarely been associated with the characterization of behaviour of animals *in situ*. Regarding invertebrates, even though laboratory studies have shown that the effects of certain neurotoxic substances are sometimes associated with behavioural changes (Scherer and McNicol, 1986), di-rect correlation between AChE activity at the individual level and re-sponses at the level of population or community remains to be established (Fig. 5). From this perspective, freshwater invertebrates may be used as models. Abundance of these organisms frequently decreases following river pollution by insecticides through an increase in drift (see, for ex-ample, Eidt, 1975; Fredeen, 1974; Muirhead-Thomson, 1978, 1987; Flannagan et al., 1979; Wallace et al., 1986, 1987; Everts et al., 1983; Kreutzweiser and Kingsbury, 1987). This phenomenon may be associ-ated, at least for living drifting animals, with an escape response (Muirhead-Thomson, 1987).

2.2 Behavioural Biomarkers

Many behavioural sequences may be more or less profoundly altered by contaminants (Table 2).

Among these, changes in avoidance and behavioural processes that regulate parental care, reproduction, and use of trophic resources may have consequences at the population level. Biomarkers may possibly be identified at this level.

2.2.1 Avoidance Behaviour

Avoidance reactions in response to potential or real danger are observed in most organisms. At this level, two major types of responses may be observed following exposure to toxicants:

- induction of escape or isolation reaction in relation to contaminants or contaminated sites;
- inhibition of escape reaction in relation to predators or parasites.

2.2.1.1 Avoidance in relation to contaminants or contaminated sites

Avoidance reaction in the presence of a contaminant is generally determined by a movement of escape or an isolation response from the external environment. Many pollutants induce an escape behaviour in the exposed animal, which leads them to drift (see above) or to search for a more suitable environment (Corta, 1967; Wentsel et al., 1977;

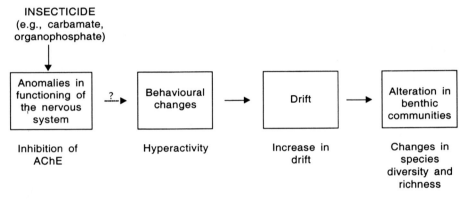

Fig. 5. Theoretical sequence of responses initiated at the biochemical, behavioural, population, and community level, by exposure to an anticholinesterase insecticide and examples of indicators of these responses in river aquatic insects (arrow: demonstrated relation; broken arrow: supposed relation). The relationship between AChE inhibition and behavioural changes has not yet been established and characterized quantitatively *in situ* (adapted from Lagadic et al., 1994).

Table 2. Possible relationships between behavioural alterations and effects on populations (after Cohn and MacPhail, 1996)

Behavioural sequence affected	Effects on populations
Avoidance behaviour (isolation, escape from predators, etc.)	Decrease in density by emigration, increase in mortality, etc.
Reproductive behaviour	Decrease in number of matings and young
Parental behaviour	Decrease in number of young attaining sexual maturity
Foraging behaviour	Increase in mortality by food shortage, alteration in growth, reduction in reproductive activity
Locomotion, migration	All of the above

McGreer, 1979). Such behaviour may have severe consequences on the evolution of the populations concerned, especially by disturbing the reproduction of adults or the settlement of young individuals (Krebs and Burns, 1977). Various devices have been designed to monitor the behaviour of vertebrates or invertebrates and to detect the appearance of anomalies (e.g., hyperactivity, abnormal swimming, decrease in resistance to current) in the presence of contaminants (Baatrup and Bailey, 1993; Gerhardt et al., 1994; Depledge et al., 1995; Gerhardt, 1995, 1996). In bivalve molluscs or barnacles, the presence of contaminants in the external environment induces a closing response of the valves of the shell or of the operculum (Davenport, 1977; Saliba and Vella, 1977; Burris et al., 1990). Isolation, if prolonged, may lead to a significant physiological stress, particularly in terms of respiration and nutrition. Devices (biocaptors) have been designed to monitor the activity of these animals in order to detect the presence of environmental contaminants (see, for example, Sloof et al., 1993; Borcherding and Volpers, 1994; Kramer, 1994). These biocaptors have various limitations, due to ignorance of the basic rhythm of animal activity and of their response to environmental factors (e.g., turbidity, hydrodynamism, presence of predators). Therefore their use is generally restricted to the laboratory.

Even though some devices may give the opportunity to study the behaviour of organisms *in situ,* there seems to be no means to quantitatively, or even qualitatively, link changes in avoidance or escape behaviour of individuals with specific biomarkers on the one hand and with the dynamics of the population on the other.

2.2.1.2 Escape Behaviour in Relation to Predators

Predators frequently exert a selective predation on the most vulnerable individuals. Contaminants sometimes deeply modify predator-prey relationships, with various consequences. For example, some substances increase the vulnerability of prey (Cooke, 1971; Goodyear, 1972; Kania and O'Hara, 1974; Galindo et al., 1985), whereas others reduce the efficacy of predators (Johanson, 1967; Witt, 1971; Woin and Larsson, 1987). The effects of certain contaminants on predator-prey relationships may sometimes vary according to the level of intoxication. Hedtke and Norris (1980) showed that low concentrations of ammonium chloride stimulated young salmons (*Oncorhynchus tshawytscha*), allowing them to escape more easily from brook trout (*Salvelinus fontinalis*), whereas higher concentrations increased the impact of predation by the latter.

The relation between quantification of the vulnerability of prey and identification of possible biomarkers remains to be established, but biomarkers related to the functioning of nervous system seem to be valuable candidates.

2.2.2 Reproductive and Parental Behaviour

In many animal species, reproductive and parental behaviour are the result of a complex set of elementary behavioural sequences such as recognition of mate, courtship, nest building, feeding of young, and protection of young against predators.

Reproductive behaviour of invertebrates, especially arthropods, is often based on mechanisms that can be disturbed by exposure to contaminants. For example, alterations in reproductive behaviour of lepidoptera exposed to sublethal concentrations of insecticides have been described (Floyd and Crowder, 1981; Linn and Roelofs, 1984, 1985; Haynes and Baker, 1985).

In birds, some field studies have shown that intoxication by neurotoxic substances may decrease the rate of survival of young through modification of the behaviour of their parents (Grue et al., 1982). The various studies devoted to birds nesting near the North American Great Lakes, particularly gulls (Fox et al., 1978; Peakall et al., 1980; Fox, 1993), suggest the existence of a correlation between the level of contamination of the environment by persistent pollutants and a change in reproductive success linked, at least partly, to behavioural problems (Ludwig and Tomoff, 1966; Peakall et al., 1980; Peakall and Fox, 1987). This hypothesis has already been made by some authors to explain the decline of populations of certain raptors or wading birds (Fyfe et al., 1976). However, it must be pointed out that if the role of behavioural changes has sometimes been suspected, the existence of a proven causal relationship with population decline remains to be demonstrated. Furthermore, associated biomarkers remain to be identified, although biomarkers of neurotoxicity appear again to be valuable candidates (Peakall, 1996).

2.2.3 Foraging Behaviour

Changes in social behaviour have been intensively studied in Hymenoptera, especially in the honey bee (*Apis mellifera*) but also in other pollinators, which are often exposed to agricultural pesticides (Johansen, 1977). Sublethal concentrations of parathion have been shown to alter the temporal division of labour between different groups of *A. mellifera* (MacKenzie and Winston, 1989). Similarly, exposure to methyl parathion disturbs the 'dance' of bees, a key behaviour for the survival of the colony because it allows 'scouts' to indicate the direction in which and distance at which a source of food may be found (Schricker and Stephen, 1970). Permethrin inhibits the search for new sources of food (Cox and Wilson, 1984). In addition to the unfavourable impact on the survival of the colony, these phenomena may also affect the pollination of certain crops.

The anomalies observed in the construction of nets by Trichoptera larvae (Decamps et al., 1973; Besch et al., 1977, 1979; Petersen and Petersen, 1983) or of webs by spiders (Johnson, 1967; Witt, 1971) are other examples of alterations in food collection by invertebrates associated with the presence of toxic substances.

2.3 Perspectives

The existence of behavioural changes in invertebrates and vertebrates induced by the presence of contaminants does not need to be demonstrated further. Even though doubts persist as to the real consequences of these changes on the dynamics of populations, some studies suggest that they are frequently significant and may sometimes cause the decline of certain species. However, due to the integrated nature of behavioural phenomena, it is very difficult to identify biomarkers that can be used to predict the appearance of these changes. Behaviour partly depends on the response of organisms to various stimuli (e.g., visual, olfactory, tactile) that generally induce a transfer of information through nerves and the triggering of an adapted response (e.g., motor, secretory). Although the neuromediators implicated in these different transfers of information are potential biomarkers (AChE, for example), the interpretation of their response is always difficult, because of the existence of confusing factors (natural external stimuli) and of their role in many physiological processes (Bocquené et al., 1997; see chapter 6). Moreover, although behaviour may itself be considered a biomarker, its study *in situ* raises many technical problems in obtaining data (measurement techniques, information, transfer and processing cost) and interpreting them (heterogeneity of populations, intervention of stimuli other than contaminants). From a fundamental point of view, causal relationships between behavioural changes and changes in the dynamics of a population remain to be proved and quantified. This should be achieved through a better understanding of biochemical determinants (molecular as well as metabolic) of individual behavioural sequences.

3. REPRODUCTION, DEVELOPMENT, GROWTH, AND ENERGY ALLOCATION

The overall consequences of the presence of contaminants on the health of ecosystems derive primarily from biochemical reactions induced at the individual level. These reactions often result in modifications at higher levels of biological organization. Molecular, cellular, or tissue changes, modifications in immune responses, or variations in the activity of certain enzymes may have consequences on survival, growth, and reproduction of individuals and, consequently, on the dynamics of populations. Ultimately, the structure of communities and functioning of ecosystems may be affected.

Reproduction and development of organisms depend essentially on physiological mechanisms regulated by hormones and on the use of available energy resources (the provision of which is related to behaviour, via the capture of prey, which depends on locomotion capacities and other behavioural events; Fig. 4). In the following sections, we will successively consider two examples of hormonal disturbance that have consequences for animal reproduction and can be proved by the measurement of biomarkers. We then consider the particular case of energy resource allocation as both a target process and a potential biomarker.

3.1 Biomarkers of Alteration of Hormonal Processes

The toxic effects of many pollutants derive from their interaction with biochemical processes implicated in the control of various physiological phenomena such as growth, reproduction, energy production, and osmoregulation (Fig. 6; Brouwer et al., 1990). In normal physiological conditions, these processes are regulated not only by hormones, but also by vitamins and other molecules that ensure the proper translation of cellular signals. Hormones generally act via two different mechanisms: interaction at the level of cell membrane followed by the release of second messengers, and interaction with nuclear receptors involved in gene regulation and expression. Interference with toxicants may occur at one or several of these levels, and the final result is usually a disturbance of a physiological process or processes controlled by these hormones.

A number of hormones have been proposed as potential biomarkers, including corticosteroids, catecholamines, steroid sexual hormones, thyroidal hormones, insulin, glucagon, and growth hormone (Mayer et al., 1992; Peakall, 1992; see also chapter 9).

3.1.1 Biomarkers and Estrogen-Mimic Compounds

Particular attention has been given to estrogen-mimic substances (such as DDT and PCB), which can interact with estrogen receptors (Nelson, 1974; Korach et al., 1988) and induce estrogen-type effects (Soto et al., 1994). These substances are known to disrupt reproduction, especially through changes in male sexual characteristics in reptiles (Bergeron et al., 1994; Guillette et al., 1994) and fish (Jobling et al., 1995). Some laboratory experiments suggest that these substances can also lead to changes in reproduction and metabolism in invertebrates (Baldwin et al., 1995; Shurin and Dodson, 1997). Among the biomarkers that can be used to prove the impact of estrogen-mimic compounds in vertebrates,

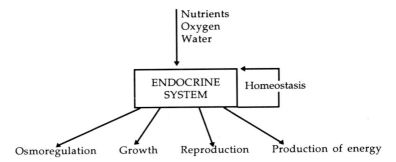

Fig. 6. Interactions between the endocrine system and the major physiological processes that can be altered by a toxic substance (from Brouwer et al., 1990)

vitellogenin appears promising, especially in oviparous species. This seric phospholipoglycoprotein is the main precursor of the reserve proteins of the egg in these animals, and some of its molecular domains (N-terminal region, especially) have been particularly well preserved in the course of evolution. Vitellogenin is synthesized and secreted by the liver in mature females in response to blood-circulating hormones. It is nearly absent in immature individuals or males. The presence of this protein in males can be used as an indicator of exposure to estrogens of endogenous or exogenous origin or to estrogen mimetics. Developed in fish for *in vitro* tests (Pelissero et al., 1991, 1993; Jobling and Sumpter, 1993; Purdom et al., 1994; Sumpter, 1995; Tyler et al., 1996), measurement of this protein has also been used successfully *in situ* to demonstrate exposure to estrogen mimetics present in effluents (Harries et al., 1996). Preparation of antibodies against the most conserved part of this protein seems to make it possible to quantify it in many species, including individuals from the natural environment (Heppell et al., 1995).

3.1.2 Biomarkers of Contamination of Aquatic Environment by Tributyltin (TBT)

In some species of vertebrates and invertebrates, intoxication by certain contaminants results in profound anatomical changes, which could have perceptible consequences at the population level, because they affect either reproductive organs ('imposex' in aquatic molluscs, see below; intersexuality in fish, see for example Howell et al., 1980; Bortone et al., 1989; Davis and Bortone, 1992; Bortone and Davis, 1994; Piferrer et al., 1994; Cody and Bortone, 1997) or organs used for feeding (menta of midge larvae, see for example Hamilton and Saether, 1971; Hare and Carter, 1976; Warwick et al., 1987; Warwick, 1988, 1990; Warwick and Tisdale, 1988; Dermott, 1991; Dickman et al., 1992; Diggs and Stewart, 1993; Bird, 1994; Bird et al., 1995; bird beaks, see for example Grier,

1968; Fox et al., 1991).midge larvae, see for example Hamilton and Saether, 1971; Hare and Carter, 1976; Warwick et al., 1987; Warwick, 1988, 1990; Warwick and Tisdale, 1988; Dermott, 1991; Dickman et al., 1992; Diggs and Stewart, 1993; Bird, 1994; Bird et al., 1995; bird beaks, see for example Grier, 1968; Fox et al., 1991).

The effects of organic derivatives of tin, especially tributyltin (TBT) used in antifouling paints on the reproductive system of prosobranch Mollusca are among the best documented cases of teratogenic effects of environmental pollutants and their impact on reproduction and the dynamics of wild populations. These compounds cause the appearance of male genital organs in females of certain marine prosobranch gastropods. This is the phenomenon of pseudohermaphrodism or imposex (Smith, 1971; Gibbs and Bryan, 1986; Gibbs et al., 1991), which affects more than 40 species of aquatic molluscs (Ellis and Pattisina, 1990). In some of them, such as *Ilyanassa obsoleta* and *Hinia reticulata*, the reproductive activity of females is hardly affected. In other species, especially in the Muricidae *Nucella lapillus* and *Ocenebra erinacea*, the structure of the oviduct can be so modified that egg-laying is hindered, which results in significant reductions in the abundance of the exposed population and may sometimes lead to its extinction (Bryan et al., 1986; Oehlmann et al., 1996). For example, *N. lapillus* considerably declined in Europe over the past 15 years (Gibbs and Bryan, 1986). The use of TBT in antifouling paints was severely restricted in France in 1982 (Ruiz et al., 1996) and in Europe, North America, Australia, and Japan in 1987 (Matthiessen et al., 1995; Langston, 1996). Even though the degree of imposex has already declined in certain populations of *N. lapillus* (Evans et al., 1991), it will probably take a long time for a complete restoration (Hawkins et al., 1994; Ruiz et al., 1996; Langston, 1996).

Many authors have proposed indices enabling quantification of imposex at the population level. A preliminary series of indices were based on biometric measurements: average female penis length (Feral, 1974), relative penis size index (RPS; Giobbs et al., 1987), and relative penis length (RPL; Stewart et al., 1992; Stroben et al., 1992). These indices were too sensitive to natural seasonal variations in the size of reproductive organs of these animals (especially in males) to be used effectively (Stroben et al., 1996). Gibbs et al. (1987) proposed an index based on the mean stage of imposex observed for a given population (VDS or vas deferens sequence index). This has been made possible by the definition of six successive levels of imposex, characterized by increasingly serious anomalies in the reproductive system. After certain improvements, this index, which exhibits little seasonal variability, has been used successfully for various species (Kohn and Almasi, 1993; Stroben et al., 1996). Reference scales have been published that allow evaluation of coastal pollution by TBT as a function of the VDS index in different species commonly found on the western coasts of Europe (Stroben et al., 1995; Fig. 7). However, though this index is particularly suitable to prove contamination by low levels of TBT (Stroben et al., 1996), it cannot be used to reveal the existence of a problem till the anomaly of development has begun to affect a significant proportion of the population. Identification of an early warning signal for this phenomenon requires a better understanding of mechanisms that lead to the occurrence of imposex.

Certain experimental studies suggest that TBT inhibits the aromatization of androgens (testosterone) into estrogens (17β-estradiol) by acting at the level of a cytochrome P450-dependent aromatase (Gibbs et al., 1991; Hawkins et al., 1994; Bettin et al., 1996; Oehlmann et al., 1996). This inhibition would be competitive (Bettin et al., 1996). In many species of prosobranchs, testosterone and 17β-estradiol concentrations show important variations during the annual cycle. However, the ratio between the concentrations of these two hormones is relatively constant (Schulte-Oehlmann et al., 1995). Measurement of this ratio in females and/or that of the activity of cytochrome P450-dependent aromatases could therefore possibly be used as early warning signals of exposure to TBT or other substances that can induce imposex and thereby cause a decline of prosobranch populations. Moreover, the targets of endocrine systems show a high level of similarity within the Animal Kingdom. Thus, biomarkers identified in prosobranch molluscs may serve as an

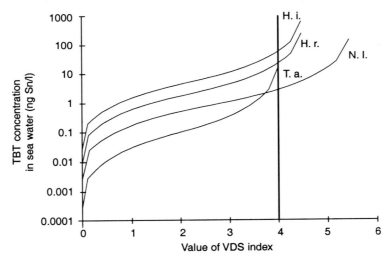

Fig. 7. Theoretical relation between the value of the VDS index measured in different species of marine prosobranchs and tributyltin concentration in sea water (TBT in ng Sn/l). The vertical line indicates the limit beyond which the fertility of individuals is altered. *H.i., Hinia incrassata; H.r., Hinia reticulata; N.l., Nucella lapillus; T.a., Trivia arctica* (from the data of Stroben et al., 1995).

early warning of the presence of hormone-mimic substances in the environment, especially mimics of male hormones (xeno-androgens), which are potentially dangerous for other species, including vertebrates.

3.2 Biomarkers and Energy Allocation for Growth and Reproduction

In any organism, growth, reproduction, and survival require energy. Energy consumption can be estimated by means of indices that can be used as biomarkers. Allocation of energy to growth and reproduction (AEGR, also known as scope for growth and scope for reproduction) is based on measurement of the quantity of energy available for growth and reproduction once the energy needs for maintaining basic metabolism have been fulfilled (Warren and Davis, 1967). This parameter may be a useful indicator of the state of health of organisms. Le Gal et al. (1997) have recently published a review on the use of such parameters as biomarkers. Most results published to date show the existence of a correlation between the presence of contaminants and alteration of the AEGR. The *in situ* use of these parameters presents some difficulties linked to the fact that their variations cannot always be related with a chemical pollution, since many other stressors (e.g., climate, food) elicit the same typé of response (Le Gal et al., 1997). The interpretation of changes in AEGR within an individual is often made difficult by lack of information about how energy is allocated between growth, reproduction, and other essential physiological processes. However, the respective roles of food and use of reserves in the satisfaction of these metabolic needs are difficult to identify (Forbes and Forbes, 1994).

Many studies have nevertheless demonstrated the usefulness of parameters implicated in the energy balance of the organism as an early warning of potential alterations within populations and communities (Bayne et al., 1985; Widdows and Johnson, 1988; Koehn and Bayne, 1989; Maltby, 1994; see also Le Gal et al., 1997). In particular, studies performed on *Gammarus pulex* enabled the identification of mechanistic links between the effects of changes in individual AEGR (effects on growth and reproduction) and the consequences at the level of populations (fitness, stress resistance), and even at the level of communities and ecosystems (Maltby, 1994).

The acquisition and allocation of energy determine the growth rate, fecundity, and survival of individuals. All these elements are essential to fitness and determine the structure and dynamics of a population. Underwood and Peterson (1988) consider that the impact of pollutants at the organism or suborganism level has more consequences on individual growth and reproduction than on the abundance of the population or on its contribution to food webs. However, it has been widely demonstrated that exposure to toxicants results in reduction of feeding in many animal species, especially in fish and aquatic invertebrates. Thus, possible secondary consequences at the food web level cannot be neglected. According to Jones et al. (1991), changes induced by sublethal concentrations of toxicants in the feeding behaviour of *Daphnia catawba* (Crustacea: Cladocera) have more rapid and pronounced effects on the algal community, its food source, than on the dynamics of the crustacean population. Similarly, the studies of Maltby (1994) on *G. pulex*, a detritivorous species that plays a key role in the processing of dead organic matter in aquatic ecosystems, indicate that a decrease in the capacity of individuals to acquire energy from food causes a reduction in the rate of incorporation of organic matter in the food webs. These phenomena could have consequences not only for the detritivorous organisms, but also on the functioning of the entire ecosystem. Other authors have actually observed accumulation of decaying plant material in polluted water courses (Wallace et al., 1982; Carpenter et al., 1983).

The reduction of feeding in the presence of pollutants results in a decrease in AEGR and, consequently, of growth. When the exposure is prolonged, development is delayed and/or fecundity decreases (Lynch, 1989). All these effects lead to a decline of individual fitness (in the Darwinian sense; Bayne et al., 1979) and may influence the structure of the population (Calow and Sibly, 1990; Calow, 1991). The organisms that succeed in adapting to long-term exposure invest less energy in reproduction. They reproduce less frequently but have more and/or larger offspring (Maltby, 1991; Daveikis and Alikhan, 1996). The energy thus saved may be used to supply processes of resistance or repair. For example, the energy needed for synthesis or functioning of certain defensive systems (e.g., metallothioneins, cytochrome P450-dependent enzymes) may deprive the normal metabolic systems of part of their energy sources, especially that which could have been used for growth and reproduction, even though such a 'downturn' of energy flows remains to be demonstrated (Forbes, 1996). The energy expenditure needed to counterbalance the effects of toxicants within an organism increases with the duration and intensity of the exposure until irreversible damage occurs in certain steps of metabolism. This theoretical linking provides the basis of a model developed by Calow and Sibly (1990), which links the metabolic effects of stress expressed as changes in AEGR of individuals with consequences for the dynamics of the population.

4. CONCLUSION—LIMITS TO THE USE OF BIOMARKERS FOR PREDICTING EFFECTS AT THE POPULATION AND COMMUNITY LEVEL

Although the existence of dose-response relationships has been demonstrated in the laboratory for many biomarkers and for many toxicants, the operational use of these parameters for evaluating the environmental impact of pollutants often gives results that are difficult to analyse. Many factors contribute to this difficulty:

— The existence of various routes of contamination and effects of various biological and physico-chemical processes may sometimes modify the level of exposure and the bioavailability of toxic substances. Therefore, even if a concentration-response relationship has been established in the laboratory, that does not mean that a similar relation could be established *in situ*. Environmental conditions (pH, level of oxygenation, etc.) considerably influence the bioavailability of contaminants and their evolution, which could lead to a significant increase or, on the contrary, decrease in the internal dose in the organism. Similar changes may affect the response of biomarkers. This problem is not specific to biomarkers that can be used to predict effects at the population and community level but it cannot be ignored and must be taken into account in any attempt to validate biomarkers *in situ*.

— Physiological, even genetic, mechanisms of adaptation to the presence of contaminants may establish rapidly in many species. These mechanisms may have pre-existed in some individuals within a natural population because of a significant phenotypic and genotypic variability among individuals. In some cases, the existence of a constant pressure of contamination (effluent discharge in a river or repeated pesticide treatments, for example) may result in selection of the individuals that possess the best faculties for adaptation. This allows them to become more easily tolerant or develop resistance more rapidly (see chapter 12).

— Differences in the duration and/or frequency of exposure to toxicants, and possibilities of synergy, or on the contrary antagonism, between the toxic substances and/or environmental factors may also cause variations in the response of individuals within a population.

— The existence of compensatory mechanisms at the population level may obscure the effects of contaminants at the individual level. Frequently, a decrease in population density results in increased reproduction of the individuals that remain (e.g., more eggs or embryos, better hatching success, increase in number of reproductive periods per year; see for example Martin, 1987; Godfray et al., 1991). These phenomena may result from a decrease in intraspecific competition for different resources (space, food, etc.) or from physiological changes (decrease in frequency of visual, tactile, or olfactory contact between individuals with consequences for the levels of certain hormones; see for example Christian, 1961; Christian and Davis, 1964; Bohlken et al., 1987). After a while, no alteration may be detected in the abundance or spatial distribution of the population concerned. However, these phenomena are frequently associated with a decrease in genetic polymorphism, which can be unfavourable if environmental conditions change (because of a probable reduction in adaptative performances; Guttman, 1994; Forbes, 1996).

There are very few examples of studies that have aimed to link or compare results of the measurement of biomarkers and those obtained for higher levels of biological organization (Adams et al., 1992; Schlenk et al., 1996). Fore et al. (1995) did not succeed in establishing relationships between the genetic diversity of *Pimephales notatus* populations and values attained by indices describing the structure of fish communities (IBI, index of biotic integrity) or of macroinvertebrate communities (ICI, invertebrate community index). Schlenk et al. (1996) examined correlations between various hepatic biomarkers measured in fish from the Arkansas River (cytochrome P450 (CYP1A) concentrations, EROD activity, concentration of mRNA encoding for a metallothionein, concentration of stress proteins (Hsp30)) and various parameters of the health of these fish and fish populations and community characteristics (abundance, species richness, species diversity). Some significant correlations have been identified between certain biomarkers (CYP1A, Hsp30) and the health of fish. Tendencies (non-significant correlations) have been observed in relations between health and characteristics of the community. Finally, correlations that are significant but difficult to explain have been found between biomarkers and the various characteristics of fish populations and communities. It is advisable to recall that the observation of a significant correlation between two parameters does not imply that there is a causal relation between them. In this perspective, the implementation of such studies and the use of adapted techniques of data analysis may give the opportunity to identify mechanistic relationships between the parameters measured at different levels of biological organization.

However, the use of biomarkers in such a context can only be considered within the framework of a multi-criteria approach based on the simultaneous measurement of parameters at increasing levels of biological organization, from the individual to the community. The predictive role that can be ascribed to certain biomarkers measured at the individual level could only be fulfilled in the perspective of the establishment of monitoring programmes based on such a multi-criteria approach. The research of quantitative mechanistic links must be encouraged because they may give the opportunity to confirm the hypotheses on which we still base too many extrapolations between levels of biological organization.

REFERENCES

Adams S.M., Crumby W.D., Greeley M.S. Jr et al. (1992). Responses of fish populations and communities to pulp mill effluents: a holistic assessment. *Ecotoxicol. Environ. Saf.*, 24:347–360.

Andersson G., Berggren H., Cronberg G. and Gelin C. (1978). Effects of planktivorous and benthivorous fish on organisms and water chemistry in eutrophic lakes. *Hydrobiologia*, 59:9–15.

Attrill M.J. and Depledge M.H. (1997). Community and population indicators of ecosystem health: targeting links between levels of biological organisation. *Aquat. Toxicol.* 38:183–197.

Baatrup E. and Bailey M. (1993). Quantitative analysis of spider locomotion employing computer-automated video tracking. *Physiol. Behav.*, 54:83–90.

Baldwin W.S., Milam D.L. and Leblanc G.A. (1995). Physiological and biochemical perturbations in *Daphnia magna* following exposure to the model environmental estrogen diethylstilbestrol. *Environ. Toxicol. Chem.*, 14:945–952.

Barbault R. (1992). *Écologie des peuplements. Structure, dynamique et évolution.* Masson, Paris.

Bayne B.L., Brown D.A., Burnes K. et al. (1985). *The Effects of Stress and Pollution on Marine Animals.* Praeger Special Studies, New York.

Bayne B.L., Moore M.N., Widdows J. et al. (1979). Measurements of the responses of individuals to environmental stress and pollution: studies with bivalve molluscs. *Philos. Trans. R. Soc. Lond. Ser. B*, 286:563–581.

Beddington J.R., Free C.A. and Lawton J.H. (1976). Concepts of stability and resilience in predator-prey models. *J. Anim. Ecol.*, 45:791–816.

Bergeron J.M., Crew D. and McLachlan J.A. (1994). PCBs as environmental estrogens: turtle sex determination as a biomarker of environmental contamination. *Environ. Health Perspect.*, 102:780–781.

Berryman A.A. (1981). *Populations Systems. A General Introduction.* Plenum Press, New York.

Besch W.K., Schreiber I. and Herbst D. (1977). Der *Hydropsyche*-Toxzitatest, erprobt an fenethcarb. *Schweiz. Z. Hydrol.*, 39:69–85.

Besch W.K., Schreiber I. and Magnin E. (1979). Influence du sulfate de cuivre sur la structure du filet des larves d'*Hydropsyche* (Insecta, Trichoptera). *Ann. Limnol.*, 15:123–138.

Bettin C., Oehlmann J. and Stroben E. (1996). TBT-induced imposex in marine neogasteropods is mediated by an increasing androgen level. *Helgoland. Meeresunters.*, 50:299–317.

Bird G.A. (1994). Use of chironomid deformities to assess environmental degradation in the Tamaska River, Quebec. *Environ. Monit. Assess.*, 30:163–175.

Bird G.A., Rosentreter M.J. and Schwartz W.J. (1995). Deformities in the menta of chironomid larvae from the Experimental Lakes Area, Ontario. *Can. J. Fish. Aquat. Sci.*, 52:2290–2295.

Blandin P. (1986). Bioindicateurs et diagnostic des systèmes écologiques. *Bull. Ecol.*, 17:215–307.

Bocquené G., Galgani F. and Walker C.H. (1997). Les cholinestérases, biomarqueurs de neurotoxicité. In: Biomarqueurs en écotoxicologie: Aspects fondamentaux. Lagadic L., Caquet Th., Amiard J.C. and Ramade F. (eds.). Masson, Paris, pp. 209–240.

Bohlken S., Joosse J. and Geraerts W.P.M. (1987). Interaction of photoperiod, grouping and isolation in female reproduction of *Lymnaea stagnalis*. *Int. J. Invert. Reprod. Dev.,* 11:45–58.

Borcherding J. and Volpers M. (1994). The 'Dreissena-Monitor'—First results on the application of this biological early warning system in the continuous monitoring of water quality. *Wat. Sci. Tech.,* 29:199–201.

Bortone S.A. and Davis W.P. (1994). Fish intersexuality as indicator of environmental stress. *BioScience,* 44:165–172.

Bortone S.A., Davis W.P. and Bundrick C.M. (1989). Morphological and behavioral characters in mosquitofish as potential bioindication of exposure to kraft mill effluent. *Bull. Environ. Contam. Toxicol.,* 43:370–377.

Brouwer A., Murk A.J. and Koeman J.H. (1990). Biochemicals and physiological approaches in ecotoxicology. *Funct. Ecol.,* 4:275–281.

Bryan G.W., Gibbs P.E., Hummerstone L.G. and Burt G.R. (1986). The decline of the gastropod *Nucella lapillus* around south-west England: evidence for the effect of tributyltin from antifouling paints. *J. Mar. Biol. Assoc. U.K.,* 66:611–640.

Burris J.A., Bamford M.S. and Steward A.J. (1990). Behavioral responses of marked snails as indicators of water quality. *Environ. Toxicol. Chem.,* 9:69–76.

Calow P. (1991). Physiological costs of combating chemical toxicants: ecological implications. *Comp. Biochem. Physiol. C,* 100:3–6.

Calow P. (1995). Ecosystem health: a critical analysis of concepts. In: Evaluating and Monitoring the Health of Large-Scale Ecosystems. Costanza R., Gaudet D.J. and Calow P. (eds.). Springer-Verlag, Heidelberg, pp. 33–42.

Calow P. and Sibly R.M. (1990). A physiological basis of population processes: ecotoxicological implications. *Funct. Ecol.,* 4:283–288.

Carpenter J., Odum W.E. and Mills A. (1983). Leaf litter decomposition in a reservoir affected by acid mine drainage. *Oikos,* 41:165–172.

Carpenter S.R. and Kitchell J.F. (1992). Trophic cascade and biomanipulation: interface of research and management: a reply to DeMelo et al. *Limnol. Oceanogr.,* 37:208–213.

Carpenter S.R. and Kitchell J.F. (1993). *The Trophic Cascade in Lakes.* Cambridge University Press, Cambridge.

Carpenter S.R., Kitchell J.F. and Hodgson J.R. (1985). Cascading trophic interactions and lake productivity. *Bioscience,* 35:634–639.

Christian J.J. (1961). Phenomena associated with population density. *Proc. Nat. Acad. Sci.,* 47:428–449.

Christian J.J. and Davis D.E. (1964). Endocrines, behavior and populations. *Science,* 146:1550–1560.

Cody R.P. and Bortone S.A. (1997). Masculinization of mosquitofish as an indicator of exposure to kraft mill effluent. *Bull. Environ. Contam. Toxicol.,* 58:429–436.

Cohen J.E. (1978). *Food Webs and Niche Space.* Monographs in Population Biology. Princeton University Press, Princeton.

Cohn J. and MacPhail R.C. (1996). Ethological and experimental approaches to behavior analysis: implications for ecotoxicology. *Environ. Health Perspect.,* 104 (Suppl. 2):299–305.

Cooke A.S. (1971). Selective predation by newts on frog tadpoles treated with DDT. *Nature,* 229:275–276.

Corta H.H. (1967). Response of *Gammarus pulex* (L.) to modified environment. II. Reactions to abnormal hydrogen ion concentration. *Crustaceana,* 3:1–10.

Costanza R., Norton B. and Haskell B. (1992). *Ecosystem Health: New Goals for Environmental Management.* Island Press, Washington.

Costanza R., Gaudet D.J. and Calow P. (1995). *Evaluating and Monitoring the Health of Large-Scale Ecosystems.* Springer-Verlag, Heidelberg.

Cox R.L. and Wilson W.T. (1984). Effects of permethrin on the behavior of individually tagged honey bee, *Apis mellifera* L. (Hymenoptera: Apidae). *Environ. Entomol.,* 13:375–378.

Daveikis V.F. and Alikhan M.A. (1996). Comparative body measurements, fecundity, oxygen uptake, and ammonia excretion in *Cambarus robustus* (Astacidae, Crustacea) from an acidic and a neutral site in northeastern Ontario, Canada. *Can. J. Zool.,* 74:1196–1203.

Davenport J. (1977). A study of the effects of copper applied continuously and discontinuously to specimens of *Mytilus edulis* (L.) exposed to steady and fluctuating salinity levels. *J. Mar. Biol. Assoc. U.K.,* 57:63–74.

Davis W.P. and Bortone S.A. (1992). Effects of kraft mill effluent on the sexuality of fishes: an environmental early warning? In: Advances in Modern Environmental Toxicology, Vol. XXI. T. Colthorn and C. Clement (eds.). Princeton Scientific Publishing Co., Princeton, pp. 113–127.

DeAngelis D.L. (1980). Energy flow, nutrient cycling, and ecosystem resilience. *Ecology,* 61:764–771.

DeAngelis D.L. (1992). *Dynamics of Nutrient Cycling and Food Webs.* Chapman and Hall, London.

Decamps H., Bech K.W. and Vobis H. (1973). Influence de produits toxiques sur la construction du filet des larves d'*Hydropsyche* (Insecta, Trichoptera). *C. R. Acad. Sci. Sér. D,* 276:375–378.

DeMelo R., France R. and McQueen D.J. (1992). Biomanipulation: hit or myth? *Limnol. Oceanogr.,* 37:192–207.

Depledge M.H., Aagaard A. and Györkös P. (1995). Assessment of trace metal toxicity using molecular, physiological and behavioural biomarkers. *Mar. Pollut. Bull.,* 31:19–27.

Dermott R.M. (1991). Deformities in larval *Procladius* spp. and dominant Chironomini from the St. Clair River. *Hydrobiologia,* 219:171–185.

Dickman M., Brindle I. and Benson M. (1992). Evidence of teratogens in sediments of the Niagara River watershed as reflected by chironomid (Diptera: Chironomidae) deformities. *J. Great Lakes Res.,* 18:467–480.

Diggins T.P. and Stewart K.M. (1993). Deformities of aquatic larval midges (Chironomidae: Diptera) in the sediments of the Buffalo River, New York. *J. Great Lakes Res.,* 19:648–659.

Drenner R.W., Smith J.D., Mummert J.R. and Lancaster H.F. (1990). Responses of a eutrophic pond community to separate and combined effects of N : P supply and planktivorous fish: a mesocosm experiment. *Hydrobiologia,* 208:161–167.

Eidt D.C. (1975). The effect of fenitrothion from large-scale forest spraying on benthos in New Brunswick headwaters streams. *Can. Entomol.,* 107:743–760.

Ellis D.V. and Pattisina A. (1990). Widespread neogasteropod imposex: a biological indicator of global TBT contamination? *Mar. Pollut. Bull.,* 21:248–253.

Evans S.M., Hutton A., Kendall M.A. and Samosir A.M. (1991). Recovery in populations of dogwhelks *Nucella lapillus* (L.) suffering from Imposex. *Mar. Pollut. Bull.,* 22:331–333.

Everts J.W., Frankenhuyzen H. van, Roman B. and Koeman J.H. (1983). Side effects of experimental pyrethroid applications for the control of tsetse flies in a riverine forest habitat (Africa). *Arch. Environ. Contam. Toxicol.,* 12:91–97.

Feral C. (1974). Étude comparée des populations d'*Ocenebra erinacea* (L.) de Granville et d'Arcachon. *Haliotis,* 4:123–134.

Flannagan J.F., Townsend B.E., De March B.G.E. et al. (1979). The effects of an experimental injection of methoxychlor on aquatic invertebrates: accumulation, standing crop, and drift. *Can. Entomol.,* 111:73–89.

Floyd J.P. and Crowder L.A. (1981). Sublethal effects of permethrin on pheromone response and mating of male pink bollworm moths. *J. Econ. Entomol.,* 74:634–638.

Forbes V.E. (1996). Chemical stress and genetic variability in invertebrate populations. *Tox. Ecotox. News,* 3:136–141.

Forbes V.E. and Forbes T.L. (1994). *Ecotoxicology in Theory and Practice.* Chapman and Hall Ecotoxicology Series 2, Chapman and Hall, London.

Foré S.A., Guttman S.I., Bailer A.J. et al. (1995). Exploratory analysis of population genetic assessment as a water quality indicator. *Ecotoxicol. Environ. Saf.,* 30:24–35.

Fox G.A. (1993). What have biomarkers told us about the effects of contaminants on the health of fish-eating birds in the Great Lakes? The theory and a literature review. *J. Great Lakes Res.,* 19:722–736.

Fox G., Gilman A.P., Peakall D.B. and Anderka F.W. (1978). Behavioral abnormalities in nesting Lake Ontario herring gulls. *J. Widl. Manage.,* 42:477–483.

Fox G.A., Collins B., Hayakawa E. et al. (1991). Reproductive outcomes of colonial fish-eating birds: a biomarker for developmental toxicants in Great Lakes food chains. II. Spatial variation in the occurrence and prevalence of bill defects in young double-crested cormorants in the Great Lakes, 1979-1987. *J. Great Lakes Res.,* 17:158–167.

Fredeen F.J.H. (1974). Tests with single injections of methoxychlor blackfly (Diptera: Simuliidae) larvicides in large rivers. *Can. Entomol.,* 106:285–305.

Frontier S. (1977). Réflexions pour une théorie des écosystèmes. *Bull. Ecol.,* 8:445–464.

Fyfe R., Risebrough R.W. and Walker W. (1976). Pollutant effects on the reproduction of the prairie falcons and merlins of the Canadian prairies. *Can. Field Nat.,* 90:346–355.

Galindo J.C., Kendall R.J., Driver C.J. and Lacher T.E. Jr. (1985). The effect of methyl parathion on susceptibility of bobwhite quail *(Colinus virginianus)* to domestic cat predation. *Behav. Neural Biol.,* 43:21–36.

Gerhardt A. (1995). Monitoring behavioural responses to metals in *Gammarus pulex* (L.) (Crustacea) with impedance conversion. *Environ. Sci. Pollut. Res.,* 2:15–23.

Gerhardt A. (1996). Behavioural early warning responses to polluted water. Performance of *Gammarus pulex* L. (Crustacea) and *Hydropsyche angustipennis* (Curtis) (Insecta) to a complex industrial effluent. *Environ. Sci. Pollut. Res.,* 3:63–70.

Gerhardt A., Clostermann M., Fridlund B. and Swensson E. (1994). Monitoring of behavioural patterns of aquatic organisms with an impedance conversion technique. *Environ. Internat.,* 20:209–219.

Gibbs P.E. and Bryan G.W. (1986). Reproductive failure in populations of the dog-whelk, *Nucella lapillus,* caused by imposex induced by tributyltin from antifouling paints. *J. Mar. Biol. Assoc. U.K.,* 66:767–777.

Gibbs P., Bryan G. and Spence S. (1991). The impact of tributyltin (TBT) pollution on the *Nucella lapillus* (Gastropoda) populations around the coast of South-East England. *Oceanol. Acta,* vol. sp. 11:257–261.

Gibbs P. Bryan G.W., Pascoe P.L. and Burt G.R. (1987). The use of the dog-whelk, *Nucella lapillus,* as an indicator of tributyltin (TBT) contamination. *J. Mar. Biol. Assoc. U.K.,* 67:507–523.

Godfray H.C.J., Partridge L. and Harvey P.H. (1991). Clutch size. *Annu. Rev. Ecol. Syst.,* 22:409–429.

Goodyear C.P. (1972). A simple technique for detecting effects of toxicants or other stresses on a predator-prey interaction. *Trans. Am. Fish. Soc.,* 101:367–370.

Greig-Smith P.W. (1991). Use of cholinesterase measurements in surveillance of wildlife poisoning in farmland. In: Cholinesterase-inhibiting Insecticides, Chemicals in Agriculture, Vol. 2. Mineau P. (ed.). Elsevier, Amsterdam, pp. 127–150.

Grier J.W. (1968). Immature bald eagle with an abnormal beak. *Bird-Banding,* 39:58–59.

Grue C.E., Powell G.V.N. and McChesney M.J. (1982). Care of nestlings by wild female starlings exposed to an organophosphate pesticide. *J. Appl. Ecol.,* 19:327–355.

Guillette L.J. Jr, Gross T.S., Masson G.R. et al. (1994). Developmental abnormalities of the gonad and abnormal sex hormone concentrations in juvenile alligators from contaminated and control lakes in Florida. *Environ. Health Perspect.,* 102:680–688.

Guttman S.I. (1994). Population genetic structure and ecotoxicology. *Environ. Health Perspect.,* 102 (suppl. 12):97–100.

Hamilton A.C. and Saether O.A. (1971). The occurrence of characteristic deformities in the chironomid larvae of several Canadian lakes. *Can. Entomol.,* 103:363–368.

Hare L. and Carter J.C.H. (1976). The distribution of *Chironomus* (s.s.? *culicini salinarius* group) larvae (Diptera: Chironomidae) in Parry Sound, Georgian Bay, with particular reference to structural deformities. *Can. J. Zool.*, 65:2129–2134.

Harries J.E., Sheahan D.A., Jobling S. et al. (1996). A survey of estrogenic activity in United Kingdom inland waters. *Environ. Toxicol. Chem.*, 15:1993–2002.

Harrison G.W. (1979). Stability under environmental stress: resistance, resilience, persistence, and variability. *Amer. Nat.*, 113:659–669.

Harte J. (1979). Ecosystem stability and the distribution of community eigenvalues. In: Theoretical Systems Ecology. Halfon E. (ed.). Academic Press, New York, pp. 453–465.

Havens K.E. (1997). Unique structural properties of pelagic food webs. *Oikos*, 78:75–80.

Hawkins S.J., Proud S.V., Spence S.K. and Southward J. (1994). From the individual to the community and beyond: water quality, stress indicators and key species in coastal systems. In: Water Quality and Stress Indicators in Marine and Freshwater Ecosystems: Linking Levels of Organisation (Individuals, Populations, Communities). Sutcliffe D.W. (ed.). Freshwater Biological Association, Ambleside, pp. 35–62.

Haynes K.F. (1988). Sublethal effects of neurotoxic insecticides on insect behaviour. *Annu. Rev. Entomol.*, 33:149–168.

Haynes K.F. and Baker T.C. (1985). Sublethal effects of permethrin on the chemical communication system of the pink bollworm moth, *Pectinophora gossypiella*. *Arch. Insect Biochem. Physiol.*, 2:283–293.

Hedtke J.L. and Norris L.A. (1980). Effects of ammonium chloride on predatory consumption rates of brook trout (*Salvelinus fontinalis*) on juvenile chinook salmon (*Oncorhynchus tshawytscha*) in laboratory streams. *Bull. Environ. Contam. Toxicol.*, 24:81–89.

Hellawell J.M. (1986). *Biological Indicators of Freshwater Pollution and Environmental Management*. Elsevier Applied Science Publishers, London.

Heppell S.A., Denslow N.D., Folmar L.C. and Sullivan C.V. (1995). Universal assay of vitellogenin as a biomarker for environmental estrogens. *Environ. Health Perspect.*, 103:9–15.

Holling C.S. (1973). Resilience and stability of ecological systems. *Annu. Rev. Ecol. Syst.*, 4:1–23.

Howell W.M., Black D.A. and Bortone S.A. (1980). Abnormal expression of secondary sex characters in a population of mosquitofish, *Gambusia affinis holbrooki*: evidence for environmentally induced masculinization. *Copeia*, 1980:43–51.

Hunter M.D. and Price P.W. (1992). Playing chutes and ladders: heterogeneity and the relative roles of bottom-up and top-down forces in natural communities. *Ecology*, 73:724–732.

Jobling S. and Sumpter J.P. (1993). Detergent component in sewage effluent are weakly oestrogenic to fish: an *in vitro* study using rainbow trout (*Oncorhynchus mykiss*) hepatocytes. *Aquat. Toxicol.*, 27:361–372.

Jobling S., Sheahan D., Osborne J.A. et al. (1995). Inhibition of testicular growth in rainbow trout (*Oncorhynchus mykiss*) exposed to estrogenic alkylphenolic chemicals. *Environ. Toxicol. Chem.*, 15:194–202.

Johansen C.A. (1977). Pesticides and pollinators. *Annu. Rev. Entomol.*, 22:177–192.

Johanson R.R. (1967). The effect of DDT on the webs of *Aranea diademata*. *Mem. Soc. Faun. Flor. Fenn.*, 43:100–104.

Jones M., Folt C. and Guarda S. (1991). Characterizing individual, population and community effects of sublethal levels of aquatic toxicants: an experimental case study using *Daphnia*. *Freshwater Biol.*, 26:35–44.

Jordan C.F., Kline J.R. and Sasscer D.S. (1972). Relative stability of mineral cycles in forest ecosystems. *Amer. Nat.*, 106:237–253.

Jørgensen S.E. (1988). *Fundamentals of Ecological Modelling*. Elsevier, Amsterdam.

Jørgensen S.E. (1992). *Integration of Ecosystem Theories: A Pattern*. Kluwer· Academic. Dordrecht.

Kania H.J. and O'Hara (1974). Behavioral alterations in a simple predator-prey system due to sublethal exposure to mercury. *Trans. Am. Fish. Soc.*, 103:134–136.

Koehn R.K. and Bayne B.L. (1989). Towards a physiological and genetical understanding of the energetics of the stress response. *Biol. J. Linn. Soc.*, 37:157–171.

Kohn A.J. and Almasi K.N. (1993). Imposex in Australian *Conus. J. Mar. Biol. Assoc. U.K.*, 21:61–89.

Korach K.S., Sarver P., Chae K., McLachlan J.A. and McKinney J.D. (1988). Estrogen receptor-binding activity of polychlorinated hydroxy-biphenyls: conformationally restricted structural probes. *Mol. Pharmacol.*, 33:120–126.

Kramer K.J.M. (1994). *Biomonitoring of Coastal Waters and Estuaries.* CRC Press, Baton Rouge.

Krebs C.T. and Burns K.A. (1977). Long-term effects of an oil spill on populations of the salt-marsh crab *Uca pugnax. Science*, 197:484–487.

Kreutzweiser D.P. and Kingsbury P.D. (1987). Permethrin treatments in Canadian forests. Part 2: Impact on stream invertebrates. *Pestic. Sci.*, 19:49–60.

Lagadic L. and Caquet Th (1996). Marqueurs biologiques de pollution. Des outils au service de l'écotoxicologie. *Phytoma*, 480:10–13.

Lagadic L., Caquet Th and Ramade F. (1994). The role of biomarkers in environmental assessment. 4. Invertebrate populations and communities. *Ecotoxicology*, 3:193–208.

Langston W.J. (1996). Recent developments in TBT ecotoxicology. *Toxicol. Ecotox. News*, 3:179–187.

Lazzaro X., Drenner R.W., Stein R.A. and Smith J.D (1992). Planktivores and plankton dynamics: effects of fish biomass and planktivore type. *Can. J. Fish. Aquat. Sci.*, 49:1466–1473.

Le Gal Y., Lagadic L., Le Bras S. and Caquet Th (1997). Charge énergétique en adénylates (CEA) et autres biomarqueurs associés au métabolisme énergétique. In: Biomarqueurs en écotoxicologie: Aspects fondamentaux. Lagadic L., Caquet Th., Amiard J.C. and Ramade F. (eds.). Masson, Paris, pp. 241–286.

Linn C.E. and Roelofs W.L. (1984). Sublethal effects of neuroactive compounds on pheromone response thresholds in male oriental fruit moths. *Arch. Insect Biochem. Physiol.*, 1:331–344.

Linn C.E. and Roelofs W.L. (1985). Multiple effects of octopamine and chlordimeform on pheromone response thresholds in the cabbage looper moth, *Trichoplusia ni. Pestic. Sci.*, 16:445–446.

Ludwig J.P. and Tomoff C.S. (1966). Reproductive success and insecticide residues in Lake Michigan herring gulls. *Jack-Pine Warbler*, 44:77–85.

Lynch M. (1989). The life history consequences of resource depression in *Daphnia pulex. Ecology*, 70:246–256.

Mackenzie K.E. and Winston M.L. (1989). Effects of sublethal exposure to diazinon on longevity and temporal division of labor in the honey bee (Hymenoptera: Apidae). *J. Econ. Entomol.*, 82:75–82.

Maltby L. (1991). Pollution as a probe of life history adaptation in *Asellus aquaticus* (Isopoda). *Oikos*, 61:11–18.

Maltby L. (1994). Stress, shredders and streams: using *Gammarus* energetics to assess water quality. In: Water Quality and Stress Indicators in Marine and Freshwater Systems: Linking Levels of Organisation. Sutcliffe D.W. (ed.). Freshwater Biological Association, Ambleside, pp. 98–110.

Martin T.E. (1987). Food as a limit on breeding birds: a life-history perspective. *Annu. Rev. Ecol. Syst.*, 18:453–487.

Martinez N. and Lawton J.H. (1995). Scale and food-web structure—from local to global. *Oikos*, 73:148–154.

Matthiessen P., Waldock R., Thain J.E. et al. (1995). Changes in periwinkle (*Littorina littorea*) populations following the ban on TBT-based antifoulings on small boats in the United Kingdom. *Ecotoxicol. Environ. Saf.*, 30:180–194.

Mauersberger P. and Straskraba M. (1987). Two approaches to generalized ecosystem modelling: thermodynamics and cybernetic. *Ecol. Model.*, 39:161–176.

May R.M. (1973). *Stability and Complexity in Model Ecosystems*. Monographs in Population Biology. Princeton University Press, Princeton.

Mayer F.L., Versteeg D.J., McKee M.J. et al. (1992). Physiological and nonspecific biomarkers. In: Biomarkers. Biochemical, Physiological, and Histological Markers of Anthropogenic Stress. Huggett R.J., Kimerle R.A., Mehrle Jr. P.M. and Bergman H.L. (eds.). SETAC Special Publication Series, Lewis Publishers, Boca Raton, pp. 5–85.

Mazumder A., McQueen D.J., Taylor W.D. and Lean D.R.S. (1988). Effects of fertilization and planktivorous fish (yellow perch) predation on size distribution of particulate phosphorus and assimilated phosphate: large enclosure experiments. *Limnol. Oceanogr.*, 33:421–430.

McGreer E.R. (1979). Sublethal effects of heavy metal contaminated sediments on the bivalve *Macoma balthica* (L.). *Mar. Pollut. Bull.*, 10:259–262.

Muirhead-Thomson R.C. (1978). Lethal and behavioural impact of chlorpyrifos methyl and temephos on selected stream macroinvertebrates: experimental studies on downstream drift. *Arch. Environ. Contam. Toxicol.*, 7:139–147.

Muirhead-Thomson R.C. (1987). *Pesticide Impact on Stream Fauna with Special Reference to Macroinvertebrates*. Cambridge University Press, Cambridge.

Munkittrick K.R. and McCarty L.S. (1995). An integrated approach to aquatic ecosystem health: top-down, bottom-up or middle-out? *J. Aquat. Ecosyst. Health*, 4:77–90.

Neel R.B., Olson J.S. (1962). *Use of Analog Computers for Simulating the Movements of Isotopes in Ecological Systems*. No. 3172. Oak Ridge National Laboratory, Oak Ridge, Tennessee.

Nelson J.A. (1974). Effects of dichlorodiphenyltrichloroethane (DDT) analogs and polychlorinated biphenyls (PCB) mixtures on 17β-[^{3}H] estradiol binding to rat uterine receptor. *Biochem. Pharmacol.*, 23:447–451.

Neubert M.G. and Caswell H. (1997). Alternatives to resilience for measuring the responses of ecological systems to perturbations. *Ecology*, 78:653–665.

Odum E.P. (1964). The new ecology. *BioScience*, 14:14–16.

Odum E.P. (1968). Energy flow in ecosystems: a historical review. *Amer. Zool.*, 8:11–18.

Oehlmann J., Fioroni P., Stroben E. and Markert B. (1996). Tributyltin (TBT) effects on *Ocinebrina aciculata* (Gastropoda: Muricidae): imposex development, sterilization, sex change and population decline. *Sci. Total Environ.*, 188:205–223.

Olla B.L., Pearson W.H. and Studholme A.L. (1980). Applicability of behavioural measures in environmental stress assessment. In: Biological Effects of Marine Pollution and the Problems of Monitoring. McIntyre A.D. and Pearce J.B. (eds.). *Rapp. P.V. Réun. Cons. Int. Explor. Mer*, 179:162–173.

Paine R.T. (1966). Food web complexity and species diversity. *Amer. Nat.*, 100:65–76.

Paine R.T. (1980). Food web: linkage, interaction strength, and community infrastructure. *J. Anim. Ecol.*, 49:667–685.

Pascoe D. and Edwards R.W. (1984). *Freshwater Biological Monitoring*. Pergamon Press, Oxford.

Patten B.C. (1966). Systems ecology: a course sequence in mathematical ecology. *BioScience*, 16:593–598.

Patten B.C. and Witkamp M. (1967). Systems analysis of ^{134}cesium kinetics in terrestrial microcosms. *Ecology*, 48:813–824.

Peakall D.B. (1992). *Animal Biomarkers as Pollution Indicators*. Chapman and Hall, London.

Peakall D.B. (1996). Disrupted patterns of behavior in natural populations as an index of ecotoxicity. *Environ. Health Perspect.*, 104 (Suppl. 2):331–335.

Peakall D.B. and Fox G.A. (1987). Toxicological investigations on pollutant-related effects in Great Lakes gulls. *Environ. Health Perspect.*, 71:187–193.

Peakall D.B. and Walker C.H. (1996). Comment on van Gestel and van Brummelen. *Ecotoxicology*, 5:227–228.

Peakall D.B., Fox G.A., Gilman A.P. et al. (1980). Reproductive success of herring gulls as an indicator of Great Lakes water quality. In: Hydrocarbons and Halogenated Hydrocarbons. Afghan B.K. and MacKay D. (eds.). Plenum Press, New York, pp. 337–344.

Pelissero C., Bennetau B., Babin P. et al. (1991). The estrogenic activity of certain phytoestrogens in the Siberian sturgeon, *Acipenser baeri. J. Steroid Biochem. Molec. Biol.*, 38:293–299.

Pelissero C., Flouriot G., Foucher J.L. et al. (1993). Vitellogenin synthesis in cultured hepatocytes: an *in vitro* test for the estrogenic potency of chemicals. *J. Steroid Biochem. Mol. Bol.*, 44:263–272.

Petersen L.B.-M. and Petersen R.C. Jr. (1983). Anomalies in hydropsychid capture nets from polluted streams. *Freshwater Biol.*, 13:185–191.

Piferrer F., Zanuy S., Carillo M. et al. (1994). Brief treatment with an aromatase inhibitor during sex differentiation causes chromosomally female salmon to develop as normal functioning males. *J. Exp. Zool.*, 270:255–262.

Pimm S.L. (1982). *Food Webs.* Chapman and Hall, London.

Pimm S.L. (1984). The complexity and stability of ecosystems. *Nature*, 307:321–326.

Pimm S.L. (1991). *The Balance of Nature.* University of Chicago Press, Chicago.

Pimm S.L. and Lawton J.H. (1977). Number of trophic levels in ecological communities. *Nature*, 268:329–331.

Power M.E. (1992). Top-down and bottom-up forces in food webs: do plants have primacy? *Ecology*, 73:733–746.

Purdom C.E., Hardiman P.A., Bye V.J. et al. (1994). Estrogenic effects of effluents from sewage treatment works. *Chem. Ecol.*, 8:275–285.

Rapport D.J. (1989). What constitutes ecosystem health? *Persp. Biol. Med.*, 33:120–132.

Rapport D.J. (1995). Ecosystem health: exploring the territory. *Ecosystem Health*, 1:5–13.

Reinersten H., Jensen A., Kokvsik J.L. et al. (1990). Effects of fish removal on the limnetic ecosystem of a eutrophic lake. *Can. J. Fish. Aquat. Sci.* 47:166–173.

Ruiz J.M., Bachelet G., Caumette P. and Donard O.F.X. (1996). Three decades of tributyltin in the coastal environment with emphasis on Arcachon Bay, France. *Environ Pollut.*, 93:195–203.

Saliba L.J. and Vella M.G. (1977). Effects of mercury on the behaviour and oxygen consumption of *Monodonta articulata. Mar. Biol.*, 43:277–282.

Schaeffer D.J. (1996). Diagnosing ecosystem health. *Ecotoxicol. Environ. Saf.*, 34:18–34.

Schaeffer D.J., Herricks E.E. and Kerster H.W. (1988). Ecosystem health. I. Measuring ecosystem health. *Environ. Mgmt.*, 12:445–455.

Schaeffer D.J., Perry J., Kerster H.W. and Cox D.K. (1985). The environmental audit. I. Concepts. *Environ. Mgmt*, 9:191–198.

Scherer E. and McNicol R.E. (1986). Behavioural responses of stream-dwelling *Acroneuria lycorias* (Ins., Plecopt.) larvae to methoxychlor and fenitrothion. *Aquat. Toxicol.*, 8:251–263.

Schlenk D., Perkins E.J., Hamilton G. et al. (1996). Correlation of hepatic biomarkers with whole animal and population-community metrics. *Can. J. Fish. Aquat. Sci.*, 53:2299–2309.

Schricker B. and Stephen W.P. (1970). The effect of sublethal doses of parathion on honey bee behavior. 1. Oral administration and the communication dance. *J. Apic. Res.*, 9:141–153.

Schulte-Oehlmann U., Bettin C., Fioroni P. et al. (1995). *Marisa cornuarietis* (Gastropoda, Prosobranchia): a potential TBT bioindicator for freshwater environments. *Ecotoxicology*, 4:372–384.

Shapiro J. and Wright D.I. (1984). Lake restoration by biomanipulation: Round Lake, Minnesota, the first two years. *Freshwater Biol.*, 14:371–383.

Shear H. (1996). The development and use of indicators to assess the state of ecosystem health in the Great Lakes. *Ecosystem Health*, 2:241–258.

Shurin J.B. and Dodson S.I. (1997). Sublethal toxic effects of cyanobacteria and nonylphenol on environmental sex determination and development in *Daphnia*. *Environ. Toxicol. Chem.*, 16:1269–1276.

Sloof W., Dezwart D. and Marquenie J.M. (1983). Detection limits of a biological monitoring system for chemical water pollution based on mussel activity. *Bull. Environ. Contam. Toxicol.*, 30:400–405.

Smith B.S. (1971). Sexuality in the American mud snail, *Nassarius obsoletus* Say. *Proc. Malacol. Soc. Lond.*, 39:377–378.

Soto A.M., Chung K.L. and Sonnenshein C. (1994). The pesticides endosulfan, toxaphene, and dieldrin have estrogenic effects on human estrogen-sensitive cells. *Environ. Health Perspect.*, 102:380–383.

Stewart C., de Mora S.J., Jones M.R.L. and Miller M.C. (1992). Imposex in New Zealand neogastropods. *Mar. Pollut. Bull.*, 24:204–209.

Stroben E., Oehlmann J. and Fioroni P. (1992). The morphological expression of imposex in *Hinia reticulata* (Gastropoda: Buccinidae): a potential indicator of tributyltin pollution. *Mar. Biol.*, 113:625–636.

Stroben E., Oehlmann J., Schulte-Oehlmann U. and Fioroni P. (1996). Seasonal variations in the genital ducts of normal and imposex-affected prosobranchs and its influence on biomonitoring indices. *Malacol. Rev.*, Suppl. 6:173–184.

Stroben E., Schulte-Oehlmann U., Fioroni P. and Oehlmann J. (1995). A comparative method for easy assessment of coastal TBT pollution by the degree of imposex in Prosobranch species. *Haliotis*, 24:1–12.

Strong D.R. (1992). Are trophic cascades all wet? Differentiation and donor-control in speciose ecosystems. *Ecology*, 73:747–754.

Sumpter J. (1995). Estrogenic surfactant-derived chemicals in the aquatic environment. *Environ. Health Perspect.*, 103:173–178.

Tansley A.G. (1935). The use and abuse of vegetational concepts and terms. *Ecology*, 16:284–307.

Tyler C.R., van der Eerden B., Jobling S. et al. (1996). Measurement of vitellogenin, a biomarker for exposure to oestrogenic chemicals, in a wide variety of cyprinid fish. *J. Comp. Physiol.*, 166B:418–426.

Underwood A.J. and Peterson C.H. (1988). Towards an ecological framework for investigating pollution. *Mar. Ecol. Prog. Ser.*, 46:227–234.

Van Dyne G.M. (1966). *Ecosystems, Systems Ecology, and Systems Ecologists*. No. 3957. Oak Ridge National Laboratory, Oak Ridge, Tennessee.

Van Gestel C.A.M. and Van Brummelen T.C. (1996). Incorporation of the biomarker concept in ecotoxicology calls for a redefinition of terms. *Ecotoxicology*, 5:217–226.

Vanni M.J. and Layne C.D. (1997). Nutrient recycling and herbivory as mechanisms in the 'top-down' effect of fish on algae in lakes. *Ecology*. 78:21–40.

Vanni M.J., Layne C.D. and Arnott S.E. (1997). 'Top-down' trophic interactions in lakes: effects of fish on nutrient dynamics. *Ecology*, 78:1–20.

Wallace J.B., Vogel D.S. and Cuffney T.F. (1986). Recovery of a headwater stream from an insecticide-induced community disturbance. *J. N. Am. Benthol. Soc.*, 5:115–126.

Wallace J.B., Webster J.R. and Cuffney T.F. (1982). Stream detritus dynamics: regulation by invertebrate consumers. *Oecologia*, 53:197–200.

Wallace J.B., Cuffney T.F., Lay C.C. and Vogel D. (1987). The influence of an ecosystem-level manipulation on prey consumption by a lotic dragonfly. *Can. J. Zool.*, 65:35–40.

Warren G.E. and Davis G.E. (1967). Laboratory studies on the feeding, bioenergetics and growth of fish. In: The Biological Basis of Freshwater Fish Production. Gerhuy S.D. (ed). Blackwell Scientific Publications. Oxford, pp. 175–214.

Warwick W.F. (1988). Morphological deformities in Chironomidae (Diptera) larvae as biological indicators of toxic stress. In: Toxic Contaminants and Ecosystem Health: A Great Lakes Focus. Evans M.S. (ed.). John Wiley & Sons, New York, pp. 281–320.

Warwick W.F. (1990). Morphological deformities in Chironomids (Diptera) larvae from the Lac St. Louis and Laprairie basins of the St. Lawrence River. *J. Great Lakes Res.,* 16:185–208.

Warwick W.F. and Tisdale N.A. (1988). Morphological deformities in *Chironomus,* and *Procladius* larvae (Diptera: Chironomidae) from two differentially stressed sites in Tobin Lake, Saskatchewan. *Can. J. Fish. Aquat. Sci.,* 45:123–144.

Warwick W.F., Fitchko J., McKee P.M. et al. The incidence of deformities in the *Chironomus* spp. from Port Harbour, Lake Ontario. *J. Great Lakes Res.,* 13:88–92.

Watt K.E.F. (1968). *Ecology and Resource Management.* Academic Press, New York.

Webster J.R., Waide J.B. and Patten B.C. (1975). Nutrient recycling and the stability of ecosystems. In: Mineral Cycling in Southeastern Ecosystems. Howell F.G., Gentry J.B. and Smith M.H. (eds.). Energy Research and Development Administration (ERDA) Symposium Series. Technical Information Center, Washington DC, pp. 1–27.

Wentsel R., McIntosh A., McCafferty W.P. et al. (1977). Avoidance response of midge larvae *(Chironomus tentans)* to sediments containing heavy metals. *Hydrobiologia,* 52:171–175.

Widdows J. and Johnson D. (1988). Physiological energetics of *Mytilus edulis:* scope for growth. *Mar. Ecol. Prog. Ser.,* 46:113–121.

Willis A.J. (1997). The ecosystem: an evolving concept viewed historically. *Funct. Ecol.,* 11:268–271.

Wilson R.C.H., Harding L.E. and Hirvonen H. (1995). Marine ecosystem monitoring network design. *Ecosystem Health,* 1:222–227.

Witt P.N. (1971). Drugs alter web-building of spiders. *Behav. Sci.,* 16:98–113.

Woin P. and Larsson P. (1987). Phthalate esters reduce predation efficiency of dragonfly larvae, *Odonata: Aeshna. Bull. Environ. Contam. Toxicol.,* 38:220–225.

Woodwell G.M. (1970). Effects of pollution on the structure and physiology of ecosystems. *Science,* 168:429–433.

Yazvenko S.B. and Rapport D.J. (1996). A framework for assessing forest ecosystem health. *Ecosystem Health,* 2:40–51.

Conclusion

Biomarkers and Evaluation of the Ecotoxicological Impact of Pollutants

L. Lagadic, J.-C. Amiard and Th. Caquet

The previous chapters illustrate the usefulness of biomarkers for monitoring environmental health. Various types of biomarkers can be used to detect the presence of pollutants in the environment and, in some cases, to evaluate their impact on living organisms. However, *in situ* use of biomarkers for evaluating the ecological impact of environmental pollutants still requires conceptual and methodological improvements. Recent introduction of biomarkers in national and international environmental quality monitoring programmes raised some questions on the actual use of these biological tools. On the basis of current knowledge, it seems possible to analyse the feedback from field use of biomarkers and to consider approaches that, in the future, should improve the assessment of environmental effects of pollutants.

1. USE OF BIOMARKERS FOR MONITORING ENVIRONMENTAL QUALITY

In the context of this book, a biomarker is defined as a change that can be observed and/or measured at the molecular, biochemical, cellular, physiological, or behavioural level, and that reveals present or past environmental exposure of an individual to at least one chemical (Lagadic et al., 1997a). A large number of individual parameters fall within this definition. Therefore, a wide range of biomarkers is available to assess nearly every type of chemical pollution occurring in the environment.

Most of the biomarkers presently identified are able to detect the presence of contaminants in a given environment through the exposure of indigenous organisms. Detection of individual exposure is usually rapid, and allows an immediate diagnostic of the state of the environment. Some biomarkers take into account the effects of pollutants on living organisms. Most often, the effects pertain to physiological functions related to individual survival, growth, and reproduction. Biomarker responses may occur some time after exposure and therefore may provide historical information on the populations that have been exposed. Usually, the cost of measurement of most biomarkers is moderate, since the techniques required are largely improved. An increasing number of biomarkers are now routinely measured.

2. LIMITATIONS TO THE USE OF BIOMARKERS IN NATURAL ENVIRONMENTS

It is now recognized that the use of biomarkers as tools for evaluating environment quality depends on the ability to (Wolfe, 1996):

- estimate the distribution of substances that are potentially toxic in the environment and in living organisms as well;
- indicate responses of organisms to the exposure to environmental contaminants;
- establish cause-and-effect relationships between the presence of contaminants and biological responses;
- evaluate the consequences of the exposure of individuals on higher levels of biological organization (populations and communities) and eventually on ecosystem health.

In controlled conditions, either laboratory or artificial ecosystems (micro- and mesocosms), most biomarkers meet these criteria and usually dose-response relations can be established. Despite fundamental knowledge presently available for a very large majority of biomarkers (McCarthy and Shugart, 1990; Huggett et al., 1992; Peakall, 1992; Peakall and Shugart, 1993; Fossi and Leonzio, 1994; Lagadic et al., 1997b), only some of them have been subjected to large-scale evaluation in natural environments, and the results are frequently questionable (see for example chapters 2, 3, and 6).

The main factors that complicate interpretation of biomarker responses in individuals sampled in natural environments are clearly identified (Depledge et al., 1995; Munkittrick and McCarty, 1995; Engel and Vaughan, 1996; Wolfe, 1996):

- *Interference with environmental factors.* In the environment, organisms are subjected to variations of both abiotic factors (e.g., temperature, illumination) and biotic factors (e.g., interspecific competition, predation) that can modify the effects of pollutants by acting either on the bioavailability of contaminants or on the biomarkers themselves. It should be noted that, apart from chemical pollution, human activities such as deforestation (or reforestation), river management, or intensive fishing may modify environmental conditions through changes of habitat conditions and/or of relationships between species.

- *Combined effects of chemicals.* Once they have been released in the environment, chemicals may interact among themselves, in the medium, and/or in organisms. In addition, compounds that naturally occur in the environment can interact with man-made chemicals. Combined action of these different types of molecules can complicate the interpretation of biomarker responses.

- *Biological characteristics of individuals.* Biological characteristics of the species used for biomarker measurements should be carefully considered for interpreting the responses. The ability of most animal species to move from one location to another may affect the reliability of detection of polluted sites. The identification of the origin of pollutants is greatly improved when the organisms used for biomarker measurements have limited mobility. In this context, sessile plants or bivalve molluscs (mussels, oysters) are often relevant, but their geographical distribution may be a limiting factor. The species used as sources of biomarkers should have wide geographic distribution areas so that different sites can be compared using the same spécies. Few species meet both these criteria.

The physiological status of the individual used for biomarker measurement also determines the significance of the response. In most organisms, physiological processes are accompanied by significant metabolic changes. In particular, reproduction and growth are responsible for biochemical changes (enzyme activity, hormone levels, energy allocation) that may interfere with biomarkers. Although the periods of growth and reproduction may be easily determined for populations that are regularly studied, more random events such as starvation, injuries, or the development of pathologies can cause stress and therefore complicate interpretation of biomarker responses.

The reaction of individuals to contaminants may vary over time with possible consequences on biomarkers mainly in terms of rapidity and/or sensitivity of response. Evolution of biomarker responses may simply be linked to the age of the individuals. It may also depend on the duration and/or on the frequency of exposure to contaminants. The more frequent and/or prolonged the exposure, the greater the possibility of adaptation of the exposed organisms. The concept of adaptation is considered here in its widest sense since it also involves repair mechanisms at the molecular, cellular, or tissue levels, processes of compensation occurring in the individual or developing within the population, and even genetic changes that can be transmitted across generations.

The results presented in the previous chapters show that, in spite of these difficulties, biomarkers exhibit a real potential for the assessment of environmental quality, and the amount of data already available for the different types of biomarkers support *in situ* validation procedures.

3. VALIDATION OF THE USE OF BIOMARKERS IN THE MONITORING OF ENVIRONMENTAL QUALITY

The validation of one or several biomarkers to follow the evolution of the quality of natural environments is not a straightforward task. It is also more difficult to implement when the areas to be monitored are large. Monitoring programmes implemented at the ecosystem scale have shown that *in situ* use of biomarkers requires some precaution, not only during procedures of sampling and measurement, but also for interpreting the results.

3.1 Methodological Aspects

The species in which biomarkers are to be measured must be chosen on the basis of a sound knowledge of their biology in the environments being studied. Therefore, biomarkers are often measured in edible species (e.g. North Sea dab, see chapters 3 and 6; mussels and oysters, chapters 2, 5, and 6), or in sedentary species with a wide distribution area (e.g. mussel and dragonet, see chapters 2 and 3).

Organism sampling must be representative of the structure and abundance of populations. In order to evaluate the extent of biological damage caused by pollutants, it is important to estimate the number of affected individuals within a population, and to determine whether a particular sex or age class has been more affected. Such information may be of great help in the implementation of *in situ* pollution reduction policies.

Biomarkers must be measured in polluted sites but also in sites where pollution is likely to occur. In this context, long-term studies are the most suitable approach to detect changes in the sensitivity of individuals over time. For such purpose, it is necessary to identify reference sites free from pollution or in which the nature and amounts of pollutants can be accurately determined (Lagadic et al., 1997c).

Sampling or caging procedures may result in non-negligible sources of error because the stress caused by such methods can interfere with pollutant exposure. Biomarker responses due to individual sampling or transplanting must be minimized and, in any case, accurately known.

Methodological rigour and knowledge of environmental conditions guarantee the reliability of the information provided by field measurements of biomarkers. The RNO experience (see chapters 2, 3, and 6) is very informative from this point of view. After initial difficulties for interpreting the first results, standardization of sampling procedures and methods of biomarker measurements, especially in the last eight years, allowed comparisons between polluted sites, thus providing a synthetic view of the evolution of the quality of the ecosystems studied.

3.2 Interpretation of Results: Significance of Biomarkers

The present tendency is frequently to extrapolate responses of biomarkers to effects on higher levels of biological organization only on the basis of conceptual mechanistic relationships. Although this approach can be justified in the context of research and development of biomarkers, it rarely meets the needs of managers of environmental quality, who are generally looking for a diagnostic of the state of the environment. Since one of the future prospects of biomarkers is their use in environmental risk assessment, suitable validation procedures should be rapidly implemented (Engel and Vaughan, 1996; Ward and Henderson, 1996; Wolfe, 1996).

3.2.1 Diagnostic Approach: Biomarkers and Environmental Impact of Pollutants

When it shows higher than natural variability, the response of a biomarker indicates that environmental changes may have affected the individuals in which the biomarker has been measured. Then, the origin of the disturbance must be determined. Various approaches may be used to detect the presence of pollutants and, in the best cases, to identify them or, on the contrary, to show that pollutants may not be responsible for biomarker response. This situation describes the most common case, because every biomarker presently used is not specific for one unique type of pollutant.

However, some biomarkers respond to a relatively restricted range of pollutants. Thus, cytochrome P4501A (or related isoforms) is classically recognized as specific for the presence of polycyclic aromatic hydrocarbons, though other compounds such as pesticides are known to cause its induction (Narbonne and Michel, 1997; Monod, 1997; see also chapters 3 and 4). Similarly, acetylcholinesterase reacts quite specifically to organophosphorous compounds and carbamates, though it also responds to metals (Bocquene et al., 1997; see also chapter 6). Even if they do not unequivocally indicate the nature of the contaminants, such biomarkers limit the analytical screening to a few chemical families, thus significantly reducing the cost of analyses.

Since biomarkers may only reveal a past exposure to toxicants (see chapter 12) and considering the specificity of response of some of them, the relevance of chemical analysis is questionable (Gillet and Monod, 1997). Basically, two distinct cases may be considered. In the first case, the exact nature of the pollutants is known and the sources of discharges are identified and even controlled. The use of biomarkers ensures that the conditions of discharge (defined using physico-chemical criteria and more rarely toxicological criteria) are respected. In particular, it gives the opportunity to ensure that discharge levels do not exceed the capacities of the receiving medium or of organisms that live in it. In the second case, biomarkers give the opportunity to detect accidental, illicit, and/or diffuse discharges. According to the intensity and extent of biomarker response, alert procedures may therefore be activated (or not) to identify the source of pollution without any information on the nature of contaminants.

3.2.2 Predictive Approach: Biomarkers and Environmental Risk Assessment

Since biomarkers can be used to detect the presence of environmental contaminants and may provide information on the effects of pollutants on organisms, they may be used for the assessment of environmental quality. Is it possible to predict the effects of pollutants on populations, communities, and ecosystems from measurements performed at the individual level? The assessment of ecological risk associated with the release of chemicals in the environment is a key problem in the context of sustainable development and long-term conservation of environmental quality. The relevance of biomarkers for environmental risk assessment has recently been under discussion and some interesting perspectives have been drawn (Blancato et al., 1996; Engel and Vaughan, 1996; Holdway, 1996; Schlenk, 1996). All the arguments that have been presented in this debate are not reviewed here in detail, but it seems interesting to recall the definition and rationale of ecological risk assessment in order to investigate the possible role of biomarkers in such an approach.

Environmental risk assessment corresponds to an estimate of the probability that an adverse effect will occur in response to the presence of one or several factors of disturbance (US EPA, 1992; Engel and Vaughan, 1996). Risk assessment for chemicals is based on four successive steps (Dary et al., 1996):

1. *Hazard identification* is performed using toxicity test data obtained for both target and non-target organisms.
2. *Establishment of dose-response relations* is used to experimentally determine the toxic activity of compounds.

3. *Exposure assessment* is based on evaluation of the nature and size of populations or ecosystems at risk, as well as the frequency, duration, and intensity of contact with the chemical
4. *Risk characterization* is performed using data from the three preceding steps. It provides a quantitative and qualitative prediction of the probability that a harmful effect will occur.

Most biomarkers fulfil the two first steps but the extrapolation of individual responses to changes in populations or communities is still difficult (see chapter 13). No *in situ* study has yet demonstrated that the effect of a chemical on a critical target in an individual was responsible for changes in the population to which it pertains. *A posteriori* studies on the effects of organochlorine pesticides (DDT, DDE) on certain bird populations (peregrine falcon in Great Britain, fish-eating birds in the North American Great Lakes) have shown that population changes were linked with effects of toxicants on individuals (Peakall, 1992). Although interesting from a scientific point of view, this approach is the opposite of the ecological risk assessment approach since there is no prediction of the effects at the population level but only a reconstruction of past phenomena.

So far, the use of biomarkers in environmental risk assessment for chemicals still depends on the identification of causal relationships or mechanistic links between the reaction of individuals to exposure and changes in populations and communities. Some studies have already initiated this research for aquatic animals (Sutcliffe, 1994), but extrapolation of effects of pollutants across different levels of biological organization still largely depends on theoretical or conceptual links.

4. CONCLUSION

The chapters in this book and the constantly growing international literature devoted to this subject demonstrate the relevance of biomarkers as tools for environmental quality assessment. Scientists, environment managers, and administrators now recognize the interest of biomarkers as diagnostic tools since they ensure rapid assessment of environmental quality through evaluation of the health of organisms.

However, the relevance of the information provided by biomarkers largely depends on the conditions of their measurement and their routine use requires many precautions. Their use to assess the evolution of environmental quality requires a reliable physicochemical and ecological characterization of ecosystems, and a sound knowledge of natural factors that may induce changes in their response in order to distinguish the signal due to a disturbance from that caused by natural fluctuations.

So far, no biomarker has been validated as a unique tool of detection of environmental pollutants and of their effects on organisms. Field monitoring programmes clearly show that the best approach is measurement of several biomarkers at different levels of biological organization within an individual (molecular, cellular and tissue level), or in an individual as a whole (physiological biomarkers).

Although many basic *in situ* studies are still necessary, the use of biomarkers as diagnostic tools of ecosystem health has already been validated. Their use in environmental health monitoring programmes is likely to increase, since numerous improvements have been made in sampling (selection of identical species, especially bioindicator species; sampling of individuals of the same age, size, or sex), interpretation (better knowledge of the effects of climate, season, and tide level on the response of biomarkers), and analytical techniques (international intercalibration studies, standardization of procedures). However, the use of biomarkers as tools for prediction of ecological risk of toxicants opens up a vast field of research and development for the next decades.

REFERENCES

Blancato J.N., Brown R.N., Dary C.C. and Saleh M.A. (eds.). (1996). *Biomarkers for Agrochemicals and Toxic Substances. Applications and Risk Assessment*. ACS Symposium Series, 643. American Chemical Society, Washington, DC.

Bocquené G., Galgani F. and Walker C.H. (1997). Les cholinestérases, biomarqueurs de neurotoxicité. In: Biomarqueurs en Écotoxicologie: Lagadic L., Caquet Th., Amiard J.C. and Ramade F. (eds.). *Aspects Fondamentaux*. Masson, Paris, pp. 209–239.

Dary C.C., Quackenboss J.J., Nauman C.H. and Hern S.C. (1996). Relationship of biomarkers of exposure to risk assessment and risk management. In: Biomarkers for Agrochemicals and Toxic Substances. Applications and Risk Assessment. Blancato J.N., Brown R.N., Dary C.C. and Saleh M.A. (eds.). ACS Symposium Series, 643. American Chemical Society, Washington, DC, pp. 2–23.

Depledge M.H., Aagaard A. and Györkös P. (1995). Assessment of trace metal toxicity using molecular, physiological and behavioural biomarkers. *Mar. Poll. Bull.*, 31:19–27.

Engel D.W. and Vaughan D.S. (1996). Biomarkers, natural variability, and risk assessment: can they coexist? *Hum. Ecol. Risk Assess.* 2:257–262.

Fossi M.C. and Leonzio C. (1994). *Nondestructive Biomarkers in Vertebrates*. Lewis Publishers, Boca Raton.

Gillet C.H. and Monod G. (1998). Chemical contamination monitored through elevation of 7-ethoxyresorufin O-deethylase (EROD) in early life stages of Arctic Charr (*Salvelinus alpinus* L.) incubated in Lake Geneva tributaries. *Mar. Environ. Res.*, 46:263–266.

Holdway D.A. (1996). The role of biomarkers in risk assessment. *Hum. Ecol. Risk Assess.* 2:263–267.

Huggett R.J., Kimerle R.A., Mehrle P.M. and Bergam H.L. (1992). *Biomarkers. Biochemical, Physiological and Histological Markers of Anthropogenic Stress*. SETAC Special Publications Series, Lewis Publishers, Boca Raton.

Lagadic L., Caquet Th. and Amiard J.C. (1997a). Biomarqueurs en écotoxicologie: Principes et définitions. In: Biomarqueurs en écotoxicologie: Aspects fondamentaux. Lagadic L., Caquet Th., Amiard J.C. and Ramade F. (eds.). Masson, Paris, pp. 1–9.

Lagadic L., Caquet Th., Amiard J.C. and Ramade F. (eds.). (1997b). *Biomarqueurs en écotoxicologie: Aspects fondamentaux.* Masson, Paris.

Lagadic L., Caquet Th. and Amiard J.C. (1997c). Intérêt d'une approche multiparamétrique pour le suivi de la qualité de l'environnement. In: Biomarqueurs en écotoxicologie: Aspects fondamentaux. Lagadic L., Caquet Th., Amiard J.C. and Ramade F. (eds.). Masson, Paris, pp. 393–401.

McCarthy J.F. and Shugart L.R. (eds.). (1990). *Biomarkers of Environmental Contamination.* Lewis Publishers, Boca Raton.

Monod G. (1997). L'induction du cytochrome P450IAl chez les poissons. In: Biomarqueurs en écotoxicologie: Aspects fondamentaux. Lagadic L., Caquet Th., Amiard J.C. and Ramade F. (eds.). Masson, Paris, pp. 33–54.

Munkittrick K.R. and McCarty L.S. (1995). An integrated approach to aquatic ecosystem health: top-down, bottom-up or middle-out? *J. Aquat. Ecosyst. Health,* 4:77–90.

Narbonne J.F. and Michel X. (1997). Systèmes de biotransformation chez les mollusques aquatiques. In: Biomarqueurs en écotoxicologie: Aspects fondamentaux. Lagadic L., Caquet Th., Amiard J.C. and Ramade F. (eds.). Masson, Paris, pp. 11–31.

Peakall D.B. (1992). *Animal Biomarkers as Pollution Indicators.* Chapman and Hall, London.

Peakall D.B. and Shugari L.R. (eds.) (1993). *Biomarkers: Research and Application in the Assessment of Environmental Health.* NATO ASI Series, serie H: Cell Biology, vol. 68. Springer Verlag, Berlin.

Schlenck D. (1996). The role of biomarkers in risk assessment. *Hum. Ecol. Risk Assess.* 2:251–256.

Sutcliffe D.W. (1994). *Water Quality and Stress Indicators in Marine and Freshwater Ecosystems: Linking Levels of Organisation (Individuals, Populations, Communities).* Freshwater Biological Association, Ambleside.

US EPA. (U. S. Environmental Protection Agency), 1992. *Framework for Ecological Risk Assessment,* Washington, DC, Risk Assessment Forum, EPA/630/R-92/001.

Ward J.B. and Henderson R.E. (1996). Identification of needs in biomarkers research. *Environ. Health Perspect.* 104 (suppl. 5):895–900.

Wolfe D.A. (1996). Insights on the utility of biomarkers for environmental impact assessment and monitoring. *Hum. Ecol. Risk Assess.* 2:245–250.

Index

A

Aberration(s), 148
  – chromosomal, 156
Abramis brama, 61
Acephate, 116, 143, 144
Acetylcholine, 120
Acetylcholinesterase (AChE), 14,
  15, 18, 19, 21, 23, 24, 120, 133, 143, 148,
  149, 151, 248–251, 252, 258, 260, 280,
  284
  – biochemical properties, 121
  – inhibition, 127–129, 133, 150, 250
  – methods of measurement, 120
  – resistant, 250, 258
  – variability, 121, 124
AChE, see Acetylcholinesterase
Acid pH, 189, 190, 193
Acid phosphatases, 237
Acipenser fulvescens, 218–219
Acyl coenzyme A retinol-acyl transferase
  (ARAT), 206
Aix sponsa, 157
ALAD, see δ-aminolevulinate dehydratase
Alaska, 140
Alca torda, 145
Aldicarb, 118, 142
Aldrin, 140, 145, 146
Alectoris graeca, 143
Alkoxycoumarin, 146
Allocation of energy in growth and repro-
  duction (AEGR), 289, 290
Alosa sp., 220
α-naphthol, 238
Aluminium (Al), 190
American eel, 220
Aminopyrene demethylase, 145
Amphibians, 139, 141

Anadara granosa, 237
Anas americana, 144
Anas platyrhynchos, 143, 146, 153
Anas rubripes, 157
Anatidae, 141
Androgen, 192, 288
Aneuploidy, 156
Anguilla anguilla, 63
Aniline, 145, 147
Anopheles arabiensis, 260
Anser anser, 151
Anser brachyrhynchus, 151
Anthracene, 12
Antifouling paints, 22, 287
AOX, 61, 63
Aphid, 251, 252, 260
Apis mellifera, 283
Apodemus flavicollis, 146, 147
Apodemus sylvaticus, 144
Arcachon basin, 28
Ardea herodias, 213–215
Arginine vasotocine (AVT), 190, 191, 199
Aroclor 1254, 147, 157
Aromatases, 288
Arsenic (As), XXII, 176, 177
Aryl hydrocarbon hydroxylase (AHH), 60,
  145–147, 152, 217
Arylsulphatase, 237
Atrazine, 134
Autophagy, 230, 232
Aythya fuligula, 146, 153
Azinphos methyl, 115, 144

B

Bank vole, see *Clethrionomys glareolus*
Barbus barbus, 61, 65
Bass, see *Dicentrarchus labrax*

List of Contributors

Amiard Jean-Claude

EP 61 du CNRS, Service d'Écotoxicologie, Faculté de Pharmacie, Université de Nantes, 1, rue Gaston Veil, F-44035 Nantes Cedex, France.

Amichot Marcel

INRA, Unité de Recherche sur la Résistance aux Insecticides, BP2078, F-06606 Antibes Cedex, France.

Bergé Jean-Baptiste

INRA, Unité de Recherche sur la Résistance aux Insecticides, BP2078, F-06606 Antibes Cedex, France.

Bocquené Gilles

IFREMER, Laboratoire d'Écotoxicologie, Rue de l'Ile d'Yeu, BP 1105, F-44311 Nantes Cedex 01, France.

Bourbonnais Diane H.

Département des Sciences Biologiques, Laboratoire TOXEN, Université du Québec à Montréal, C.P. 8888, Succursale Centre-Ville, Montréal, Québec, Canada, H3C 3P8.

Budzinski Hélène

URA 348 CNRS, Université Bordeaux I, 351 cours de la Libération, F-33405 Talence Cedex, France.

Burgeot Thierry

IFREMER, Laboratoire d'Écotoxicologie, Rue de l'Ile d'Yeu, BP 1105, F-44311 Nantes Cedex 01, France.

Caquet Thierry

Laboratoire d'Écologie et de Zoologie, URA CNRS 2154, Bt 442, Université de Paris-Sud, F-91405 Orsay Cedex, France.

Clijsters Herman

Limburgs Universitair Centrum, Dep. SBG, Universitaire Campus, B-3590 Diepen-beek, Belgique.

Cosson Richard P.

EP 61 du CNRS, Service d'Écotoxicologie, Faculté de Pharmacie, Université de Nantes, 1, rue Gaston Veil, F-44035 Nantes Cedex, France.

Cuany André

INRA, Unité de Recherche sur la Résistance aux Insecticides, BP2078, F-06606 Antibes Cedex, France.

Daubèze Michèle

Laboratoire de Toxicologie Alimentaire, ISTAB/Université Bordeaux I, Avenue des Facultés, F-33405 Talence Cedex, France.

Flammarion Patrick

Laboratoire d'Écotoxicologie, Division Biologie des Écosystèmes Aquatiques, CEMAGREF, 3 bis Quai Chauveau, CP 220, F-69336 Lyon Cedex 09, France.

Fouchécourt Marie-Odile

École Nationale Vétérinaire de Lyon, Unité Associée de Toxicologie Métabolique et d'Écotoxicologie, INRA-ENVL, BP 83, F-69280 Marcy l'Étoile, France.

Galgani François

IFREMER, Laboratoire d'Écotoxicologie, Rue de l'Ile d'Yeu, BP 1105, F-44311 Nantes Cedex 01, France.

Garric Jeanne

Laboratoire d'Écotoxicologie, Division Biologie des Écosystèmes Aquatiques, CEMAGREF, 3 bis Quai Chauveau, CP 220, F-69336 Lyon Cedex 09, France.

Garrigues Philippe

URA 348 CNRS, Université Bordeaux I, 351 cours de la Libération, F-33405 Talence Cedex, France.

Hontela Alice

Département des Sciences Biologiques, Laboratoire TOXEN, Université du Québec à Montréal, C.P. 8888, Succursale Centre-Ville, Montréal, Québec, Canada, H3C 3P8.

Lafaurie Marc

Laboratoire de Toxicologie Marine, Faculté de Médecine, Université de Nice-Sofia Antipolis, avenue de Valombrose, F-06107 Nice Cedex 02, France.

Lagadic Laurent

INRA, Unité d'Écotoxicologie Aquatique, 65, rue de St.-Brieuc, F-35042 Rennes Cedex, France.

Mench Michel

INRA, Unité d'Agronomie, Centre de Recherches de Bordeaux, B.P. 81, F-33883 Villenave d'Ornon Cedex, France.

Michel Xavier

Laboratoire de Toxicologie Alimentaire, ISTAB/Université Bordeaux I, Avenue des Facultés, F-33405 Talence Cedex, France.

Mocquot Bernard

INRA, Unité d'Agronomie, Centre de Recherches de Bordeaux, B.P. 81, F-33883 Villenave d'Ornon Cedex, France.

Monod Gilles

INRA, Unité d'Écotoxicologie Aquatique, IFR 43 : Biologie et Écologie du Poisson, 65, rue de St.-Brieuc, F-35042 Rennes Cedex, France.

Mora Pascal

Laboratoire de Toxicologie Alimentaire, ISTAB/Université Bordeaux I, Avenue des Facultés, F-33405 Talence Cedex, France.

Narbonne Jean-François

Laboratoire de Toxicologie Alimentaire, ISTAB/Université Bordeaux I, Avenue des Facultés, F-33405 Talence Cedex, France.

Pasteur Nicole

ISEM, Laboratoire Génétique et Évolution, USTL, F-34095 Montpellier Cedex 05, France.

Pauron David

INRA, Unité de Recherche sur la Résistance aux Insecticides, BP2078, F-06606 Antibes Cedex, France.

Pellerin-Massicote Jocelyne

Groupe de Recherche en Environnement Côtier, Département d'Océanographie, Université du Québec à Rimouski, 310 allée des Ursulines, Rimouski, Québec, Canada, G5L 3A1.

Ramade François

Laboratoire d'Écologie et de Zoologie, URA CNRS 2154, Bt 442, Université de Paris-Sud, F-91405 Orsay Cedex, France.

Raymond Michel

ISEM, Laboratoire Génétique et Évolution, USTL, F-34095 Montpellier Cedex 05, France.

Ribera Daniel

Association pour le Développement, l'Étude et le Conseil en Toxicologie (ADEC-Tox), 120 rue Quintin, F-33000 Bordeaux, France.

Rivière Jean-Louis

INRA, Centre de Versailles, Unité de Phytopharmacie et Médiateurs Chimiques, Route de St Cyr, F-78026 Versailles Cedex, France.

Spear Philip A.

Département des sciences biologiques, Laboratoire TOXEN, Université du Québec à Montréal, C.P. 8888, Succursale Centre-Ville, Montréal, Québec, Canada, H3C 3P8.

Tremblay Réjean

Groupe de Recherche en Environnement Côtier, Département d'Océanographie, Université du Québec à Rimouski, 310 allée des Ursulines, Rimouski, Québec, Canada, G5L 3A1.

Vangronsveld Jaco

Limburgs Universitair Centrum, Dep. SBG, Universitaire Campus, B-3590 Diepenbeek, Belgique.

Walker Colin H.

University of Reading, School of Animal and Microbial Sciences, Whiteknights, PO Box 228, Reading RG6 2AJ, Grande-Bretagne.